Managing Projects with Smart Technologies

With a focus on project managers (PMs) in the construction industry, this book addresses the impact of smart technology applications on project management and examines how technologically competent PMs can be developed for successfully managing and delivering projects with smart technologies.

The book assesses the changes to the knowledge and skillsets required to manage projects with smart technologies; develops a Technological Competency Framework to improve PM competency when managing projects with smart technologies; and develops a Knowledge-Based Technological Competency Analytics and Innovations System to assess and improve the technological competency of PMs and provide recommendations to improve their competency.

Managing Projects with Smart Technologies is ideal for PMs and academics in the areas of construction project management, engineering, architecture, and infrastructure and anyone involved in the technical training of professionals in these areas.

Bon-Gang Hwang is a professor in the Department of the Built Environment, College of Design and Engineering, National University of Singapore.

Jasmine Ngo is a PhD degree holder in the Department of the Built Environment, National University of Singapore, Singapore.

Hanjing Zhu is a PhD candidate in the Department of the Built Environment, College of Design and Engineering, National University of Singapore.

Spon Research

Publishes a stream of advanced books for built environment researchers and professionals from one of the world's leading publishers. The ISSN for the Spon Research programme is ISSN 1940–7653 and the ISSN for the Spon Research E-book programme is ISSN 1940–8005.

Work Stress Induced Chronic Diseases in Construction
Discoveries Using Data Analytics
Imriyas Kamardeen

Life-Cycle Greenhouse Gas Emissions of Commercial Buildings
An Analysis for Green-Building Implementation Using A Green Star
Rating System
Cuong N. N. Tran, Vivian W. Y. Tam and Khoa N. Le

Data-driven BIM for Energy Efficient Building Design
Saeed Banihashemi, Hamed Golizadeh and Farzad Pour Rahimian

Successful Development of Green Building Projects
Tayyab Ahmad

BIM and Construction Health and Safety
Uncovering, Adoption and Implementation
Hamed Golizadeh, Saeed Banihashemi, Carol Hon and Robin Drogemuller

Digitalisation in Construction
Recent Trends and Advances
Edited by Chansik Park, Farzad Pour Rahimian, Nashwan Dawood, Akeem Pedro, Dongmin Lee, Rahat Hussain and Mehrtash Soltani

Managing Projects with Smart Technologies
Developing Technological Competency for Project Managers
Bon-Gang Hwang, Jasmine Ngo and Hanjing Zhu

For more information about this series, please visit: www.routledge.com

Managing Projects with Smart Technologies

Developing Technological
Competency for Project Managers

**Bon-Gang Hwang, Jasmine Ngo and
Hanjing Zhu**

Routledge
Taylor & Francis Group
LONDON AND NEW YORK

First published 2024
by Routledge
4 Park Square, Milton Park, Abingdon, Oxon OX14 4RN

and by Routledge
605 Third Avenue, New York, NY 10158

Routledge is an imprint of the Taylor & Francis Group, an informa business

British Library Cataloguing-in-Publication Data
A catalogue record for this book is available from the British Library

ISBN: 978-1-032-60364-3 (hbk)
ISBN: 978-1-032-61149-5 (pbk)
ISBN: 978-1-003-46223-1 (ebk)

DOI: 10.1201/9781003462231

Typeset in Times New Roman
by Apex CoVantage, LLC

Contents

13 Conclusions and Recommendations

About the Authors

Professor Bon-Gang Hwang is a professor in the Department of the Built Environment, National University of Singapore, Singapore. He has spent over 25 years conducting research in construction project management, publishing over 150 high-quality papers in this field. His influential work in advancing the digital transformation within Singapore's construction industry has been recognised with the Outstanding Project Manager Award in both 2020 and 2022.

Dr Jasmine Ngo is a PhD degree holder in the Department of the Built Environment, National University of Singapore, Singapore. She focuses on project management within the digitally transforming construction industry. She has contributed to over 10 research papers, with her co-authored works recognised with the Outstanding Project Manager Award.

Hanjing Zhu is a PhD candidate in the Department of the Built Environment, National University of Singapore. She focuses on digital transformation and construction safety research. Her contributions to various papers and projects are reflected in her co-authored work receiving the 2022 Outstanding Project Manager Award.

Acronyms and Abbreviations

3D	Three-dimensional
4IR	Fourth Industrial Revolution
AHP	Analytic Hierarchy Process
AI	Artificial Intelligence
AIP	Aggregation of Individual Priorities
AtIP-GMM	Average-to-Interpolated Geometric Mean Method
AM	Additive Manufacturing
AR	Augmented Reality
ATT	Attitude Towards Technology
AV	Autonomous Vehicles
AVE	Average Variance Extracted
BCA	Building and Construction Authority
BD	Big Data
BDA	Big Data Analytics
BDE	Big Data Engineering
BEI	Behavioural Event Interview
BI	Behavioural Intention
BIM	Building Information Modelling
CAD	Computer-Aided Design
CC	Cloud Computing
CICs	Cognitive Intelligence Competencies
CII	Construction Industry Institute
CPM	Construction Project Management
CPS	Cyber-Physical System
CR	Composite Reliability
CTA	Confirmatory Tetrad Analysis
DLT	Distributed Ledger Technology
DSS	Decision Support System
EFA	Exploratory Factor Analysis
EICs	Emotional Intelligence Competencies
GPS	Global-Positioning System
GUI	Graphical User Interface
HTMT	Heterotrait–Monotrait

ICT	Information and Communication Technology
IDT	Innovation Diffusion Theory
IMDA	Infocomm Media Development Authority
IoT	Internet-of-Things
IPMA	International Project Management Association
I-PMA	Importance-Performance Matrix Analysis
IR	Industrial Revolution
IS	Information System
IT	Information Technology
KBTCAIS	Knowledge-Based Technological Competency Analytics and Innovations System
KMO	Kaiser-Meyer-Olkin
LM	Linear Regression Model
LEED	Leadership in Energy and Environmental Design
MAPE	Mean Absolute Percentage Error
MGA	Multigroup Analysis
MM	Motivation Model
MPCU	Model of PC Utilisation
MPE	Mean Percentage Error
PBC	Perceived Behavioural Control
PCA	Principal Component Analysis
PE	Percentage Error
PESTLE	Political, Economic, Social, Technological, Legal and Environmental
PEU	Perceived Ease of Use
PIT	Personal Innovativeness in Technologies
PIIT	Personal Innovativeness in IT
PLS–SEM	Partial Least Squares–Structural Equation Modelling
PM	Project Manager
PMBOK	Project Management Body of Knowledge
PMCD	Project Manager Competency Development
PMI	Project Management Institute
PMP	Project Management Professionals
PPVC	Prefabricated Prefinished Volumetric Construction
PU	Perceived Usefulness
RFID	Radio Frequency Identification
RMSE	Root Mean Square Error
SARSA	State-Action-Reward-State-Action
SCT	Social Cognitive Theory
SEM	Structural Equation Modelling
SICs	Social Intelligence Competencies
SME	Small and Medium-Sized Enterprises
TAFF	Technology Affect
TAM	Technology Acceptance Model
TANX	Technology Anxiety

TCF	Technological Competency Framework
TIB	Theory of Interpersonal Behaviour
TPB	Theory of Planned Behaviour
TR	Technology Readiness
TRA	Theory of Reasoned Action
TRI	Technology Readiness Index
TSE	Technology Self-Efficacy
UAV	Unmanned Aerial Vehicles
UTAUT	Unified Theory of Acceptance and Use of Technology
UWB	Ultra-Wide Band
VBA	Visual Basic for Applications
VIF	Variance Inflation Factor
VR	Virtual Reality
WEF	World Economic Forum

1   Introduction

1.1   Research Motivation

It is inarguable that all industries are being disrupted by rapidly advancing technologies, leading to transformations in organisational processes and individual work processes (Dombrowski & Wagner, 2014; Edirisinghe, 2019; Lee et al., 2018; Lu, 2017). With the transformations in organisational processes and individual work processes, it has been widely recognised that a 4IR has emerged (Schwab, 2016). 4IR can be represented by the convergence of the physical and digital paradigms, enabling the digitalisation, automation and integration of entire value chains (Kagermann et al., 2013; Schwab, 2016). The applications of key 4IR technologies, including the CPS, IoT, BD, AI, Robotics, AV, AM, 3D imaging, AR, VR, and DLT, enable self-monitor, self-organise and self-execute work tasks, and the technologies are often termed 'smart technologies' (Dallasega et al., 2018; Ghobakhloo, 2018; Jabbour et al., 2018; Kamble et al., 2018; Lu, 2017; J. M. Müller et al., 2018; Oesterreich & Teuteberg, 2016; Pereira & Romero, 2017; Stock et al., 2018). These technologies build on and amplify one another, resulting in the acceleration of the pace of change (Schwab, 2016; Whyte et al., 2016; World Economic Forum, 2018b).

A growing demand for digital transformation has also been witnessed in the construction industry under the 4IR (Jia et al., 2019). However, the construction industry is known to be a laggard in technology adoption (Sepasgozar et al., 2016). The slow technology adoption is largely due to the nature of the industry, leading to the general resistance towards change, lack of training and development, complex and unpredictable production environment, lack of interoperability across the value chain, large number of small- and medium-sized enterprises with limited capabilities to invest in new technologies and fragmentation across specialisations and lack of skilled professionals (Chen et al., 2018; Dainty et al., 2001; Dallasega et al., 2018; Edirisinghe, 2019; Froese, 2010; Niu et al., 2017; Oesterreich & Teuteberg, 2016; Sepasgozar et al., 2016). On top of that, the ways buildings and cities are designed, constructed and managed and behave and interact with users are being transformed, as new technologies are being incorporated into buildings and cities (Costa et al., 2015). The changes result in uncertainties in work processes and may require different knowledge and skillsets to be competent in the management of

DOI: 10.1201/9781003462231-1

construction projects. The interdisciplinary knowledge and skills, problem-solving skills and interpersonal skills will increasingly be required in response to the shifts in industrial processes, business models and work environments (Dombrowski & Wagner, 2014; Magruk, 2016; Pereira & Romero, 2017; Schwab, 2016). It is crucial that this change be proactively managed in order to reap the associated benefits, instead of increasing the risk of widening skill gaps, inequality and polarisation (Brynjolfsson & McAfee, 2014; Schwab, 2016; World Economic Forum, 2018b).

Specifically, PMs have been found to directly and indirectly influence project performance, leading a project team to achieve the project's objectives (R. Müller & Turner, 2007; Nixon et al., 2012; Zhang et al., 2013). As projects grow increasingly complex and smart technologies continue to proliferate across the construction industry, it is crucial for PMs, the key contributors to project success, to possess the competencies to fully leverage these innovations (Ahmed & Anantatmula, 2017; Davis, 2011; Edum-Fotwe & McCaffer, 2000; Loufrani-Fedida & Missonier, 2015; Tabassi et al., 2016; Trivellas & Drimoussis, 2013; Zhang et al., 2013). The importance of PMs' competencies is also evident with the development of competency frameworks tailored for their role, such as the PM's Competency Development Framework (Project Management Institute, 2017) and the IPMA Competence Baseline (International Project Management Association, 2015).

Although numerous studies focus on smart technologies in the 4IR, there is a lack of research exploring the impact of key smart technologies on CPM knowledge, skillsets and the technological competency required to manage projects with smart technologies. Existing studies typically focus on the applications of the smart technologies in specific areas of construction projects, or on improving the performance of the smart technologies. Apart from that, studies that investigate the competency required of workers in managing smart technologies are typically conducted in other industries such as education, manufacturing and healthcare. Therefore, there exists a knowledge gap in the technological competencies of PMs in managing projects with smart technologies, especially in the era of the 4IR. This study fills the gap through the establishment of a theoretical framework of technological competency for PMs and the development of a tool that can assess and improve PMs' technological competency.

1.2   Research Scope

This book focuses on PMs in Singapore's construction industry in the era of 4IR. Singapore has been ranked as one of the top countries that are technologically ready and competitive globally in 2018, alongside Australia and Sweden (World Economic Forum, 2018a). However, the technology adoption rate of Singapore's construction industry remains low. This is in contrast with Australia's construction industry, which was found to be moderately investing in technologies (Leviäkangas et al., 2017) and in Sweden's construction industry, which was found to be highly innovative (European Commission, 2018). With the disparity between the technological readiness of Singapore as a country and the construction industry specifically, it presents a huge opportunity to further increase the technology adoption rate and maximise the potential of technologies in the construction industry.

'Construction industry' in this book is limited to the processes involved, from initialisation of a project to the handover of the end-product to the client, as these are the main responsibilities of the PM. The operations and maintenance of infrastructure and buildings are not considered in this study, as these are mainly managed by facilities managers.

1.3   Research Aim and Objectives

The objectives of this work are to

a) assess the changes to the knowledge and skillsets required to manage projects with smart technologies,
b) develop a TCF for PMs to be competent in managing projects with smart technologies, and
c) develop a KBTCAIS to assess and improve the technological competency of PMs.

1.4   Research Significance

This book contributes to the existing body of knowledge, in both theory and in practice. Firstly, while there have been many studies on the applications of smart technologies in projects, a limited literature studies the impact of smart technologies on CPM knowledge and skillsets. This book also develops the theoretical framework of technological competency, which extends the existing theories of competency. In particular, as technologies are expected to continually proliferate every aspect of human life and get increasingly integrated into work processes, technological competency may be able to provide more accurate measurement and prediction of one's job performance.

Apart from the contribution to theory, this book contributes to practice, by providing valuable reference to the industry in navigating the disruption and in the training and development of technologically competent PMs who are able to consistently deliver successful projects through the identification of the changes in the required CPM knowledge and skillsets to manage projects with smart technologies. Furthermore, the development of the KBTCAIS can be used by PMs to assess their technological competency level and receive recommendations to improve their level of technological competency. The KBTCAIS can also be used by organisations as a reference for training and development of PMs, as a tool for recruitment and resource allocation. By accurately assessing the technological competency level of individual PMs, organisations can optimally assign PMs to projects of varying technological complexity, ultimately improving the overall performance of the projects, organisations and industry. On a broader level, improvements in PM's technological competency can also aid in reducing the skills disparity among PMs in managing projects with smart technologies. This can facilitate the digital transformation of the construction industry, ultimately improving the performance of the construction industry through the increased adoption of smart technologies.

Although the book focuses on PMs in Singapore's construction industry, the process of assessing the impact of smart technologies on knowledge and skillsets can be extended to other industries globally. Subsequently, the TCF and KBTCAIS for other specific professions and industries may be developed with relevant modifications of the domain-related factors.

1.5 Structure of This Book

The book is structured into 13 chapters:

Chapter 1 gives a brief overview of the smart technologies in the 4IR and their impact, the construction industry and PM's competencies, the research scope, questions, aims and objectives, approach and significance of the research.

Chapter 2 introduces the key smart technologies in the 4IR, including CPS, IoT, BD, AI, Robotics, AV, AM, 3D imaging, AR, VR and DLT.

Chapter 3 provides the common applications of smart technologies in projects and the changes to project management processes.

Chapter 4 introduces the competency theory, common competency models and the assessment method of competency levels and personal characteristics that result in superior work performance.

Chapter 5 introduces the evolution of CPM along with the advancements in technologies. This chapter also explores PM competencies, including the knowledge, skills and common PM competencies.

Chapter 6 explores the theories of technology adoption and acceptance, including the Technology Acceptance Model, its variations and theoretical underpinnings, Innovation Diffusion Theory, Social Cognitive Theory, Model of PC Utilisation, Motivation Model, Unified Theory of Acceptance and Use of Technology, Personal Innovativeness in IT and the Technology Readiness Index.

Chapter 7 proposes a conceptual model for PMs' technological competency, which aims to combine the theory of competency and theories of technology adoption and acceptance to form the TCF. The hypotheses in this book are also provided in Chapter 7.

Chapter 8 details the research methods to be adopted in the research, including the overall research approach, research design, data collection methods, data analysis methods and the validation of the findings. This chapter also presents the profile of the survey respondents and interviewees.

Chapter 9 provides comprehensive data analysis regarding the existing implementation status of smart technologies in Singapore's construction industry, survey respondents' attitude towards technologies and self-assessed technological competency level, the importance and changes in project management knowledge and skills when using smart technologies, the applicability, impact and significance of the factors affecting one's attitude towards technologies and the differences in the importance of knowledge in project management, skills and factors affecting attitude towards technologies in contributing to one's technological competency to manage projects with smart technologies.

Chapter 10 details the development and validation processes of the TCF.

Chapter 11 provides case studies that investigate the most common smart technology applications in projects, common challenges faced by organisations during the implementation of new technologies, changes to the knowledge, skills and personal attributes required of PMs to manage projects with smart technologies and effective strategies to facilitate technology adoption.

Chapter 12 details the development of the KBTCAIS, including the objectives, architecture, development process, demonstration of the KBTCAIS and validation of the tool.

Chapter 13 provides the conclusion of the study, contributions to knowledge and practice, the limitations of the study and recommendations for practice and future studies.

References

Ahmed, R., & Anantatmula, V. S. (2017). Empirical study of project managers leadership competence and project performance. *Engineering Management Journal, 29*(3), 189–205. https://doi.org/10.1080/10429247.2017.1343005

Brynjolfsson, E., & McAfee, A. (2014). *The second machine age: Work, progress, and prosperity in a time of brilliant technologies* (p. 306). Norton & Company.

Chen, Q., García de Soto, B., & Adey, B. T. (2018). Construction automation: Research areas, industry concerns and suggestions for advancement. *Automation in Construction, 94*, 22–38. https://doi.org/10.1016/j.autcon.2018.05.028

Costa, A. A., Lopes, P. M., Antunes, A., Cabral, I., Grilo, A., & Rodrigues, F. M. (2015). 3I Buildings: Intelligent, interactive and immersive buildings. *Procedia Engineering, 123*, 7–14. https://doi.org/10.1016/j.proeng.2015.10.051

Dainty, A. R. J., Briscoe, G. H., & Millett, S. J. (2001). New perspectives on construction supply chain integration. *Supply Chain Management, 6*(4), 163–173. https://doi.org/10.1108/13598540110402700

Dallasega, P., Rauch, E., & Linder, C. (2018). Industry 4.0 as an enabler of proximity for construction supply chains: A systematic literature review. *Computers in Industry, 99*, 205–225. https://doi.org/10.1016/j.compind.2018.03.039

Davis, S. A. (2011). Investigating the impact of project managers' emotional intelligence on their interpersonal competence. *Project Management Journal, 42*(4), 37–57. https://doi.org/10.1002/pmj.20247

Dombrowski, U., & Wagner, T. (2014). Mental strain as field of action in the 4th industrial revolution. *Procedia CIRP, 17*, 100–105. https://doi.org/10.1016/j.procir.2014.01.077

Edirisinghe, R. (2019). Digital skin of the construction site: Smart sensor technologies towards the future smart construction site. *Engineering, Construction and Architectural Management, 26*(2), 184–223. https://doi.org/10.1108/ECAM-04-2017-0066

Edum-Fotwe, F. T., & McCaffer, R. (2000). Developing project management competency: Perspectives from the construction industry. *International Journal of Project Management, 18*, 111–124.

European Commission. (2018). *European construction sector observatory: Country profile Sweden.* https://ec.europa.eu/docsroom/documents/30682/attachments/1/translations/en/renditions/native

Froese, T. M. (2010). The impact of emerging information technology on project management for construction. *Automation in Construction, 19*(5), 531–538. https://doi.org/10.1016/j.autcon.2009.11.004

Ghobakhloo, M. (2018). The future of manufacturing industry: A strategic roadmap toward Industry 4.0. *Journal of Manufacturing Technology Management, 29*(6), 910–936. https://doi.org/10.1108/JMTM-02-2018-0057

International Project Management Association. (2015). *IPMA individual competence baseline for project, programme and portfolio management.* International Project Management Association.

Jabbour, A. B. L. de S., Jabbour, C. J. C., Foropon, C., & Filho, M. G. (2018). When titans meet – Can Industry 4.0 revolutionise the environmentally-sustainable manufacturing wave? The role of critical success factors. *Technological Forecasting and Social Change, 132,* 18–25. https://doi.org/10.1016/j.techfore.2018.01.017

Jia, M., Komeily, A., Wang, Y., & Srinivasan, R. S. (2019). Adopting Internet of Things for the development of smart buildings: A review of enabling technologies and applications. *Automation in Construction, 101,* 111–126. https://doi.org/10.1016/j.autcon.2019.01.023

Kagermann, H., Wahlster, W., & Helbig, J. (2013). Recommendations for implementing the strategic initiative Industrie 4.0: Securing the future of German manufacturing industry. Final Report of the Industrie 4.0 Working Group. Forschungsunion (p. 82).

Kamble, S. S., Gunasekaran, A., & Gawankar, S. A. (2018). Sustainable Industry 4.0 framework: A systematic literature review identifying the current trends and future perspectives. *Process Safety and Environmental Protection, 117,* 408–425. https://doi.org/10.1016/j.psep.2018.05.009

Lee, M., Yun, J. J., Pyka, A., Won, D., Kodama, F., Schiuma, G., Park, H., Jeon, J., Park, K., Jung, K., Yan, M.-R., Lee, S., & Zhao, X. (2018). How to respond to the fourth industrial revolution, or the second information technology revolution? Dynamic new combinations between technology, market, and society through open innovation. *Journal of Open Innovation: Technology, Market, and Complexity, 4*(21). https://doi.org/10.3390/joitmc4030021

Leviäkangas, P., Paik, S. M., & Moon, S. (2017). Keeping up with the pace of digitization: The case of the Australian construction industry. *Technology in Society, 50,* 33–43. https://doi.org/10.1016/j.techsoc.2017.04.003

Loufrani-Fedida, S., & Missonier, S. (2015). The project manager cannot be a hero anymore! Understanding critical competencies in project-based organizations from a multilevel approach. *International Journal of Project Management, 33,* 1220–1235. https://doi.org/10.1016/j.ijproman.2015.02.010

Lu, Y. (2017). Industry 4.0: A survey on technologies, applications and open research issues. *Journal of Industrial Information Integration, 6,* 1–10. https://doi.org/10.1016/J.JII.2017.04.005

Magruk, A. (2016). Uncertainty in the sphere of the Industry 4.0 – Potential areas to research. *Business, Management and Education, 14*(2), 275–291. https://doi.org/10.3846/bme.2016.332

Müller, J. M., Kiel, D., & Voigt, K. I. (2018). What drives the implementation of Industry 4.0? The role of opportunities and challenges in the context of sustainability. *Sustainability, 10*(1), 247–270. https://doi.org/10.3390/su10010247

Müller, R., & Turner, J. R. (2007). Matching the project manager's leadership style to project type. *International Journal of Project Management, 25*(1), 21–32. https://doi.org/10.1016/j.ijproman.2006.04.003

Niu, Y., Lu, W., Liu, D., Chen, K., Anumba, C., & Huang, G. G. (2017). An SCO-enabled logistics and supply chain – management system in construction. *Journal of Construction Engineering and Management, 143*(3), 04016103. https://doi.org/10.1061/(ASCE)CO.1943-7862.0001232

Nixon, P., Harrington, M., & Parker, D. (2012). Leadership performance is significant to project success or failure: A critical analysis. *International Journal of Productivity and Performance Management, 61*(2), 204–216. https://doi.org/10.1108/17410401211194699

Oesterreich, T. D., & Teuteberg, F. (2016). Understanding the implications of digitisation and automation in the context of Industry 4.0: A triangulation approach and elements of a research agenda for the construction industry. *Computers in Industry, 83,* 121–139. https://doi.org/10.1016/j.compind.2016.09.006

Pereira, A. C., & Romero, F. (2017). A review of the meanings and the implications of the Industry 4.0 concept. *Procedia Manufacturing, 13,* 1206–1214. https://doi.org/10.1016/j.promfg.2017.09.032

Project Management Institute. (2017). *Project manager competency development framework*. Project Management Institute.

Schwab, K. (2016). *The fourth industrial revolution*. PENGUIN GROUP.

Sepasgozar, S. M. E., Loosemore, M., & Davis, S. R. (2016). Conceptualising information and equipment technology adoption in construction: A critical review of existing research. *Architectural Management, 23*, 158–176. https://doi.org/10.1108/ECAM-05-2015-0083

Stock, T., Obenaus, M., Kunz, S., & Kohl, H. (2018). Industry 4.0 as enabler for a sustainable development: A qualitative assessment of its ecological and social potential. *Process Safety and Environmental Protection, 118*, 254–267. https://doi.org/10.1016/j.psep.2018.06.026

Tabassi, A. A., Roufechaei, K. M., Ramli, M., Bakar, A. H. A., Ismail, R., & Pakir, A. H. K. (2016). Leadership competences of sustainable construction project managers. *Journal of Cleaner Production, 124*, 339–349. https://doi.org/10.1016/j.jclepro.2016.02.076

Trivellas, P., & Drimoussis, C. (2013). Investigating leadership styles, behavioural and managerial competency profiles of successful project managers in Greece. *Procedia – Social and Behavioral Sciences, 73*, 692–700. https://doi.org/10.1016/j.sbspro.2013.02.107

Whyte, J., Stasis, A., & Lindkvist, C. (2016). Managing change in the delivery of complex projects: Configuration management, asset information and "big data". *International Journal of Project Management, 34*(2), 339–351. https://doi.org/10.1016/j.ijproman.2015.02.006

World Economic Forum. (2018a). *Readiness for the future of production report*. World Economic Forum.

World Economic Forum. (2018b). *The future of jobs* (p. 147). World Economic Forum.

Zhang, F., Zuo, J., & Zillante, G. (2013). Identification and evaluation of the key social competencies for Chinese construction project managers. *International Journal of Project Management, 31*(5), 748–759. https://doi.org/10.1016/j.ijproman.2012.10.011

2 Smart Technologies in the Fourth Industrial Revolution

2.1 Introduction

The advent of the Fourth Industrial Revolution (4IR) has brought about digital transformation across various industries (Schwab, 2016). 'Industrial revolution' refers to a set of technologies that amplify productivity through deep changes in entire systems (Pereira & Romero, 2017; Schwab, 2016; Stock et al., 2018). This is evident in the previous three IRs, in which the first IR introduced mechanisation and steam engine from the 18th to the 19th centuries; mass production and electricity in the second IR, spanning from the late 19th to the early 20th century; and automation, computers and electronics in the third IR, which started in the late 20th century (Devezas et al., 2017; Y. Lu, 2017; Pereira & Romero, 2017; Schwab, 2016). The technologies in the 4IR are characterised by their ability to self-monitor, self-organise and self-execute.

The representative smart technology in the 4IR is the CPS, where every entity can be identified and can continuously interact and exchange information amongst themselves (Kamble et al., 2018; Y. Lu, 2017; Posada et al., 2015; Wan et al., 2016). This leads to more flexible, self-organising and autonomous processes (Kamble et al., 2018; Liu & Xu, 2017; Y. Lu, 2017). Other smart technologies in the 4IR include the IoT, BD, AI, Robotics, AV, AM, 3D imaging, AR, VR and DLT (Dallasega et al., 2018; Ghobakhloo, 2018; Jabbour et al., 2018; Kamble et al., 2018; Y. Lu, 2017; Müller et al., 2018; Oesterreich & Teuteberg, 2016; Pereira & Romero, 2017; Stock et al., 2018). CPS generates huge volume, variety and velocity of data, leading to the need for BD analytics, which further enables other technologies. Automation and optimisation of work processes also require the use of AI, robotics, AV, AM, 3D imaging and DLT. Furthermore, as collaboration is critical in the 4IR, AR and VR can improve the visualisation of 3D models and communication processes. With data and networking as the core to 4IR, cyber-security becomes essential, giving rise to the demand for DLT.

2.2 Cyber-Physical System

CPS is formed by the integration of physical systems to a processing unit through the use of embedded systems to control, monitor and coordinate the physical systems (Ghobakhloo, 2018; Stock et al., 2018). An embedded system is a combination

DOI: 10.1201/9781003462231-2

of hardware and software components, connected to the system environment via sensors and actuators (Stock et al., 2018).

CPS can be characterised by the measurement and monitoring of physical processes using data collected through sensors; controlling of the physical processes through actuators; creation of a virtual copy of the system environment, also known as the digital twin; and processing, evaluation and storage of the acquired data that interacts between the system environment and the digital twin, interconnected through digital networks and providing various human–machine interfaces according to the user's needs (Y. Lu, 2017; Stock et al., 2018). The two main functional components of CPS are (i) the real-time data acquisition from the physical paradigm and information feedback from the cyber paradigm through advanced connectivity and (ii) intelligent data management, analytics and computational capability that make up the cyber paradigm (Lee et al., 2015; Monostori et al., 2016).

CPS is seen as the fundamental technology of 4IR (Kagermann et al., 2013; Monostori et al., 2016). Some of the prominent examples of CPS include AV, robot, intelligent buildings and smart factories (Monostori et al., 2016). In order for CPS to be effective, horizontal integration through value networks, vertical integration and networked manufacturing systems and end-to-end digital integration across entire value chains must be implemented (Kagermann et al., 2013; Monostori et al., 2016).

2.3   Internet-of-Things

IoT is commonly associated with CPS, consisting of a network of interconnected CPS or devices through real-time networks (Ghobakhloo, 2018; Kamble et al., 2018; Y. Lu, 2017; Stock et al., 2018). While IoT typically perform the same functions as CPS, collecting real-time data of the physical environment and continuously interacting and exchanging information among devices, one key difference between IoT and CPS is in their origins. In particular, CPS emerged from the systems engineering and control perspective, focusing on 'a system of collaborating computational elements controlling physical entities' (Greer et al., 2019). The systems are networked using software components and use shared knowledge and information from processes to independently control logistics and production systems (Greer et al., 2019). On the other hand, 'IoT' is an umbrella term encompassing the different aspects of the 'extension of the Internet and the Web into the physical realm, by means of the widespread deployment of spatially distributed devices with embedded identification, sensing and/or actuation capabilities' (Greer et al., 2019). Greer et al. (2019) further analysed the factors that differentiate CPS and IoT and found that current CPS and IoT concepts are increasingly converging. Hence, CPS and IoT are used to refer to 'systems-of-systems that consist of engineered, physical systems integrated with networking, data, and computational systems linked via transducers and interact with humans' (Greer et al., 2019). Similarly, this study takes the view that the concepts of CPS and IoT have converged and are used to refer to systems-of-systems that collect information on the physical environment and send them to the digital twin for processing and analysis and for controlling of physical processes through actuators.

2.4 Big Data

BD is associated with the 3Vs – volume, velocity and variety – where the size and complexity of the datasets have exceeded the capability of conventional database software tools to capture, store, manage and analyse (Gill et al., 2009). Volume refers to the magnitude of the data; velocity refers to the rate of data generation and data delivery; and variety refers to the structural heterogeneity in a dataset (Beyer & Laney, 2012; Chen et al., 2012; Gandomi & Haider, 2015).

BD encompasses technologies that can extract value through the analysis of the large volume, variety and velocity of data to provide decision-making support and drive process automation (Bilal et al., 2016; De Mauro et al., 2016; Gandomi & Haider, 2015; Ghobakhloo, 2018; Schwab, 2016). BD can be categorised into BDE and BDA. BDE is concerned with the supporting functions of BD, such as data storage and processing for analytics, while BDA is concerned with pattern identification and extraction of knowledge through the application of analytical techniques on large datasets to generate value (Bilal et al., 2016; Matsunaga et al., 2014; Provost & Fawcett, 2013). BDA can be differentiated from traditional data analysis based on the volume and types of data that are analysed. Traditional data analysis can be applied on structured data, while BDA can be applied on structured, semi-structured and unstructured data (IBM, 2013). Structured data refers to data that is clearly defined, organised in a standardised way and stored in relational databases, while unstructured data refers to data that contains information but does not have an explicit structure, such as images, emails, websites and reports (Baars & Kemper, 2008; Losee, 2006).

The BD lifecycle includes data ingestion, data storage, data processing, data analysis and data visualisation (Ellingwood, 2016). Data ingestion is the feeding of data into the data-processing system. Data may be processed in batches, in streams or in a hybrid form for analysis. Batch processing handles a large dataset collected over a period of time and results in a batch result, while stream processing handles real-time data resulting in real-time data outputs (Vaseekaran, 2017). A hybrid data processing framework uses a combination of batch and stream data processing.

2.5 Artificial Intelligence

BD plays a pivotal role in empowering AI, where machines replicate human intelligence processes, including learning and decision-making. Predictive and prescriptive analytics on vast amounts of data facilitate process optimisation and enable machines to adapt autonomously (Lee et al., 2015). The ability to adapt and learn from experience is a core aspect of human intelligence (Skilton & Hovsepian, 2018b). Modern AI systems are able to autonomously make decisions based on the stimuli in order to achieve their pre-determined objectives through machine learning (Skilton & Hovsepian, 2018a).

Machine learning can be categorised into supervised learning, unsupervised learning, reinforcement learning and deep learning (Skilton & Hovsepian, 2018b). Supervised learning aims to predict an outcome based on a given set of inputs (Hastie et al., 2001). The main techniques under supervised learning include classification and regression. Classification aims to categorise data into collections with similar characteristics to emulate decisions automatically, while regression aims to extrapolate trends present in existing datasets to predict future outcomes (Bilal et al., 2016; Skilton & Hovsepian, 2018b). Unsupervised learning describes association and patterns among a set of input measures and classifies them into clusters. Common techniques associated with unsupervised learning include clustering, natural language processing, dimensionality reduction and information retrieval (Skilton & Hovsepian, 2018b). Reinforcement learning can maximise performance by taking appropriate action according to the inputs and feedback of its previous action (Skilton & Hovsepian, 2018b). Common reinforcement learning technologies include Markov decision processes, Q-learning and SARSA (Huang, 2018). Finally, deep learning uses multiple layers to progressively recognise features from the raw data that would be used by the next layer in the network (Skilton & Hovsepian, 2018b). Deep learning is typically used on complex datasets and for image and speech recognition (Allam & Dhunny, 2019; Luan et al., 2020).

AI requires large volumes of high-quality data in order to achieve accuracy and improve decision-making (Luan et al., 2020). On the other hand, BD utilises AI methods to analyse the large volume, variety and velocity of data (Bilal et al., 2016). Hence, we take the view that AI is a subset of BD.

2.6 Robotic Systems

Robotic systems are machines that are designed and programmed to execute one or more tasks with little or no human intervention. Robotic systems are examples of CPS and IoT as they collect information about the surrounding physical environment through heterogeneous sensors; process and analyse the collected real-time information through advanced computing devices; automatically make planning and control decisions and continuously actuate the corresponding mechanical components (Zhu et al., 2018). There have been several variations of robots, such as humanoid robots, industrial robots, telepresence robots, lifting robots, cobots and robotic systems. Although there is a great variation in robotic systems, robotic systems can be characterised by their manoeuvrability, location awareness via sensors and artificial intelligence (Hudson et al., 2019).

The operations of robotic systems can be classified into three main components, namely perception, planning and control (Kato et al., 2015; Pendleton et al., 2017). Perception is concerned with the development of a contextual understanding of the environment, including the spatial information of obstacles, detection of markings, categorisation of data based on their semantic meaning and the localisation of its position with respect to the environment (Kato et al., 2015; Pendleton et al., 2017). Planning is the process of making meaningful decisions to achieve the goals of the robot, which involves moving from a start point to a goal location or to complete a pre-determined set of tasks while avoiding obstacles and optimising its route and processes (Pendleton et al., 2017). Control is concerned with the execution of the planned actions and encompasses the processes of path tracking and trajectory tracking (Kato et al., 2015; Pendleton et al., 2017). Control requires a feedback system in which the measured system response is compared with the desired behaviour to trigger actions to reduce the discrepancies between the system response and the desired behaviour in order to achieve the goals of the planned actions (Pendleton et al., 2017). Apart from perception, planning and control, robots should be able to cooperate and coordinate with one another.

2.7   Autonomous Vehicles

Similar to robotic systems, AVs are machines programmed to perform tasks with little or no human intervention. Although AVs may not typically be taken as robots, AVs possess the three characteristics of robotic systems and operate using perception, planning and control (Rosique et al., 2019). Hence, we take the view that AVs are a subset of robotic systems and both robotic systems and AVs will be referred to as machines designed and programmed to execute tasks with little or no human intervention.

2.8   Augmented Reality

AR is defined as the display of a real-world scene and virtual information into the user's view (Chi et al., 2013; Ghobakhloo, 2018; Golparvar-Fard et al., 2009). According to Chi et al. (2013), there are four phases required to develop an AR application. The first phase involves the setting up of the access to data for end users over the cloud; the second phase describes the setting up of the computing processes, which mainly involves the localisation technologies and algorithms that are required for identifying the positions of the subjects in the applications and to provide appropriate reactions back to the users; the third phase involves the tangible aspect of AR, specifically the portable and mobile devices; the final phase is concerned with the presentation of information to end users, specifically the provision of a natural user interface, which is intuitive for users to operate and control the AR application (Chi et al., 2013; X. Li et al., 2018).

AR requires inputs in the form of visual, audio, motorial, location and/or orientation to provide information on objects, motion, speech, text and positions. Through the computing algorithms, the outputs of AR may come in the form of visual, audio, touch, smell and taste to provide text, sounds, motion or haptic feedback, flavour and fragrances to enhance user experience and understanding (X. Li et al., 2018).

2.9   Virtual Reality

VR refers to the display of an environment generated from a model to allow for user experience (Flavián et al., 2019). While both VR and AR display digital information to users, one key difference between AR and VR is the context in which the user receives the digital information (Flavián et al., 2019). Specifically, AR displays digital information for user experience onto a real-world scene, providing digital information related to the real-time environment the user is in, whereas VR displays digital information from a 3D model in which the user may or may not be in real time. In other words, VR simulates a 3D model based on the immersion of a user in the virtual environment through the interaction and feedback from the system (Arnaldi et al., 2017; Neo et al., 2021). VR is typically used to simulate designs, provide a safe space for learning to occur and improve comprehension through the interactive feedback provided (Arnaldi et al., 2017). As AR and VR have commonly been discussed as a group of technologies that enable user interaction supported within a virtual environment in the construction industry, we view AR and VR similarly (X. Li et al., 2018; Manuel et al., 2020; Noghabaei et al., 2020).

2.10   Additive Manufacturing

AM is the process of building up successive layers of materials based on a CAD model to make objects (Craveiro et al., 2019; Ghobakhloo, 2018; Kamble et al., 2018), as opposed to traditional manufacturing processes or subtractive manufacturing, where objects are constructed by successively cutting material from a solid block to achieve the desired form. There are several types of AM processes that utilise various tools and materials to build the object. The materials used in AM must be compatible with the AM process selected, while displaying acceptable service properties in order to perform successfully in the application (Bourell et al., 2017). For example, vat photo-polymerisation and material jetting require the use of photosensitive thermosets; any feedstock that can be formed into powder or sheet can be used for binder jetting and sheet lamination; material extrusion uses amorphous polymers with a large viscous softening temperature range; powder bed fusion typically uses materials with a large difference in melt temperature on heating and crystallisation temperature on cooling; and directed energy deposition uses metallic powder or wire feedstock (Bourell et al., 2017).

2.11   3D Imaging

3D imaging refers to the collection of 3D geometric as-built information using technologies such as laser scanners or photogrammetry. 3D imaging consists of the capturing of the 3D geometry of surfaces and the generation of point clouds with 3D coordinates processed through algorithms to form the 3D model (Álvares et al., 2018). Photogrammetry is concerned with processing images to reconstruct the objects in 3D, while laser scanning uses laser to generate the point clouds with the 3D coordinates of each point from scanning of the scene (Álvares et al., 2018; Baltsavias, 1999). From the point clouds collected, data processing consists of three phases: (i) scan editing and verification to identify and remove erroneous data, (ii) registering the point cloud to an appropriate coordinate system and (iii) to combining the individual point clouds into one point cloud for the entire area of interest by matching the overlapping portions of each point cloud and removal of duplicated data in overlapping areas (Whitworth et al., 2011). The resulting 3D model is then used for the various applications.

2.12   Distributed Ledger Technology

DLT, more commonly known as blockchain, is an open, distributed, peer-validated, transparent, write-only and time-stamped ledger, using community validation to synchronise the ledgers across multiple users (Aste et al., 2017; Grover et al.,

2019). It is a data structure and consists of an ordered list of blocks that contain transactions, logged and added in chronological order that is permanent (Hald & Kinra, 2019; Q. Lu & Xu, 2017). DLT supports public and private blockchain models (Dai & Vasarhelyi, 2017; Grover et al., 2019; Hald & Kinra, 2019). Public blockchain, or permission-less blockchain, is completely decentralised and can be accessed by any Internet user, while private blockchain, or permissioned blockchain, is only accessible by authorised users with pre-determined access rights (Hald & Kinra, 2019).

DLT consists of nodes which form the peer-to-peer network that stores the data in a decentralised manner across the network, a cryptography system that encrypts and decrypts the transaction and a consensus mechanism that validates the transaction before it is added into the chain and stored across all the nodes (Hald & Kinra, 2019; J. Li et al., 2019; Ølnes et al., 2017). Each block includes a hash value, which contains the contents of the previous block, which is inherited from the first block in the chain (Hald & Kinra, 2019). This ensures that new information cannot be forged retrospectively by changing a prior entry in the blockchain as these would cause changes in the subsequent blocks, breaking the chain (Aste et al., 2017; Grover et al., 2019; Hald & Kinra, 2019). These features of DLT provide immutability, transparency, integrity, auditability, reliability and security of the data (J. Li et al., 2019; Ølnes et al., 2017).

2.13   Enabling Technologies of Smart Technologies

The enabling technologies of smart technologies include sensors, virtual models, mobile devices, communication networks, data management, cloud computing, user interface and actuators, as shown in Table 2.1.

Sensors provide real-time information of the target components, including the location of the tagged component and the environmental conditions of the sensed areas and enable CPS and IoT, AR, AV and robotics, and 3D imaging. Sensors

Table 2.1 Enabling Technologies of Smart Technologies

	Sensors	Virtual Models	Mobile Devices	Data Management	Cloud Computing	Communication Networks	User Interface	Actuators
CPS	✓	✓	✓	✓	✓	✓	✓	✓
IoT	✓	✓	✓	✓	✓	✓	✓	✓
BD				✓	✓		✓	
AI				✓	✓		✓	
AR	✓	✓	✓	✓	✓	✓	✓	
VR	✓	✓		✓	✓		✓	
AV	✓	✓	✓	✓	✓	✓	✓	✓
Robotics	✓	✓	✓	✓	✓	✓		✓
AM		✓						
3D imaging	✓			✓				
DLT				✓	✓	✓		

can be categorised into position sensors, environmental sensors and image sensors. Position sensors, such as RFID, UWB and GPS, provide location information of the tagged components or data collection devices. The location information enables the trackability and traceability of the tagged components and provides lifecycle information in CPS and IoT (Zhong et al., 2017). Position sensors allow for the identification of the spatial information in AR to determine the locations where the information from the virtual models should be superimposed (Chi et al., 2013). In AV and robotics, location information enables the navigation of the pre-determined route (Pendleton et al., 2017). Location information for 3D imaging helps to determine the coordinates of the point cloud generated (Álvares et al., 2018). Environmental sensors such as temperature sensors, pressure sensors and wind sensors capture the conditions of the physical paradigm in order to (i) optimise and monitor work processes in CPS and IoT and (ii) navigate safely for AVs and robots. Image sensors include cameras and video cameras, which provide rich information on the physical environment, and can be used in CPS and IoT, AR, AV and robotics, and 3D imaging (Jia et al., 2019).

Virtual model is the platform in which the replica of the physical paradigm is developed digitally for the visualisation and storage of the information collected through sensors and can serve as a benchmark for the comparison between the planned and actual performance of work processes in CPS and IoT (Akanmu & Anumba, 2015). Virtual model also provides the information which is to be superimposed according to a user's position in AR and the environment to be displayed for the user's experience in VR (Chi et al., 2013; X. Li et al., 2018). Virtual models enable AV and robotics by providing the network information for the planning of the route to be taken (Pendleton et al., 2017). They are also essential in AM as they provide the information on the final form of the object that is to be manufactured (Craveiro et al., 2019).

Mobile devices serve as touch points for the collection of data and provision of access to information from the cloud. Mobile devices in CPS and IoT serve as agents to acquire real-time information on the physical conditions and performance data, which will be sent to the cyber paradigm to be analysed and optimised (Oesterreich & Teuteberg, 2016). In CPS, IoT, AR and VR, mobile devices provide access to information from the virtual model for users and to support collaboration and communication among stakeholders (Chi et al., 2013; Oesterreich & Teuteberg, 2016). AV and robots play the role as devices that execute the pre-determined program.

Data is the cornerstone of the 4IR and hence data management plays a critical role in enabling the smart technologies (Marcus, 2015). Data management includes data acquisition, data storage, data processing and data analysis. CPS and IoT generate large volumes of data, which require BD to manage the huge dataset to extract meaningful information (Oesterreich & Teuteberg, 2016). It is critical to ensure the quality of the input data to efficiently utilise the data and to provide accurate information for decision making (Cai & Zhu, 2015). CPS and IoT rely on the accuracy and quality of the analysis of the data collected for optimisation and lifecycle management (Cai & Zhu, 2015; Oesterreich & Teuteberg,

2016). AR and VR require data management to determine what information should be displayed according to the user's location (Chi et al., 2013); AV and robotics require the real-time processing of data collected in order to complete the pre-assigned tasks safely (Kato et al., 2015; Pendleton et al., 2017); and data collected from 3D imaging must be processed in order to generate an accurate model (Whitworth et al., 2011).

Due to the need to acquire, store, process and analyse a vast amount of data and to provide access to information in real time from any location in CPS, IoT, BD, AI, AR, VR, AV and robotics, cloud computing is a critical enabler of these smart technologies (Stock et al., 2018). Cloud computing provides computing capacity, storage, database, processing and other services on demand over the Internet. As such, communication networks are also critical in enabling the control of physical artefacts remotely and the exchange and access of information across devices (Kamble et al., 2018; Oesterreich & Teuteberg, 2016).

User interface is concerned with data visualisation and user touch points. In CPS, IoT, BD and AI, the analysis of data must be visualised in a manner that users can understand in order to support decision making (Lee et al., 2015; Masiane et al., 2019). This translates into the requirement to tailor the display of the analysed data according to each user's needs. In AR and VR, user interfaces must be intuitive for users to control the application so that risks of operating AR and VR will be reduced (Chi et al., 2013).

Finally, actuators play a critical role in executing actions in the physical artefacts for CPS, IoT, AV and robotics according to real-time data collected and processed (Y. Lu, 2017; Stock et al., 2018).

2.14 Summary

The emergence of the 4IR has led to the disruption of business models, organisational processes, systems, and culture through the introduction of smart technologies. This chapter investigated 11 key smart technologies, namely CPS, IoT, BD, AI, AR, VR, AV, robotics, AM, 3D imaging and DLT as well as their enabling technologies. The enabling technologies driving the advancement of smart technologies encompass sensor technologies, virtual modelling, improvements in mobile devices, data management technologies, cloud computing, communication networking technologies, enhancements in user interface and actuators. The comprehensive literature review in this chapter paves the way for studying the applications of these technologies in the construction industry and the updated technological competencies of PMs managing projects enhanced by smart technologies.

References

Akanmu, A., & Anumba, C. J. (2015). Cyber-physical systems integration of building information models and the physical construction. *Engineering, Construction and Architectural Management, 22*(5), 516–535. https://doi.org/10.1108/ECAM-07-2014-0097

Allam, Z., & Dhunny, Z. A. (2019). On big data, artificial intelligence and smart cities. *Cities, 89*, 80–91. https://doi.org/10.1016/j.cities.2019.01.032

Álvares, J. S., Costa, D. B., & Melo, R. R. S. de. (2018). Exploratory study of using unmanned aerial system imagery for construction site 3D mapping. *Construction Innovation*, *18*(3), 301–320. https://doi.org/10.1108/CI-05-2017-0049

Arnaldi, B., Guitton, P., & Moreau, G. (2017). Virtual reality and augmented reality: Myths and realities. In *Virtual reality and augmented reality: Myths and realities*. https://doi.org/10.1002/9781119341031

Aste, T., Tasca, P., & Di Matteo, T. (2017). Blockchain technologies: The foreseeable impact on society and industry. *Computer*, *50*(9), 18–28. https://doi.org/10.1109/MC.2017.3571064

Baars, H., & Kemper, H.-G. (2008). Management support with structured and unstructured data – an integrated business intelligence framework. *Information Systems Management*, *25*(2), 132–148. https://doi.org/10.1080/10580530801941058

Baltsavias, E. P. (1999). A comparison between photogrammetry and laser scanning. *ISPRS Journal of Photogrammetry and Remote Sensing*, *54*(2), 83–94. https://doi.org/10.1016/S0924-2716(99)00014-3

Beyer, M. A., & Laney, D. (2012). *The importance of 'big data': A definition*. Gartner. www.gartner.com/en/documents/2057415/the-importance-of-big-data-a-definition

Bilal, M., Oyedele, L. O., Qadir, J., Munir, K., Ajayi, S. O., Akinade, O. O., Owolabi, H. A., Alaka, H. A., & Pasha, M. (2016). Big Data in the construction industry: A review of present status, opportunities, and future trends. *Advanced Engineering Informatics*, *30*, 500–521. https://doi.org/10.1016/j.aei.2016.07.001

Bourell, D., Kruth, J. P., Leu, M., Levy, G., Rosen, D., Beese, A. M., & Clare, A. (2017). Materials for additive manufacturing. *CIRP Annals*, *66*(2), 659–681. https://doi.org/10.1016/j.cirp.2017.05.009

Cai, L., & Zhu, Y. (2015). The challenges of data quality and data quality assessment in the big data era. *Data Science Journal*, *14*(2), 1–10. https://doi.org/10.5334/dsj-2015-002

Chen, H., Chiang, R. H. L., & Storey, V. C. (2012). Business intelligence and analytics: From big data to big impact. *MIS Quarterly*, *36*(4), 1165. https://doi.org/10.2307/41703503

Chi, H.-L., Kang, S.-C., & Wang, X. (2013). Research trends and opportunities of augmented reality applications in architecture, engineering, and construction. *Automation in Construction*, *33*, 116–122. https://doi.org/10.1016/j.autcon.2012.12.017

Craveiro, F., Duarte, J. P., Bartolo, H., & Bartolo, P. J. (2019). Additive manufacturing as an enabling technology for digital construction: A perspective on Construction 4.0. *Automation in Construction*, *103*, 251–267. https://doi.org/10.1016/j.autcon.2019.03.011

Dai, J., & Vasarhelyi, M. A. (2017). Toward blockchain-based accounting and assurance. *Journal of Information Systems*, *31*(3), 5–21. https://doi.org/10.2308/isys-51804

Dallasega, P., Rauch, E., & Linder, C. (2018). Industry 4.0 as an enabler of proximity for construction supply chains: A systematic literature review. *Computers in Industry*, *99*, 205–225. https://doi.org/10.1016/j.compind.2018.03.039

De Mauro, A., Greco, M., & Grimaldi, M. (2016). A formal definition of big data based on its essential features. *Library Review*, *65*, 122–135. https://doi.org/10.1108/LR-06-2015-0061

Devezas, T., Leitão, J., & Sarygulov, A. (2017). Introduction. In *Industry 4.0. studies on entrepreneurship, structural change and industrial dynamics* (pp. 1–10). Springer. https://doi.org/10.1007/978-3-319-49604-7_1

Ellingwood, J. (2016). An introduction to big data concepts and terminology. *DigitalOcean*. www.digitalocean.com/community/tutorials/an-introduction-to-big-data-concepts-and-terminology

Flavián, C., Ibáñez-Sánchez, S., & Orús, C. (2019). The impact of virtual, augmented and mixed reality technologies on the customer experience. *Journal of Business Research*, *100*, 547–560. https://doi.org/10.1016/J.JBUSRES.2018.10.050

Gandomi, A., & Haider, M. (2015). Beyond the hype: Big data concepts, methods, and analytics. *International Journal of Information Management*, *35*(2), 137–144. https://doi.org/10.1016/j.ijinfomgt.2014.10.007

Ghobakhloo, M. (2018). The future of manufacturing industry: A strategic roadmap toward Industry 4.0. *Journal of Manufacturing Technology Management*, *29*(6), 910–936. https://doi.org/10.1108/JMTM-02-2018-0057

Gill, S. K., Nguyen, P., & Koren, G. (2009). Adherence and tolerability of iron-containing prenatal multivitamins in pregnant women with pre-existing gastrointestinal conditions. *Journal of Obstetrics and Gynaecology, 29*(7), 594–598. https://doi.org/10.1080/01443610903114527

Golparvar-Fard, M., Peña-Mora, F., & Savarese, S. (2009). D4AR – A 4-Dimensional augmented reality model for automating construction progress monitoring data collection, processing and communication. *Journal of Information Technology in Construction, 14*(13), 129–153.

Greer, C., Burns, M. J., Wollman, D., & Griffor, E. (2019). *Cyber-physical systems and Internet of Things.* https://doi.org/10.6028/NIST.SP.1900-202

Grover, P., Kar, A. K., & Janssen, M. (2019). Diffusion of blockchain technology: Insights from academic literature and social media analytics. *Journal of Enterprise Information Management.* https://doi.org/10.1108/JEIM-06-2018-0132

Hald, K. S., & Kinra, A. (2019). How the blockchain enables and constrains supply chain performance. *International Journal of Physical Distribution and Logistics Management, 49*(4), 376–397. https://doi.org/10.1108/IJPDLM-02-2019-0063

Hastie, T., Friedman, J., & Tibshirani, R. (2001). *The elements of statistical learning.* Springer. https://doi.org/10.1007/978-0-387-21606-5

Huang, K.-H. (2018). *Introduction to various reinforcement learning algorithms. Part I (Q-Learning, SARSA, DQN, DDPG).* https://towardsdatascience.com/introduction-to-various-reinforcement-learning-algorithms-i-q-learning-sarsa-dqn-ddpg-72a5e0cb6287

Hudson, J., Orviska, M., & Hunady, J. (2019). People's attitudes to autonomous vehicles. *Transportation Research Part A: Policy and Practice, 121*, 164–176. https://doi.org/10.1016/j.tra.2018.08.018

IBM. (2013). *What is big data? – Bringing big data to the enterprise.* http://www-01.ibm.com/software/data/bigdata/

Jabbour, A. B. L. de S., Jabbour, C. J. C., Foropon, C., & Filho, M. G. (2018). When titans meet – Can industry 4.0 revolutionise the environmentally-sustainable manufacturing wave? The role of critical success factors. *Technological Forecasting and Social Change, 132*, 18–25. https://doi.org/10.1016/j.techfore.2018.01.017

Jia, M., Komeily, A., Wang, Y., & Srinivasan, R. S. (2019). Adopting Internet of Things for the development of smart buildings: A review of enabling technologies and applications. *Automation in Construction, 101*, 111–126. https://doi.org/10.1016/j.autcon.2019.01.023

Kagermann, H., Wahlster, W., & Helbig, J. (2013). Recommendations for implementing the strategic initiative Industrie 4.0: Securing the future of German manufacturing industry. Final Report of the Industrie 4.0 Working Group. Forschungsunion (p. 82).

Kamble, S. S., Gunasekaran, A., & Gawankar, S. A. (2018). Sustainable Industry 4.0 framework: A systematic literature review identifying the current trends and future perspectives. *Process Safety and Environmental Protection, 117*, 408–425. https://doi.org/10.1016/j.psep.2018.05.009

Kato, S., Takeuchi, E., Ishiguro, Y., Ninomiya, Y., Takeda, K., & Hamada, T. (2015). An open approach to autonomous vehicles. *IEEE Micro, 35*(6), 60–68. https://doi.org/10.1109/MM.2015.133

Lee, J., Bagheri, B., & Kao, H.-A. (2015). A cyber-physical systems architecture for Industry 4.0-based manufacturing systems. *Manufacturing Letters, 3*, 18–23. https://doi.org/j.mfglet.2014.12.001

Li, J., Greenwood, D., & Kassem, M. (2019). Blockchain in the built environment and construction industry: A systematic review, conceptual models and practical use cases. *Automation in Construction, 102*, 288–307. https://doi.org/10.1016/j.autcon.2019.02.005

Li, X., Yi, W., Chi, H.-L., Wang, X., & Chan, A. P. C. (2018). A critical review of virtual and augmented reality (VR/AR) applications in construction safety. *Automation in Construction, 86*, 150–162. https://doi.org/10.1016/j.autcon.2017.11.003

Liu, Y., & Xu, X. (2017). Industry 4.0 and cloud manufacturing: A comparative analysis. *Journal of Manufacturing Science and Engineering, 139*(3), 034701. https://doi.org/10.1115/1.4034667

Losee, R. M. (2006). Browsing mixed structured and unstructured data. *Information Processing & Management, 42*(2), 440–452. https://doi.org/10.1016/j.ipm.2005.02.001

Lu, Q., & Xu, X. (2017). Adaptable blockchain-based systems: A case study for product traceability. *IEEE Software, 34*(6), 21–27. https://doi.org/10.1109/MS.2017.4121227

Lu, Y. (2017). Industry 4.0: A survey on technologies, applications and open research issues. *Journal of Industrial Information Integration, 6,* 1–10. https://doi.org/10.1016/J.JII.2017.04.005

Luan, H., Geczy, P., Lai, H., Gobert, J., Yang, S. J. H., Ogata, H., Baltes, J., Guerra, R., Li, P., & Tsai, C. C. (2020). Challenges and future directions of big data and artificial intelligence in education. *Frontiers in Psychology, 11,* 2748. https://doi.org/10.3389/FPSYG.2020.580820/BIBTEX

Manuel, J., Delgado, D., Asce, A. M., Oyedele, L., Beach, T., & Demian, P. (2020). Augmented and virtual reality in construction: Drivers and limitations for industry adoption. *Journal of Construction Engineering and Management, 146*(7), 04020079. https://doi.org/10.1061/(ASCE)CO.1943-7862.0001844

Marcus, A. (2015). Jobs and the fourth industrial revolution. *World Economic Forum.* www.weforum.org/agenda/2015/12/data-and-the-fourth-industrial-revolution/

Masiane, M. M., Driscoll, A., Feng, W., Wenskovitch, J., & North, C. (2019). Towards insight-driven sampling for big data visualisation. *Behaviour & Information Technology,* 1–20. https://doi.org/10.1080/0144929X.2019.1616223

Matsunaga, F. T., Brancher, J. D., & Busto, R. M. (2014). Data mining applications and techniques: A systematic review. *Revista Eletrônica Argentina-Brasil de Tecnologias Da Informação e Da Comunicação, 1*(2). https://doi.org/10.5281/zenodo.59454

Monostori, L., Kádár, B., Bauernhansl, T., Kondoh, S., Kumara, S., Reinhart, G., Sauer, O., Schuh, G., Sihn, W., & Ueda, K. (2016). Cyber-physical systems in manufacturing. *CIRP Annals, 65*(2), 621–641. https://doi.org/10.1016/j.cirp.2016.06.005

Müller, J. M., Kiel, D., & Voigt, K. I. (2018). What drives the implementation of Industry 4.0? The role of opportunities and challenges in the context of sustainability. *Sustainability, 10*(1), 247–270. https://doi.org/10.3390/su10010247

Neo, J. R. J., Won, A. S., & Shepley, M. M. (2021). Designing immersive virtual environments for human behavior research. *Frontiers in Virtual Reality, 5.* https://doi.org/10.3389/FRVIR.2021.603750

Noghabaei, M., Heydarian, A., Balali, V., & Han, K. (2020). Trend analysis on adoption of virtual and augmented reality in the architecture, engineering, and construction industry. *Data, 5*(1). https://doi.org/10.3390/DATA5010026

Oesterreich, T. D., & Teuteberg, F. (2016). Understanding the implications of digitisation and automation in the context of Industry 4.0: A triangulation approach and elements of a research agenda for the construction industry. *Computers in Industry, 83,* 121–139. https://doi.org/10.1016/j.compind.2016.09.006

Ølnes, S., Ubacht, J., & Janssen, M. (2017). Blockchain in government: Benefits and implications of distributed ledger technology for information sharing. *Government Information Quarterly, 34*(3), 355–364. https://doi.org/10.1016/j.giq.2017.09.007

Pendleton, S. D., Andersen, H., Du, X., Shen, X., Meghjani, M., Eng, H., Rus, D., & Ang, M. H. (2017). *Perception, planning, control, and coordination for autonomous vehicles.* https://doi.org/10.3390/machines5010006

Pereira, A. C., & Romero, F. (2017). A review of the meanings and the implications of the Industry 4.0 concept. *Procedia Manufacturing, 13,* 1206–1214. https://doi.org/10.1016/j.promfg.2017.09.032

Posada, J., Toro, C., Barandiaran, I., Oyarzun, D., Stricker, D., de Amicis, R., Pinto, E. B., Eisert, P., Dollner, J., & Vallarino, I. (2015). Visual computing as a key enabling technology for Industrie 4.0 and industrial internet. *IEEE Computer Graphics and Applications, 35*(2), 26–40. https://doi.org/10.1109/MCG.2015.45

Provost, F., & Fawcett, T. (2013). *Data science for business: What you need to know about data mining and data-analytic thinking.* O'Reilly.

Rosique, F., Navarro, P. J., Fernández, C., & Padilla, A. (2019). A systematic review of perception system and simulators for autonomous vehicles research. *Sensors, 19*(3), 648. https://doi.org/10.3390/S19030648

Schwab, K. (2016). *The fourth industrial revolution.* PENGUIN GROUP.

Skilton, M., & Hovsepian, F. (2018a). Intelligent agents. In *The 4th industrial revolution* (pp. 99–119). Springer International Publishing. https://doi.org/10.1007/978-3-319-62479-2_4

Skilton, M., & Hovsepian, F. (2018b). Machine learning. In *The 4th industrial revolution* (pp. 121–157). Springer International Publishing. https://doi.org/10.1007/978-3-319-62479-2_5

Stock, T., Obenaus, M., Kunz, S., & Kohl, H. (2018). Industry 4.0 as enabler for a sustainable development: A qualitative assessment of its ecological and social potential. *Process Safety and Environmental Protection, 118*, 254–267. https://doi.org/10.1016/j.psep.2018.06.026

Vaseekaran, G. (2017). *Big data battle: Batch processing vs Stream processing.* Medium. https://medium.com/@gowthamy/big-data-battle-batch-processing-vs-stream-processing-5d94600d8103

Wan, J., Yi, M., Li, D., Zhang, C., Wang, S., & Zhou, K. (2016). Mobile services for customization manufacturing systems: An example of Industry 4.0. *IEEE Access, 4*, 8977–8986. https://doi.org/10.1109/ACCESS.2016.2631152

Whitworth, M., Anderson, I., & Hunter, G. (2011). Geomorphological assessment of complex landslide systems using field reconnaissance and terrestrial laser scanning. In *Geomorphological mapping* (Vol. 15, pp. 459–474). Elsevier. https://doi.org/10.1016/B978-0-444-53446-0.00017-3

Zhong, R. Y., Peng, Y., Xue, F., Fang, J., Zou, W., Luo, H., Thomas Ng, S., Lu, W., Shen, G. Q. P., & Huang, G. Q. (2017). Prefabricated construction enabled by the Internet-of-Things. *Automation in Construction, 76*, 59–70. https://doi.org/10.1016/j.autcon.2017.01.006

Zhu, Q., Sangiovanni-Vincentelli, A., Hu, S., & Li, X. (2018). Design automation for cyber-physical systems. *Proceedings of the IEEE, 106*(9), 1479–1483. https://doi.org/10.1109/JPROC.2018.2865229

3 Applications of Key Smart Technologies in Projects

3.1 Introduction

The construction industry has witnessed unprecedented technological advancements during the past years, and the 4IR has brought forth a paradigm shift with the emergence of smart technologies. As the construction landscape undergoes profound digital transformation, it becomes imperative to understand and harness the full potential of these smart technologies in CPM.

These cutting-edge innovations encompass CPS, IoT, BD, AI, AV, Robotics, AR, VR, AM, 3D imaging and DLT. These technologies build on and amplify one another, resulting in significant improvements in CPM performance. CPS, in conjunction with IoT, enables real-time monitoring and control of project performance throughout its lifecycle. BD plays a pivotal role by integrating with CPS and IoT to automate data processing and analysis of sensor-generated real-time data. This facilitates autonomous control of physical processes and can be further augmented by AI to learn from historical data for predictive and estimative purposes. Additionally, AI aids in generating automated and optimised design and planning strategies, as well as seamless document classification and information retrieval.

The integration of AV and Robotics in construction processes automates fabrication and on-site assembly, significantly benefiting building construction. Furthermore, human–robot collaborations in AV and Robotics influence project planning and execution, impacting project schedule and cost. These technologies also contribute to automatic data collection, particularly through the acquisition of 3D images for computer vision analysis.

AR and VR technologies are employed for simulating and visualising construction environments, supporting virtual safety training, project schedule management, resource planning and project progress monitoring. These visual aids substantially enhance user comprehension of project progress and site conditions (Portman et al., 2015).

Furthermore, AM allows for customisation in CPM through the printing of building components and automated manufacturing of complex elements. DLT enables automated contract monitoring and execution, while providing records of changes made in BIM models, thus enhancing accountability and transparency. The forthcoming sections specify applications of these smart technologies in the context of construction projects.

DOI: 10.1201/9781003462231-3

3.2   Cyber-Physical System and Internet-of-Things

3.2.1   *Real-Time Project Monitoring and Control*

CPS and IoT can be applied in real-time project monitoring and control along the construction supply chain and on site (Edirisinghe, 2019; Oesterreich & Teuteberg, 2016; Zhong et al., 2017). As the data is collected from sensors, cameras and RFID tags, status or progress information can be sent to the digital twin automatically in real time for data processing and analysis for monitoring and control, decision-making and automated execution (Edirisinghe, 2019).

Edirisinghe (2019) reviewed the studies on context-aware digital skin of the construction site and identified the following applications of CPS and IoT: (i) labour tracking, (ii) supply chain management, (iii) mobile equipment operation tracking, (iv) schedule and progress monitoring and (v) safety management. Similar applications were also identified in Tang et al. (2019)'s review on the integration of BIM and IoT devices.

In particular, CPS and IoT can be used to locate and track construction workers within the construction site, which can be further extended to monitor the productivity of workers (Behzadan et al., 2008; T. Cheng et al., 2013; H. Jiang et al., 2015). In the applications of CPS and IoT for supply chain management, material and equipment tracking can be conducted to reduce construction waste (H. Li et al., 2005); for asset inventory tracking and allocation to improve utilisation and efficiency (Goodrum et al., 2006); to reduce the time required to locate specific precast components in the storage yard for just-in-time deliveries to construction sites (Ergen et al., 2007) and to improve visibility and traceability of the prefabricated components (Zhong et al., 2017).

CPS and IoT can also track and monitor construction equipment to automate equipment operation processes (Roberts et al., 1999); track and analyse construction work progress and productivity (Pradhananga & Teizer, 2013; Sacks et al., 2005; Yang et al., 2014) and prevent collisions (Oloufa et al., 2003; Ray & Teizer, 2013; S. Zhang et al., 2015). With the automated tracking of resources' activities and productivity, monitoring and control of project schedule can also be achieved (M.-Y. Cheng & Chen, 2002; Chin et al., 2008). Finally, real-time monitoring and control can be applied to track near-miss incidents (Teizer & Cheng, 2015; W. Wu et al., 2010); to monitor if construction workers are complying with personal protection equipment requirements (Barro-Torres et al., 2012; Kelm et al., 2013); for safety monitoring during the placement of concrete (Moon et al., 2015); monitoring of hazardous areas in the construction site to detect fall accidents (Carbonari et al., 2011; Lee et al., 2009) and monitoring of construction site conditions to provide early warning in underground construction sites (Ding et al., 2013) and confined spaces (Riaz et al., 2014).

3.2.2   *Integrated Data Platform*

A key feature of CPS and IoT is the automated data processing and analysis of real-time data collected through the sensors and the autonomous control of the physical processes. Hence, it is common for BD to be integrated with CPS and IoT for the processing and analysis of the data in real-time. The project data collected from the sensors is stored in the CPS and IoT platform, serving as an integrated project data platform for project stakeholders (Kochovski & Stankovski, 2018; Oesterreich & Teuteberg, 2016). At the same time, project stakeholders can access updated and integrated project information from any location for viewing, managing, sharing and collaboration in real time (Oesterreich & Teuteberg, 2016). The processing and analysis of integrated project data provides a more holistic view of the project, enabling context-aware decision-making capabilities and intelligent autonomy (Jia et al., 2019).

3.3 Big Data and Artificial Intelligence

Bilal et al. (2016) reviewed the use of BD in the construction industry including the common tools and techniques, applications of BD in the construction industry, the research issues, possible future directions and the challenges faced in the implementation of BD. The common applications of BD in CPM can be categorised into the following: (i) cause analysis, prediction and estimation; (ii) automated and optimised design and planning and (iii) document classification and information retrieval.

3.3.1 *Cause Analysis, Prediction and Estimation*

BD and AI can be applied in CPM through the analysis of historical project data to identify and investigate the causes of project delays (Aibinu & Jagboro, 2002; H. Kim et al., 2008; Soibelman & Kim, 2002), project cost overruns (Soibelman & Kim, 2002; Williams & Gong, 2014), project quality issues (Soibelman & Kim, 2002), occupational injuries (C.-W. Cheng et al., 2012; Liao & Perng, 2008) and schedule and workspace conflicts (Y.-J. Chen et al., 2011), which make up project risks. With BD and AI, their patterns and probabilities of occurrences can be analysed and extended to predict the occurrences of these project risks. Prescribed remedial solutions can also be provided (Bilal, Oyedele, Qadir et al., 2016). Apart from the prediction of the project risks, the analysis results can be used to estimate project duration (Siu et al., 2013; Soibelman et al., 2008) and project costs (Elfaki et al., 2014) and for project tender evaluations (Olsson & Bull-Berg, 2015; Y. Zhang et al., 2015). Other applications include the use of BD for construction waste management in terms of construction waste generation estimation, benchmarking and waste management performance (Bilal, Oyedele, Akinade et al., 2016; Ekanayake & Ofori, 2004; Lu et al., 2015), estimation of optimal mark up for tender prices (Moselhi et al., 1991) and prediction of construction firms' failure (Alaka et al., 2018).

3.3.2 *Automated and Optimised Design and Planning*

BD and AI can be used to synthesise historical project data, specified project requirements, targeted users and sustainability and environmental needs to automatically generate designs that are able to fulfil the wider requirements (Bilal, Oyedele, Qadir et al., 2016). Apart from that, BD and AI can be used for clash detection (L. Wang & Leite, 2013), for generation of optimised project schedule and allocation of labour automatically to reduce workspace conflicts and increase productivity (Y.-J. Chen et al., 2011; H. Liu et al., 2015).

3.3.3   Document Classification and Information Retrieval

Finally, BD and AI can be used to classify documents and retrieve information automatically from project documents such as post-project review documents, contract documents, project progress reports, construction drawings, site images and videos (Bilal, Oyedele, Qadir et al., 2016; S. Liu et al., 2008; Soibelman et al., 2008; Ur-Rahman & Harding, 2012). Classification and information retrieval from post-project review documents enable learning from past projects to improve future project planning and execution (Carrillo et al., 2011; Choudhary et al., 2009). The classification of site images and videos enables the identification of workers' and equipment actions towards project safety and to predict site injuries (Gong et al., 2011; S. U. Han et al., 2012), to reduce the time required to document and index site images (Bilal, Oyedele, Qadir et al., 2016) and to improve PMs' understanding of project progress (Sun et al., 2020). Automated information retrieval from project documents also enable the automation of regulatory and contractual compliance system (Salama & El-Gohary, 2016).

3.4   Autonomous Vehicles and Robotics

3.4.1   Automated Construction

Robotics can be applied in construction projects to automate construction work such as bricklaying and surface finishing and can replace human workers for simple and repetitive tasks or dangerous works (Bock, 2015; Castro-Lacouture et al., 2007; Q. Chen et al., 2018; De Soto et al., 2018, 2019; Oesterreich & Teuteberg, 2016; M. Skibniewski & Hendrickson, 1988; Warszawski & Rosenfeld, 1994). Robotics can also be used to automate construction fabrication and assembly both onsite and offsite (Bock, 2015; De Soto et al., 2018, 2019; Willmann et al., 2016). Specifically, robotics may be used together with AM in the case of automated construction fabrication (or additive construction) and assembly onsite. However, due to the current limitations faced in terms of site space, equipment cost and logistics (De Soto et al., 2018), the use of robotics with AM for automated construction fabrication and assembly are mostly conducted offsite. Khoshnevis (2004) explored the application of robotic construction fabrication using contour crafting to construct a house. Willmann et al. (2016) presented the prototype application of an automated timber roof assembly using robotic assembly. De Soto et al. (2018) presented the application of an in-situ fabricator mobile robot designed for additive construction onsite. In automated construction fabrication and assembly offsite, prefabrication processes would be carried out by robots instead of human workers (Kasperzyk et al., 2017; Linner & Bock, 2012).

While these applications affect construction work processes, CPM will also be impacted by these applications due to the need to manage the changes in work processes. In addition, the automation of construction tasks requires significant adjustments to the project schedule and resources (M. J. Skibniewski, 1988). De Soto et al. (2019) emphasised the transformation of the current roles in the construction sectors brought about by construction robots. In particular, robotic construction fabrication is expected to change the construction planning and execution phase significantly due to the introduction of human–robot collaborations (De Soto et al., 2019). From the case study, PMs' tasks in the planning phase remain largely similar, although the coordination among the different project participants will shift towards new roles that will be required to manage the new tasks, such as digital fabrication. In the execution phase, PMs' tasks remain largely similar, except for the shift of their workload with the improved availability and reliability of the information. Efforts to monitor and control project schedule and cost may shift towards the coordination among programmers and collaboration among stakeholders, in the case of onsite construction fabrication.

The impacts to CPM from robotic prefabrication are expected to be similar to those resulting from the adoption of PPVC. This includes the need to coordinate between onsite demand and offsite construction, cooperation among stakeholders, changes to quality management processes, management of human resources and logistical processes including longer lead times and changes to site layout (Hwang et al., 2018; L. Jiang et al., 2018; Luo et al., 2015; Yunus & Yang, 2012). (Khoshnevis, 2004)

3.4.2   *Automated Data Collection*

AV and robotics are more commonly used in CPM for automated data collection. They are typically used in conjunction with 3D imaging tools such as laser scanners or photogrammetry, and as data collection tools for CPS or IoT systems. AV and robotics can be used to create a 3D map of construction sites to assist in project management tasks such as progress monitoring through the comparison of as-built and as-planned building models, surveys of construction sites such as topographic surveys and measurements for construction planning, quality control and inspection (Álvares et al., 2018; Ham et al., 2016; K. K. Han & Golparvar-Fard, 2017; S. Kim & Irizarry, 2016; Oesterreich & Teuteberg, 2016).

3.5   Augmented Reality and Virtual Reality

AR and VR may be considered to be one of the most widely applied smart technologies in CPM due to the increasing use and emphasis on BIM (Gu & London, 2010). Apart from BIM, other applications of AR and VR in CPM can be categorised into simulation and information retrieval and communication. AR and VR have been applied to project safety management, project scheduling and resource management, project communication management and project stakeholder management.

3.5.1   *Simulation*

AR and VR have been applied to simulate the construction environment for (i) construction safety training, (ii) project schedule management and resource planning,

(iii) project progress monitoring and (iv) client walkthrough. AR and VR are used as tools of communication in a visual form to improve user comprehension (Portman et al., 2015).

In construction safety training, workers are trained to recognise construction hazards and to visualise the construction environment and work processes to improve user comprehension of the dynamic conditions faced in a construction work site (Fang et al., 2014; Guo et al., 2017; X. Li et al., 2018; Park & Kim, 2013; Sacks et al., 2015; X. Wang & Dunston, 2007). Furthermore, the workers can be trained to operate the equipment in a virtual construction environment to replicate the working environment to improve comprehension of the training (Fang et al., 2014; H. Li et al., 2012).

In project schedule management and resource planning, the project schedule and planned resources are added to 3D models to identify potential workspace conflicts and to determine the optimal schedule and site layout (Ahmed, 2019; Golparvar-Fard et al., 2009; Kamat et al., 2011; Park & Kim, 2013). When used in combination with data collection tools such as laser scanners or construction site images, identification of the discrepancies between the as-built and as-planned models are enabled, allowing PMs to monitor construction progress (Ahmed, 2019; Golparvar-Fard et al., 2009; Ham et al., 2016; B. Kim et al., 2012).

Simulation of the building model also enables walkthrough of the designed building, to support constructability analysis, client involvement, collaboration among project stakeholders and decision making (Boton, 2018; Mobach, 2008; Oesterreich & Teuteberg, 2016; Portman et al., 2015; Y. Wang et al., 2014).

3.5.2   *Information Retrieval and Communication*

AR can be combined with BIM and applied to visualise BIM data onto the physical context of the construction activity and enables real-time communication onsite (Chi et al., 2013; X. Li et al., 2018; Oesterreich & Teuteberg, 2016; Y. Wang et al., 2014). This allows users to retrieve building models and building information onsite. For example, a Smart Helmet has been proposed by Daqri, which enables 3D visual overlays of model information onto the worker's field of vision, provides access to task procedures for the assigned task, and can connect a worker to a colleague to ask for advice and task guidance (Greenhalgh et al., 2016). Other devices such as Smart Glasses have also been proposed for the same functions (Oesterreich & Teuteberg, 2016). AR and VR can also be used in conjunction with mobile computing to support communication among project stakeholders (Oesterreich & Teuteberg, 2016; Y. Wang et al., 2014).

3.6   Additive Manufacturing

AM can be applied in construction projects in terms of the printing of building components and automated manufacturing of complex components, enabling mass customisation (Kothman & Faber, 2016; Oesterreich & Teuteberg, 2016; Tay et al., 2017; P. Wu et al., 2016). Similar to the applications to automate construction processes, the applications impact CPM in terms of the need to manage the changes in construction work processes, which have been introduced in Section 3.3.1. In addition, mass customisation increases the complexity of projects and requires PMs to have the ability to identify the client's needs and to translate them into solutions or designs to fulfil their needs (Brunoe & Nielsen, 2016; Nahmens & Bindroo, 2011).

3.7   3D Imaging

As highlighted in Section 3.4.2, 3D imaging tools such as laser scanners and photogrammetry can be applied in CPM to automate progress tracking, survey of construction sites, quality control and inspections. 3D imaging can capture the as-built conditions of projects, and may be converted into a 3D model using algorithms to be compared with the as-planned building model to determine the project progress automatically (Bosché et al., 2015; El-Omari & Moselhi, 2008; Golparvar-Fard et al., 2009; Omar & Nehdi, 2016; Turkan et al., 2012). Automated progress tracking can also be combined with the project schedule and cost data to develop an automated earned value tracking system (Turkan et al., 2013).

3.8   Distributed Ledger Technology

Li et al. (2019) reviewed the applications of DLT in the built-environment and construction industry. In CPM, DLT can be applied through smart contracts and as a distributed ledger that maintains all transaction records. Furthermore, the Institution of Civil Engineers has published a report that examines the potential of DLT in the construction industry in December 2018, demonstrating the emerging importance of DLT in the industry (Institution of Civil Engineers, 2018).

3.8.1   *Smart Contracts*

According to Li et al. (2019), smart contracts can be considered one of the most important aspects of DLT for the construction industry. Smart contracts are self-executing codes that will automatically execute the terms of a contract when the conditions set are met (J. Li et al., 2019; McNamara & Sepasgozar, 2018). As such, contract execution can be automated. In terms of CPM, to track if the conditions are being met, the integration of DLT with BIM and CPS or IoT may be necessary (J. Li et al., 2019). Automated monitoring of project progress may be achieved using CPS or IoT tracking devices, which automatically collect data from the construction site (Heiskanen, 2017). With the automation of contract monitoring and execution, integrated BIM processes can be achieved and payment processes within a project will be automated and improved in terms of the fulfilment of the requirements and on time payment (J. Li et al., 2019; McNamara & Sepasgozar, 2018).

3.8.2   *Maintenance of Transaction Records*

When DLT is combined with BIM, some of the challenges that hinder BIM adoption can be overcome due to the availability of a real-time and immutable record of data with timestamps, which increases reliability, integrity and transparency of the data (J. Li et al., 2019; Safa et al., 2019; Turk & Klinc, 2017). Furthermore, the ownership and rights of the information contained within the BIM model can be made explicit and transparent to all project parties, increasing trust (J. Li et al., 2019). With the records of changes made in the BIM model combined with DLT, the responsibility for modifications and errors can be identified readily (McNamara & Sepasgozar, 2018; Safa et al., 2019; Turk & Klinc, 2017). At the same time, as project information will be consolidated and updated within the model, project stakeholders can have access to the updated information as and when required (Safa et al., 2019). DLT also ensures security of the project data (McNamara & Sepasgozar, 2018).

3.9   Summary

This chapter introduced the common applications of the key smart technologies in projects, as shown in Table 3.1. The common applications of key smart technologies in projects demonstrated the potential to improve CPM performance and subsequently project performance.

Table 3.1 Common Applications of the Key Smart Technologies in Projects

Smart Technology	Applications in Projects	Examples
CPS/IoT	Real-time project monitoring and control	Labour tracking within the construction site to improve safety and productivity
		Supply chain management to reduce construction wastes, to improve asset utilisation and efficiency, to locate specific precast components and to improve visibility and traceability of the prefabricated components
		Tracking and monitoring of construction equipment to automate equipment operation processes, tracking and analysis of progress and productivity and prevention of collisions
		Safety monitoring such as tracking of near-miss incidents, compliance in personal protection equipment requirements, monitoring of hazardous areas on site and site conditions
	Integrated data platform	Allow stakeholders to access project information from any location for viewing, managing, sharing and collaboration in real-time
BD/AI	Cause analysis, prediction and estimation	Identify causes of project delays, cost overruns, quality issues, workspace conflicts to predict probability of occurrences and estimate duration/cost required for project to be completed
	Automated and optimised design and planning	Generative designs that synthesise historical project data, project requirements, needs of users and sustainability requirements
		Automatically generate optimised schedule and allocation of workers to minimise workspace conflicts
	Document classification and information retrieval	Enable learning from post-project review documents
		Enable identification of workers' and equipment actions towards site safety
		For project progress monitoring and control
		Automated regulatory and contractual compliance system

(Continued)

Table 3.1 (Continued)

Smart Technology	Applications in Projects	Examples
AV/ Robotics	Automated construction	Automate construction work onsite such as bricklaying
		Automate construction fabrication and assembly both onsite and offsite
	Automated data collection	Automate data collection for project tracking
AR/VR	Simulation	Construction safety training to improve worker's comprehension
		Project schedule and resource planning to identify potential workspace conflicts and determine optimal schedule and site layout, and to monitor construction progress
		Walkthrough of the designed building to support constructability analysis, client involvement and collaboration among project stakeholders and decision making
	Information retrieval and communication	Access to building information onsite to support collaboration and communication
AM	Printing of building components	Print building components based on a computer-aided design which enables mass customisation
3D imaging	Capture as-built conditions of projects automatically	Capture as-built conditions of projects to determine project progress automatically
DLT	Smart contracts	Automated contract monitoring and execution
	Maintenance of transaction records	Records of changes made in BIM model to improve accountability and transparency

References

Ahmed, S. (2019). A review on using opportunities of augmented reality and virtual reality in construction project management. *Technology and Management in Construction, 11*, 1839–1852. https://doi.org/10.2478/otmcj-2018-0012

Aibinu, A. A., & Jagboro, G. O. (2002). The effects of construction delays on project delivery in Nigerian construction industry. *International Journal of Project Management, 20*(8), 593–599. https://doi.org/10.1016/S0263-7863(02)00028-5

Alaka, H., Oyedele, L., Owolabi, H., Akinade, O., Bilal, M., & Ajayi, S. (2018). A big data analytics approach for construction firms failure prediction models. *IEEE Transactions on Engineering Management*, 1–10. https://doi.org/10.1109/TEM.2018.2856376

Álvares, J. S., Costa, D. B., & Melo, R. R. S. de. (2018). Exploratory study of using unmanned aerial system imagery for construction site 3D mapping. *Construction Innovation, 18*(3), 301–320. https://doi.org/10.1108/CI-05-2017-0049

Barro-Torres, S., Fernández-Caramés, T. M., Pérez-Iglesias, H. J., & Escudero, C. J. (2012). Real-time personal protective equipment monitoring system. *Computer Communications, 36*(1), 42–50. https://doi.org/10.1016/j.comcom.2012.01.005

Behzadan, A. H., Aziz, Z., Anumba, C. J., & Kamat, V. R. (2008). Ubiquitous location tracking for context-specific information delivery on construction sites. *Automation in Construction, 17*(6), 737–748. https://doi.org/10.1016/j.autcon.2008.02.002

Bilal, M., Oyedele, L. O., Akinade, O. O., Ajayi, S. O., Alaka, H. A., Owolabi, H. A., Qadir, J., Pasha, M., & Bello, S. A. (2016). Big data architecture for construction waste analytics (CWA): A conceptual framework. *Journal of Building Engineering, 6*, 144–156. https://doi.org/10.1016/j.jobe.2016.03.002

Bilal, M., Oyedele, L. O., Qadir, J., Munir, K., Ajayi, S. O., Akinade, O. O., Owolabi, H. A., Alaka, H. A., & Pasha, M. (2016). Big Data in the construction industry: A review of present status, opportunities, and future trends. *Advanced Engineering Informatics, 30*, 500–521. https://doi.org/10.1016/j.aei.2016.07.001

Bock, T. (2015). The future of construction automation: Technological disruption and the upcoming ubiquity of robotics. *Automation in Construction, 59*, 113–121. https://doi.org/10.1016/j.autcon.2015.07.022

Bosché, F., Ahmed, M., Turkan, Y., Haas, C. T., & Haas, R. (2015). The value of integrating Scan-to-BIM and Scan-vs-BIM techniques for construction monitoring using laser scanning and BIM: The case of cylindrical MEP components. *Automation in Construction, 49*, 201–213. https://doi.org/10.1016/j.autcon.2014.05.014

Boton, C. (2018). Supporting constructability analysis meetings with Immersive Virtual Reality-based collaborative BIM 4D simulation. *Automation in Construction, 96*, 1–15. https://doi.org/10.1016/j.autcon.2018.08.020

Brunoe, T. D., & Nielsen, K. (2016). Complexity management in mass customization SMEs. *Procedia CIRP, 51*, 38–43. https://doi.org/10.1016/j.procir.2016.05.099

Carbonari, A., Giretti, A., & Naticchia, B. (2011). A proactive system for real-time safety management in construction sites. *Automation in Construction, 20*(6), 686–698. https://doi.org/10.1016/j.autcon.2011.04.019

Carrillo, P., Harding, J., & Choudhary, A. (2011). Knowledge discovery from post-project reviews. *Construction Management and Economics, 29*(7), 713–723. https://doi.org/10.1080/01446193.2011.588953

Castro-Lacouture, D., Maynard, C., Bryson, L., Williams, R., & Bosscher, P. (2007). Concrete paving productivity improvement using a multi-task autonomous robot. *24th International Symposium on Automation & Robotics in Construction (ISARC 2007)*, 223–228.

Chen, Q., García de Soto, B., & Adey, B. T. (2018). Construction automation: Research areas, industry concerns and suggestions for advancement. *Automation in Construction, 94*, 22–38. https://doi.org/10.1016/j.autcon.2018.05.028

Chen, Y.-J., Feng, C.-W., Wang, Y.-R., & Wu, H.-M. (2011). Using BIM model and genetic algorithms to optimize the crew assignment for construction project planning. *International Journal of Technology, 2*(3).

Cheng, C.-W., Leu, S.-S., Cheng, Y.-M., Wu, T.-C., & Lin, C.-C. (2012). Applying data mining techniques to explore factors contributing to occupational injuries in Taiwan's construction industry. *Intelligent Speed Adaptation + Construction Projects, 48*, 214–222. https://doi.org/10.1016/j.aap.2011.04.014

Cheng, M.-Y., & Chen, J.-C. (2002). Integrating barcode and GIS for monitoring construction progress. *Automation in Construction, 11*(1), 23–33. https://doi.org/10.1016/S0926-5805(01)00043-7

Cheng, T., Teizer, J., Migliaccio, G. C., & Gatti, U. C. (2013). Automated task-level activity analysis through fusion of real time location sensors and worker's thoracic posture data. *Automation in Construction, 29*, 24–39. https://doi.org/10.1016/j.autcon.2012.08.003

Chi, H.-L., Kang, S.-C., & Wang, X. (2013). Research trends and opportunities of augmented reality applications in architecture, engineering, and construction. *Automation in Construction, 33*, 116–122. https://doi.org/10.1016/j.autcon.2012.12.017

Chin, S., Yoon, S., Choi, C., & Cho, C. (2008). RFID+4D CAD for progress management of structural steel works in high-rise buildings. *Journal of Computing in Civil Engineering, 22*(2), 74–89. https://doi.org/10.1061/(ASCE)0887-3801(2008)22:2(74)

Choudhary, A. K., Oluikpe, P. I., Harding, J. A., & Carrillo, P. M. (2009). The needs and benefits of text mining applications on post-project reviews. *Computers in Industry, 60*(9), 728–740. https://doi.org/10.1016/j.compind.2009.05.006

De Soto, G. B., Agustí-Juan, I., Hunhevicz, J., Joss, S., Graser, K., Habert, G., & Adey, B. T. (2018). Productivity of digital fabrication in construction: Cost and time analysis of a robotically built wall. *Automation in Construction, 92,* 297–311. https://doi.org/10.1016/j.autcon.2018.04.004

De Soto, G. B., Agustí-Juan, I., Joss, S., & Hunhevicz, J. (2019). Implications of Construction 4.0 to the workforce and organizational structures. *International Journal of Construction Management,* 1–13. https://doi.org/10.1080/15623599.2019.1616414

Ding, L. Y., Zhou, C., Deng, Q. X., Luo, H. B., Ye, X. W., Ni, Y. Q., & Guo, P. (2013). Real-time safety early warning system for cross passage construction in Yangtze Riverbed Metro Tunnel based on the internet of things. *Automation in Construction, 36,* 25–37. https://doi.org/10.1016/j.autcon.2013.08.017

Edirisinghe, R. (2019). Digital skin of the construction site: Smart sensor technologies towards the future smart construction site. *Engineering, Construction and Architectural Management, 26*(2), 184–223. https://doi.org/10.1108/ECAM-04-2017-0066

Ekanayake, L. L., & Ofori, G. (2004). Building waste assessment score: Design-based tool. *Building and Environment, 39*(7), 851–861. https://doi.org/10.1016/j.buildenv.2004.01.007

Elfaki, A. O., Alatawi, S., & Abushandi, E. (2014). Using intelligent techniques in construction project cost estimation: 10-year survey. *Advances in Civil Engineering, 2014,* 1–11. https://doi.org/10.1155/2014/107926

El-Omari, S., & Moselhi, O. (2008). Integrating 3D laser scanning and photogrammetry for progress measurement of construction work. *Automation in Construction, 18*(1), 1–9. https://doi.org/10.1016/j.autcon.2008.05.006

Ergen, E., Akinci, B., & Sacks, R. (2007). Tracking and locating components in a precast storage yard utilizing radio frequency identification technology and GPS. *Automation in Construction, 16*(3), 354–367. https://doi.org/10.1016/j.autcon.2006.07.004

Fang, Y., Teizer, J., & Marks, E. (2014). A framework for developing an as-built virtual environment to advance training of crane operators, 31–40. https://doi.org/10.1061/9780784413517.004

Golparvar-Fard, M., Peña-Mora, F., & Savarese, S. (2009). D4AR – A 4-Dimensional augmented reality model for automating construction progress monitoring data collection, processing and communication. *Journal of Information Technology in Construction, 14*(13), 129–153.

Gong, J., Caldas, C. H., & Gordon, C. (2011). Learning and classifying actions of construction workers and equipment using bag-of-video-feature-words and Bayesian network models. *Advances and Challenges in Computing in Civil and Building Engineering, 25*(4), 771–782. https://doi.org/10.1016/j.aei.2011.06.002

Goodrum, P. M., McLaren, M. A., & Durfee, A. (2006). The application of active radio frequency identification technology for tool tracking on construction job sites. *Automation in Construction, 15*(3), 292–302. https://doi.org/10.1016/j.autcon.2005.06.004

Greenhalgh, P., Mullins, B., Grunnet-Jepsen, A., & Bhowmik, A. K. (2016). Industrial deployment of a full-featured head-mounted augmented-reality system and the incorporation of a 3D-sensing platform. *SID Symposium Digest of Technical Papers, 47*(1), 448–451. https://doi.org/10.1002/sdtp.10704

Gu, N., & London, K. (2010). Understanding and facilitating BIM adoption in the AEC industry. *Automation in Construction, 19,* 988–999. https://doi.org/10.1016/j.autcon.2010.09.002

Guo, H., Yu, Y., & Skitmore, M. (2017). Visualization technology-based construction safety management: A review. *Automation in Construction, 73,* 135–144. https://doi.org/10.1016/j.autcon.2016.10.004

Ham, Y., Han, K. K., Lin, J. J., & Golparvar-Fard, M. (2016). Visual monitoring of civil infrastructure systems via camera-equipped Unmanned Aerial Vehicles (UAVs): A review of related works. *Visualization in Engineering, 4*(1), 1. https://doi.org/10.1186/s40327-015-0029-z

Han, K. K., & Golparvar-Fard, M. (2017). Potential of big visual data and building information modeling for construction performance analytics: An exploratory study. *Automation in Construction, 73,* 184–198. https://doi.org/10.1016/j.autcon.2016.11.004

Han, S. U., Lee, S. H., & Peña-Mora, F. (2012). A machine-learning classification approach to automatic detection of workers' actions for behavior-based safety analysis. *Computing in Civil Engineering, 2012*, 65–72. https://doi.org/10.1061/9780784412343.0009

Heiskanen, A. (2017). The technology of trust: How the Internet of Things and blockchain could usher in a new era of construction productivity. *Construction Research and Innovation, 8*(2), 66–70. https://doi.org/10.1080/20450249.2017.1337349

Hwang, B. G., Shan, M., & Looi, K. Y. (2018). Key constraints and mitigation strategies for prefabricated prefinished volumetric construction. *Journal of Cleaner Production, 183*, 183–193. https://doi.org/10.1016/j.jclepro.2018.02.136

Institution of Civil Engineers. (2018). *Blockchain technology in the construction industry: Digital transformation for high productivity* (p. 52). Institution of Civil Engineers.

Jia, M., Komeily, A., Wang, Y., & Srinivasan, R. S. (2019). Adopting Internet of Things for the development of smart buildings: A review of enabling technologies and applications. *Automation in Construction, 101*, 111–126. https://doi.org/10.1016/j.autcon.2019.01.023

Jiang, H., Lin, P., Qiang, M., & Fan, Q. (2015). A labor consumption measurement system based on real-time tracking technology for dam construction site. *Automation in Construction, 52*, 1–15. https://doi.org/10.1016/j.autcon.2015.02.004

Jiang, L., Li, Z., Li, L., & Gao, Y. (2018). Constraints on the promotion of prefabricated construction in China. *Sustainability, 10*(7). https://doi.org/10.3390/su10072516

Kamat, V. R., Golparvar-Fard, M., Savarese, S., Fischer, M., Peña-Mora, F., & Martinez, J. C. (2011). Research in visualization techniques for field construction. *Journal of Construction Engineering and Management, 137*(10), 853–862. https://doi.org/10.1061/(ASCE)CO.1943-7862.0000262

Kasperzyk, C., Kim, M.-K., & Brilakis, I. (2017). Automated re-prefabrication system for buildings using robotics. *Automation in Construction, 83*, 184–195. https://doi.org/10.1016/j.autcon.2017.08.002

Kelm, A., Laußat, L., Meins-Becker, A., Platz, D., Khazaee, M. J., Costin, A. M., Helmus, M., & Teizer, J. (2013). Mobile passive Radio Frequency Identification (RFID) portal for automated and rapid control of Personal Protective Equipment (PPE) on construction sites. *Automation in Construction, 36*, 38–52. https://doi.org/10.1016/j.autcon.2013.08.009

Khoshnevis, B. (2004). Automated construction by contour crafting – Related robotics and information technologies. *The Best of ISARC 2002, 13*(1), 5–19. https://doi.org/10.1016/j.autcon.2003.08.012

Kim, B., Kim, H., & Kim, C. (2012). Interactive modeler for construction equipment operation using augmented reality. *Journal of Computing in Civil Engineering, 26*(3), 331–341. https://doi.org/10.1061/(ASCE)CP.1943-5487.0000137

Kim, H., Soibelman, L., & Grobler, F. (2008). Factor selection for delay analysis using knowledge discovery in databases. *Automation in Construction, 17*(5), 550–560. https://doi.org/10.1016/j.autcon.2007.10.001

Kim, S., & Irizarry, J. (2016). Lessons learned from unmanned aerial system-based 3D mapping experiments. *52 ASC Annual International Conference Proceedings* (pp. 1–8). http://ascpro0.ascweb.org/archives/cd/2016/paper/CPRT157002016.pdf

Kochovski, P., & Stankovski, V. (2018). Supporting smart construction with dependable edge computing infrastructures and applications. *Automation in Construction, 85*, 182–192. https://doi.org/10.1016/j.autcon.2017.10.008

Kothman, I., & Faber, N. (2016). How 3D printing technology changes the rules of the game: Insights from the construction sector. *Journal of Manufacturing Technology Management, 27*(7), 932–943. https://doi.org/10.1108/JMTM-01-2016-0010

Lee, U.-K., Kim, J.-H., Cho, H., & Kang, K.-I. (2009). Development of a mobile safety monitoring system for construction sites. *Automation in Construction, 18*(3), 258–264. https://doi.org/10.1016/j.autcon.2008.08.002

Li, H., Chan, G., & Skitmore, M. (2012). Multiuser virtual safety training system for tower crane dismantlement. *Journal of Computing in Civil Engineering, 26*(5), 638–647. https://doi.org/10.1061/(ASCE)CP.1943-5487.0000170

Li, H., Chen, Z., Yong, L., & Kong, S. C. W. (2005). Application of integrated GPS and GIS technology for reducing construction waste and improving construction efficiency. *Automation in Construction, 14*(3), 323–331. https://doi.org/10.1016/j.autcon.2004.08.007

Li, J., Greenwood, D., & Kassem, M. (2019). Blockchain in the built environment and construction industry: A systematic review, conceptual models and practical use cases. *Automation in Construction, 102*, 288–307. https://doi.org/10.1016/j.autcon.2019.02.005

Li, X., Yi, W., Chi, H.-L., Wang, X., & Chan, A. P. C. (2018). A critical review of virtual and augmented reality (VR/AR) applications in construction safety. *Automation in Construction, 86*, 150–162. https://doi.org/10.1016/j.autcon.2017.11.003

Liao, C.-W., & Perng, Y.-H. (2008). Data mining for occupational injuries in the Taiwan construction industry. *Safety Science, 46*(7), 1091–1102. https://doi.org/10.1016/j.ssci.2007.04.007

Linner, T., & Bock, T. (2012). Evolution of large-scale industrialisation and service innovation in Japanese prefabrication industry. *Construction Innovation, 12*(2), 156–178. https://doi.org/10.1108/14714171211215921

Liu, H., Al-Hussein, M., & Lu, M. (2015). BIM-based integrated approach for detailed construction scheduling under resource constraints. *Automation in Construction, 53*, 29–43. https://doi.org/10.1016/j.autcon.2015.03.008

Liu, S., McMahon, C. A., & Culley, S. J. (2008). A review of structured document retrieval (SDR) technology to improve information access performance in engineering document management. *Computers in Industry, 59*(1), 3–16. https://doi.org/10.1016/j.compind.2007.08.001

Lu, W., Chen, X., Peng, Y., & Shen, L. (2015). Benchmarking construction waste management performance using big data. *Resources, Conservation & Recycling, 105*, 49–58. https://doi.org/10.1016/j.resconrec.2015.10.013

Luo, L. Z., Mao, C., Shen, L. Y., & Li, Z. D. (2015). Risk factors affecting practitioners' attitudes toward the implementation of an industrialized building system a case study from China. *Engineering, Construction and Architectural Management, 22*(6), 622–643. https://doi.org/10.1108/ECAM-04-2014-0048

McNamara, A., & Sepasgozar, S. (2018, September 26). Barriers and drivers of Intelligent Contract implementation in construction. *42nd AUBEA Conference 2018: Educating Building Professionals for the Future in the Globalised World.* https://www.researchgate.net/publication/329050104_Barriers_and_drivers_of_Intelligent_Contract_implementation_in_construction

Mobach, M. P. (2008). Do virtual worlds create better real worlds? *Virtual Reality, 12*(3), 163–179. https://doi.org/10.1007/s10055-008-0081-2

Moon, S., Choi, E., & Yang, B. (2015). Holistic integration based on USN technology for monitoring safety during concrete placement. *Automation in Construction, 57*, 112–119. https://doi.org/10.1016/j.autcon.2015.05.001

Moselhi, O., Hegazy, T., & Fazio, P. (1991). Neural networks as tools in construction. *Journal of Construction Engineering and Management, 117*(4), 606–625. https://doi.org/10.1061/(ASCE)0733-9364(1991)117:4(606)

Nahmens, I., & Bindroo, V. (2011). Is customization fruitful in industrialized homebuilding industry? *Journal of Construction Engineering and Management, 137*(12), 1027–1035. https://doi.org/10.1061/(ASCE)CO.1943-7862.0000396

Oesterreich, T. D., & Teuteberg, F. (2016). Understanding the implications of digitisation and automation in the context of Industry 4.0: A triangulation approach and elements of a research agenda for the construction industry. *Computers in Industry, 83*, 121–139. https://doi.org/10.1016/j.compind.2016.09.006

Oloufa, A. A., Ikeda, M., & Oda, H. (2003). Situational awareness of construction equipment using GPS, wireless and web technologies. *Automation in Construction, 12*(6), 737–748. https://doi.org/10.1016/S0926-5805(03)00057-8

Olsson, N. O. E., & Bull-Berg, H. (2015). Use of big data in project evaluations. *International Journal of Managing Projects in Business, 8*(3), 491–512. https://doi.org/10.1108/IJMPB-09-2014-0063

Omar, T., & Nehdi, M. L. (2016). Data acquisition technologies for construction progress tracking. *Automation in Construction, 70*, 143–155. https://doi.org/10.1016/j.autcon.2016.06.016

Park, C.-S., & Kim, H.-J. (2013). A framework for construction safety management and visualization system. *Automation in Construction, 33*, 95–103. https://doi.org/10.1016/j.autcon.2012.09.012

Portman, M. E., Natapov, A., & Fisher-Gewirtzman, D. (2015). To go where no man has gone before: Virtual reality in architecture, landscape architecture and environmental planning. *Computers, Environment and Urban Systems, 54*, 376–384. https://doi.org/10.1016/j.compenvurbsys.2015.05.001

Pradhananga, N., & Teizer, J. (2013). Automatic spatio-temporal analysis of construction site equipment operations using GPS data. *Automation in Construction, 29*, 107–122. https://doi.org/10.1016/j.autcon.2012.09.004

Ray, S. J., & Teizer, J. (2013). Computing 3D blind spots of construction equipment: Implementation and evaluation of an automated measurement and visualization method utilizing range point cloud data. *Automation in Construction, 36*, 95–107. https://doi.org/10.1016/j.autcon.2013.08.007

Riaz, Z., Arslan, M., Kiani, A. K., & Azhar, S. (2014). CoSMoS: A BIM and wireless sensor based integrated solution for worker safety in confined spaces. *Automation in Construction, 45*, 96–106. https://doi.org/10.1016/j.autcon.2014.05.010

Roberts, G. W., Dodson, A. H., & Ashkenazi, V. (1999). Global positioning system aided autonomous construction plant control and guidance. *Automation in Construction, 8*(5), 589–595. https://doi.org/10.1016/S0926-5805(99)00008-4

Sacks, R., Navon, R., Brodetskaia, I., & Shapira, A. (2005). Feasibility of automated monitoring of lifting equipment in support of project control. *Journal of Construction Engineering and Management, 131*(5), 604–614. https://doi.org/10.1061/(ASCE)0733-9364(2005)131:5(604)

Sacks, R., Whyte, J., Swissa, D., Raviv, G., Zhou, W., & Shapira, A. (2015). Safety by design: Dialogues between designers and builders using virtual reality. *Construction Management and Economics, 33*(1), 55–72. https://doi.org/10.1080/01446193.2015.1029504

Safa, M., Baeza, S., & Weeks, K. (2019). Incorporating blockchain technology in construction management. *Strategic Direction, 35*(10), 1–3. https://doi.org/10.1108/sd-03-2019-0062

Salama, D. M., & El-Gohary, N. M. (2016). Semantic text classification for supporting automated compliance checking in construction. *Journal of Computing in Civil Engineering, 30*(1), 4014106. https://doi.org/10.1061/(ASCE)CP.1943-5487.0000301

Siu, M. F., Ekyalimpa, R., Lu, M., & Abourizk, S. (2013). Applying regression analysis to predict and classify construction cycle time. *Computing in Civil Engineering – Proceedings of the 2013 ASCE International Workshop on Computing in Civil Engineering* (pp. 669–676). https://doi.org/10.1061/9780784413029.084

Skibniewski, M., & Hendrickson, C. (1988). Analysis of robotic surface finishing work on construction site. *Journal of Construction Engineering and Management, 114*(1), 53–68. https://doi.org/10.1061/(ASCE)0733-9364(1988)114:1(53)

Skibniewski, M. J. (1988). Framework for decision-making on implementing robotics in construction. *Journal of Computing in Civil Engineering, 2*(2), 188–201. https://doi.org/10.1061/(ASCE)0887-3801(1988)2:2(188)

Soibelman, L., & Kim, H. (2002). Data preparation process for construction knowledge generation through knowledge discovery in databases. *Journal of Computing in Civil Engineering, 16*(1), 39–48. https://doi.org/10.1061/(ASCE)0887-3801(2002)16:1(39)

Soibelman, L., Wu, J., Caldas, C., Brilakis, I., & Lin, K.-Y. (2008). Management and analysis of unstructured construction data types. *Intelligent Computing in Engineering and Architecture, 22*(1), 15–27. https://doi.org/10.1016/j.aei.2007.08.011

Sun, J., Lei, K., Cao, L., Zhong, B., Wei, Y., Li, J., & Yang, Z. (2020). Text visualization for construction document information management. *Automation in Construction, 111*, 103048. https://doi.org/10.1016/j.autcon.2019.103048

Tang, S., Shelden, D. R., Eastman, C. M., Pishdad-Bozorgi, P., & Gao, X. (2019). A review of building information modeling (BIM) and the internet of things (IoT) devices integration: Present status and future trends. *Automation in Construction, 101,* 127–139. https://doi.org/10.1016/J.AUTCON.2019.01.020

Tay, Y. W. D., Panda, B., Paul, S. C., Noor Mohamed, N. A., Tan, M. J., & Leong, K. F. (2017). 3D printing trends in building and construction industry: A review. *Virtual and Physical Prototyping, 12*(3), 261–276. https://doi.org/10.1080/17452759.2017.1326724

Teizer, J., & Cheng, T. (2015). Proximity hazard indicator for workers-on-foot near miss interactions with construction equipment and geo-referenced hazard areas. *Automation in Construction, 60,* 58–73. https://doi.org/10.1016/j.autcon.2015.09.003

Turk, Ž., & Klinc, R. (2017). Potentials of blockchain technology for construction management. *Procedia Engineering, 196,* 638–645. https://doi.org/10.1016/j.proeng.2017.08.052

Turkan, Y., Bosché, F., Haas, C. T., & Haas, R. (2012). Automated progress tracking using 4D schedule and 3D sensing technologies. *Automation in Construction, 22,* 414–421. https://doi.org/10.1016/j.autcon.2011.10.003

Turkan, Y., Bosché, F., Haas, C. T., & Haas, R. (2013). Toward automated earned value tracking using 3D imaging tools. *Journal of Construction Engineering and Management, 139*(4), 423–433. https://doi.org/10.1061/(ASCE)CO.1943-7862.0000629

Ur-Rahman, N., & Harding, J. A. (2012). Textual data mining for industrial knowledge management and text classification: A business oriented approach. *Expert Systems with Applications, 39*(5), 4729–4739. https://doi.org/10.1016/j.eswa.2011.09.124

Wang, L., & Leite, F. (2013). Knowledge discovery of spatial conflict resolution philosophies in BIM-enabled MEP design coordination using data mining techniques: A proof-of-concept. *Computing in Civil Engineering – Proceedings of the 2013 ASCE International Workshop on Computing in Civil Engineering* (pp. 419–426). https://doi.org/10.1061/9780784413029.053

Wang, X., & Dunston, P. (2007). Design, strategies, and issues towards an augmented reality-based construction training platform. *Journal of Information Technology in Construction (ITcon), 12*(25), 363–380. https://www.itcon.org/2007/25

Wang, Y., Wang, X., Truijens, M., Hou, L., & Zhou, Y. (2014). Integrating augmented reality with building information modeling: Onsite construction process controlling for liquefied natural gas industry. *Automation in Construction, 40,* 96–105. https://doi.org/10.1016/j.autcon.2013.12.003

Warszawski, A., & Rosenfeld, Y. (1994). Robot for interior-finishing works in building: Feasibility analysis. *Journal of Construction Engineering and Management, 120*(1), 132–151. https://doi.org/10.1061/(ASCE)0733-9364(1994)120:1(132)

Williams, T. P., & Gong, J. (2014). Predicting construction cost overruns using text mining, numerical data and ensemble classifiers. *Automation in Construction, 43,* 23–29. https://doi.org/10.1016/j.autcon.2014.02.014

Willmann, J., Knauss, M., Bonwetsch, T., Apolinarska, A. A., Gramazio, F., & Kohler, M. (2016). Robotic timber construction – Expanding additive fabrication to new dimensions. *Automation in Construction, 61,* 16–23. https://doi.org/10.1016/j.autcon.2015.09.011

Wu, P., Wang, J., & Wang, X. (2016). A critical review of the use of 3-D printing in the construction industry. *Automation in Construction, 68,* 21–31. https://doi.org/10.1016/j.autcon.2016.04.005

Wu, W., Yang, H., Chew, D. A. S., Yang, S., Gibb, A. G. F., & Li, Q. (2010). Towards an autonomous real-time tracking system of near-miss accidents on construction sites. *Automation in Construction, 19*(2), 134–141. https://doi.org/10.1016/j.autcon.2009.11.017

Yang, J., Vela, P., Teizer, J., & Shi, Z. (2014). Vision-based tower crane tracking for understanding construction activity. *Journal of Computing in Civil Engineering, 28*(1), 103–112. https://doi.org/10.1061/(ASCE)CP.1943-5487.0000242

Yunus, R., & Yang, J. (2012). Critical sustainability factors in industrialised building systems. *Construction Innovation, 12*(4), 447–463. https://doi.org/10.1108/14714171211272216

Zhang, S., Teizer, J., Pradhananga, N., & Eastman, C. M. (2015). Workforce location track-ing to model, visualize and analyze workspace requirements in building information mod-els for construction safety planning. *Automation in Construction, 60,* 74–86. https://doi.org/10.1016/j.autcon.2015.09.009

Zhang, Y., Luo, H., & He, Y. (2015). A system for tender price evaluation of construction project based on big data. *Procedia Engineering, 123,* 606–614. https://doi.org/10.1016/j.proeng.2015.10.114

Zhong, R. Y., Peng, Y., Xue, F., Fang, J., Zou, W., Luo, H., Thomas Ng, S., Lu, W., Shen, G. Q. P., & Huang, G. Q. (2017). Prefabricated construction enabled by the Internet-of-Things. *Automation in Construction, 76,* 59–70. https://doi.org/10.1016/j.autcon.2017.01.006

4 Competency Theory

4.1 Introduction

The evolving dynamics of the smart technologies necessitate effective strategies for talent acquisition, development and management. One prominent tool in this regard, gaining momentum within human resource management, is competency. Competency can guide recruitment and digital transformation within organisations by illuminating the knowledge, skill and personal attributes of exceptional performers (Boyatzis, 2008; Dainty et al., 2004; Markus et al., 2005; Moore et al., 2002). It is noted that 'competency' is often confused with 'competence' in most literature, despite their distinct implications (Ahadzie et al., 2008; Dainty et al., 2004; Moore et al., 2002). Competence refers to the skills and knowledge required to perform a job well, and is typically work related (Moore et al., 2002; Woodruffe, 1993) while *competency* refers to the sets of behaviour the person must display in order to perform the work tasks with competence and is regarded as 'person related' (Ahadzie et al., 2008; Boyatzis, 1982; Moore et al., 2002; Spencer & Spencer, 1993; Woodruffe, 1993).

Spencer and Spencer (1993) argued that competency is a better predictor of performance in complex jobs due to a 'restricted range effect' where almost everyone has similar levels of high intelligence. This is also supported by Boyatzis (1982). Furthermore, competency models are found to be dynamic such that although the way tasks are carried out is likely to change, core motivational, interpersonal and cognitive competencies that predict success remain the same over time (Spencer & Spencer, 1993).

In this chapter, the focus is on the concept of competency, exploring existing competency models and examining key factors contributing to an individual's competency. Various approaches to assessing competency levels are also discussed. Moreover, the chapter identifies specific personality traits that influence an individual's competency.

4.2 Definition of Competency

Competency models are widely used as a human resource management tool to guide the recruitment and promotion of individuals in organisations (Boyatzis, 2008; Dainty et al., 2004; Markus et al., 2005; Moore et al., 2002). One of the earliest works on modern competency theory was introduced in McClelland's work in

DOI: 10.1201/9781003462231-4

1973: 'Testing for Competence rather than for Intelligence'. The article argued that aptitude and intelligence tests could not predict an individual's job performance (McClelland, 1973). Instead, job-related factors that distinguish a superior performer should form the basis of the test to predict proficiency on the job (McClelland, 1973). The need to analyse the behaviours of superior performers of a job was also emphasised in the study (McClelland, 1973).

In Boyatzis (1982), competency is defined as 'an underlying characteristic of a person which results in effective and/or superior performance in a job'. Competency broadly refers to two factors: (i) the job aspects that have to be performed competently and (ii) the set of behaviour patterns the individual must display to perform job tasks and functions with the required level of competence (Moore et al., 2002; Woodruffe, 1993). Spencer and Spencer (1993) further extended the definition of competency to include the causal relation of the underlying characteristics to job performance.

Competency and competence are two distinct concepts. Competency refers to the underlying characteristics or behavioural patterns that enable the individual to be competent, and is 'person related' (Moore et al., 2002; Sanghi, 2012; Woodruffe, 1993). On the other hand, competence refers to the standards of the job tasks to be achieved, and is 'work related' (Moore et al., 2002; Sanghi, 2012; Woodruffe, 1993). According to Woodruffe (1993), the distinction lies in the aspects of the job in which the individual is competent (i.e. competence), and the aspects of the person which enables him or her to be competent (i.e. competency). Despite the distinctions between the terms, it is noted that 'competence' and 'competency' have been used rather interchangeably (Dainty et al., 2004; Moore et al., 2002).

In complex jobs, competencies were found to be more important in the prediction of superior performance than competence due to the 'restricted range effect' (Spencer & Spencer, 1993). Restricted range effect describes the similarity in competence levels of individuals in the job (Spencer & Spencer, 1993). As such, competencies play a distinguishing role between superior performers in these jobs (Spencer & Spencer, 1993). Furthermore, the underlying core competencies that predict success remain the same over time although the way job tasks are carried out constantly evolves (Spencer & Spencer, 1993).

Hence, this study will focus on the competency of PMs with the following definition: the underlying characteristics of an individual that is causally related to superior project management performance.

4.3   Approaches to Competency

Since the concept of competency was popularised within the field of human resource management, several competency models have been proposed. This section details three common models of competency.

4.3.1   *Knowledge, Skills and Personal Attributes*

According to Spencer and Spencer (1993), competency can be categorised into five components: (i) knowledge; (ii) skill; (iii) self-concept; (iv) traits and (v) motives.

This is recognised as an attribute-based competency model (Crawford, 2005). The definition of each component is as follows:

- Knowledge: information an individual has in specific areas
- Skill: the ability to perform a certain physical or mental task
- Self-concept: an individual's attitudes, values or self-image
- Traits: physical characteristics and consistent responses to situations or information
- Motives: the things an individual consistently thinks about or wants that cause action

Knowledge and skill are more visible and are relatively easy to develop while self-concept, traits and motives are more hidden and more difficult to assess and develop (Spencer & Spencer, 1993). It was further highlighted that a behaviour without intent does not define competency (Spencer & Spencer, 1993). Accordingly, 'intent' is influenced by motives, traits, self-concept and knowledge, which drives 'action' or the behaviour that uses one's skills to perform the task, resulting in the 'outcome' or the resultant job performance (Spencer & Spencer, 1993).

Competencies are further categorised into 'threshold competencies' and 'differentiating competencies' according to the job performance criterion they predict. Threshold competencies are essential characteristics that individuals in the job need to have to fulfil the job tasks while differentiating competencies distinguish superior performers from average performers (Spencer & Spencer, 1993). Threshold competencies involve expertise and experience, knowledge and basic cognitive competencies such as memory and deductive reasoning while differentiating competencies involve more advanced CICs, EICs and SIs (Boyatzis, 2008, 2009).

- CI: the ability to think or analyse information and situations that leads to effective or superior performance
- Emotional intelligence: the ability to recognise, understand and use emotional information about oneself that leads to effective or superior performance
- SI: the ability to recognise, understand and use emotional information about others that leads to effective or superior performance

4.3.2 Input, Personal and Output Competencies

Building upon the five characteristics of competency proposed by Spencer and Spencer (1993), Crawford (2005) developed a framework to reconcile the attribute-based competency model with performance-based competency standards, providing a basis for the identification and measurement of competences against standards. In this framework, components include (i) input competencies, (ii) personal competencies and (iii) output competencies.

- Input competencies: the knowledge, understanding, skills and abilities that an individual brings to the job
- Personal competencies: the core personality characteristics underlying an individual's capability to do a job, such as attitude and personality traits

- Output competencies: the demonstrable performance or the ability to perform activities within an occupational area to the levels of performance expected in employment

However, Crawford (2005) highlighted that no statistically significant relationship was found between output competencies and perceived performance of individuals.

4.3.3   *Task and Contextual Performance Behaviours*

Another approach to predicting job performance utilises the task performance behaviours and contextual behaviours to understand job performance of managerial roles (Ahadzie et al., 2008; W. C. Borman & Motowidlo, 1997; W. Borman & Motowidlo, 1993; Conway, 1999).

- Task performance behaviours: the effectiveness of individuals in performing activities that contribute to the technical function of the organisation, are job specific and are typically related to the knowledge, skills and abilities of individuals.
- Contextual performance behaviours: non-job-specific behaviours that contribute to organisational effectiveness but are not formally recognised as part of the job. Contextual performance behaviours shape the organisational, social, and psychological context for the technical function to operate and serve as the catalyst for task activities and processes. These are typically associated with the individuals' predisposition and volition.

According to previous studies, task performance behaviours account for about 50–55 per cent of the variance in the managerial performance domains while contextual performance behaviours account for about 30 per cent of the variance in the managerial performance domains (Ahadzie et al., 2008; W. Borman & Motowidlo, 1993). Variance in task performance can typically be explained by the differences in knowledge, skills and abilities, whereas variance in contextual performance is due to the differences in volition and personality (Conway, 1996).

As knowledge, skills and personal attributes are the underlying factors of all the competency models (Succar et al., 2013), this study will focus on these factors in PMs in managing smart technologies in projects to continuously deliver successful projects in the 4IR.

4.4   Assessment of Competency Levels

To assess the job competency levels of individuals and to distinguish superior performers from average performers, McClelland (1973) proposed six principles of job competency testing:

- Criterion sampling refers to identifying effectiveness criteria for a job and subsequently identifying a group of superior performers and a group of average performers according to the effectiveness criteria.

- Test should reflect changes in what has been learnt over time.
- Scales for each characteristic tested should be made public and explicit.
- Tests should assess competencies involved based on the clusters of life outcomes beyond occupational outcomes and include social outcomes such as leadership, interpersonal skills and communication skills.
- Tests should involve both operant and respondent behaviours: respondent behaviours are elicited by stimuli while operant behaviours are responses by an individual that are not triggered by any clearly defined stimuli.
- Tests should sample operant thought patterns to get maximum generalisability to various action outcomes instead of specific skills with little general predictive power.

These six principles form the basis of the job competency assessment methodology, which includes six steps according to Spencer and Spencer (1993):

- Step 1: Define performance effective criteria – to identify the criteria that define superior or effective performance in the job to be studied and can include 'hard' outcomes such as sales and productivity measures, supervisor nominations and peer ratings.
- Step 2: Identify a criterion sample – to identify a group of superior performers and a group of average performers based on the criteria or ratings identified in Step 1.
- Step 3: Collect data – data for the development of competency models to be collected, through behavioural event interviews, expert panels, survey, computerised expert systems, job task analysis or observation.
- Step 4: Analyse collected data and develop a competency model – data from all sources and methods are analysed to identify the personality and skill competencies that distinguish superior from average performers through hypothesis generation, thematic analysis or concept formation.
- Step 5: Validate the competency model – the competency model developed can be validated using a second criterion sample through concurrent cross-validation, concurrent construct validation or predictive validation.
- Step 6: Prepare applications of the competency model – the competency model can be used design selection interviews, tests and assessment centres for selection, career pathing, performance management, succession planning, training and development, compensation and management information systems.

The McBer Job Competency Assessment Methodology serves as a basis in competency studies and has been used in studies such as Cheng et al. (2005a), Murphy (2014), Ajayi et al. (2016), Dainty et al. (2004) and Calhoun et al. (2008). These studies have demonstrated the validity of the McBer Job Competency Assessment Methodology. The behavioural event interview is also widely recognised as the most effective way of identifying the underlying behaviours of superior performance in management roles (Cheng et al., 2005b; Ajayi et al., 2016; Calhoun et al., 2008; Murphy, 2014).

Accordingly, the McBer Competency Dictionary was developed, which con-solidates the common competencies based on the BEIs conducted (Spencer & Spencer, 1993). The competencies provided in the McBer Competency Diction-ary are clustered based on their underlying intents. Spencer and Spencer (1993) provided the scales for 21 most common competencies, which serve as refer-ences for the development of a competency model for each job type in general. Scales are typically derived based on the following dimensions: (i) intensity and completeness of action taken to execute an intent, (ii) size of impact, (iii) com-plexity, (iv) amount of effort and (v) unique dimensions (Ryan et al., 2012; Spen-cer & Spencer, 1993). Combinations of various competencies at different levels are required for different job roles and hence need to be adapted according to the job requirements (Alvarenga et al., 2019; M. I. Cheng et al., 2005b; Spencer & Spencer, 1993).

4.5   Personal Characteristics

Based on extensive research, the McBer Competency Dictionary identified 12 personal characteristics that distinguished the superior managers from average performing managers (Dainty et al., 2004). This was further confirmed in Dainty et al. (2004), who investigated the competency profile of superior construction PMs. The 12 personal characteristics that consistently distinguished superior managers from average-performing managers were found to be achievement orientation, initiative, information seeking, focus on client's needs, impact and influence, directiveness/assertiveness, teamwork and co-operation, team leader-ship, analytical thinking, conceptual thinking, self-control and flexibility (Dainty et al., 2004; Spencer & Spencer, 1993). Achievement orientation refers 'the man-ager's concern working towards a standard of excellence'; initiative refers to the taking of proactive actions to avert problems to enhance job results and avoid problems; information seeking refers to an underlying curiosity or desire to know more about things, people or issues and not accepting situations at face value; focus on client's needs refers to the desire to meet the needs of the clients; impact and influence refers to the ability to influence, persuade, and convince people to support their agenda; directiveness/assertiveness refers to the ability to use personal power or the power of the individual's position for the long term good of the project; teamwork refers to the ability of the manager to work coopera-tively with the project team; team leadership refers to the ability to lead groups to achieve project objectives; analytical thinking refers to the ability to break down problems systematically; conceptual thinking refers to the ability to iden-tify patterns in complex situations; self-control refers to the ability to keep emo-tions under control; and flexibility refers to the manager's ability to adapt one's thinking, attitude or behaviour to changing situation (Dainty et al., 2004; Spencer & Spencer, 1993). As these underlying personal characteristics were found to consistently distinguish superior managers from average-performing managers, these personal characteristics are also theorised to affect PMs' level of techno-logical competency level.

4.6 Summary

This chapter presents the common competency models, along with the methodology for assessing individual competency levels and identifying the consistent personal characteristics that distinguish superior managers from average performers. In this study, competency is defined as the underlying characteristics of an individual causally related to superior project management performance. As technology continues to permeate every aspect of human life, competence in managing these technologies becomes indispensable. However, the adoption of these technologies also necessitates a shift in how tasks are performed. Therefore, we aim to develop the concept of technological competency to predict the success of PMs in managing projects with smart technologies, as detailed in Chapter 7.

References

Ahadzie, D. K., Proverbs, D. G., & Olomolaiye, P. (2008). Towards developing competency-based measures for construction project managers: Should contextual behaviours be distinguished from task behaviours? *International Journal of Project Management, 26,* 631–645. https://doi.org/10.1016/j.ijproman.2007.09.011

Ajayi, S. O., Oyedele, L. O., Kadiri, K. O., Akinade, O. O., Bilal, M., Owolabi, H. A., & Alaka, H. A. (2016). Competency-based measures for designing out construction waste: Task and contextual attributes. *Engineering, Construction and Architectural Management, 23*(4), 464–490. https://doi.org/10.1108/ECAM-06-2015-0095

Alvarenga, J. C., Branco, R. R., Guedes, A. L. A., Soares, C. A. P., & E Silva, W. D. S. (2019). The project manager core competencies to project success. *International Journal of Managing Projects in Business.* https://doi.org/10.1108/IJMPB-12-2018-0274

Borman, W. C., & Motowidlo, S. J. (1993). Expanding the criterion domain to include elements of contextual performance. *Personnel Selection in Organizations* (pp. 71–98)

Borman, W. C., & Motowidlo, S. J. (1997). Task performance and contextual performance: The meaning for personnel selection research. *Human Performance, 10*(2), 99–109. https://doi.org/10.1207/s15327043hup1002_3

Boyatzis, R. E. (1982). *The competent manager: A model for effective performance.* Wiley.

Boyatzis, R. E. (2008). Competencies in the 21st century. *Journal of Management Development, 27*(1), 5–12. https://doi.org/10.1108/02621710810840730

Boyatzis, R. E. (2009). Competencies as a behavioral approach to emotional intelligence. *Journal of Management Development, 28*(9), 749–770. https://doi.org/10.1108/02621710910987647

Calhoun, J. G., Dollett, L., Sinioris, M. E., Wainio, J. A., Butler, P. W., Griffith, J. R., Patullo, A., & Warden, G. L. (2008). Development of an interprofessional competency model for healthcare leadership. *Journal of Healthcare Management, 53*(6), 375–389. https://doi.org/10.1097/00115514-200811000-00006

Cheng, M. I., Dainty, A. R. J., & Moore, D. R. (2005a). Towards a multidimensional competency-based managerial performance framework: A hybrid approach. *Journal of Managerial Psychology, 20*(5), 380–396. https://doi.org/10.1108/02683940510602941

Cheng, M. I., Dainty, A. R. J., & Moore, D. R. (2005b). What makes a good project manager? *Human Resource Management Journal, 15*(1), 25–37. https://doi.org/10.1111/j.1748-8583.2005.tb00138.x

Conway, J. M. (1996). Additional construct validity evidence for the task/contextual performance distinction. *Human Performance, 9*(4), 309–329. https://doi.org/10.1207/s15327043hup0904_1

Conway, J. M. (1999). Distinguishing contextual performance from task performance for managerial jobs. *Journal of Applied Psychology, 84*(1), 3–13. https://doi.org/10.1037/0021-9010.84.1.3

Crawford, L. (2005). Senior management perceptions of project management competence. *International Journal of Project Management, 23*(1), 7–16. https://doi.org/10.1016/j.ijproman.2004.06.005

Dainty, A. R. J., Cheng, M. I., & Moore, D. R. (2004). A competency-based performance model for construction project managers. *Construction Management and Economics, 22*(8), 877–886. https://doi.org/10.1080/0144619042000202726

Markus, L. H., Cooper-Thomas, H. D., & Allpress, K. N. (2005). Confounded by competencies? An evaluation of the evolution and use of competency models. *New Zealand Journal of Psychology, 34*(2), 117.

McClelland, D. C. (1973). Testing for competence rather than for "intelligence". *The American Psychologist, 28*(1), 1–14. https://doi.org/10.1037/h0034092

Moore, D. R., Cheng, M. I., & Dainty, A. R. J. (2002). Competence, competency and competencies: Performance assessment in organisations. *Work Study, 51*(6), 314–319. https://doi.org/10.1108/00438020210441876

Murphy, M. E. (2014). Implementing innovation: A stakeholder competency-based approach for BIM. *Construction Innovation, 14*(4), 433–452. https://doi.org/10.1108/CI-01-2014-0011

Ryan, G., Spencer, L. M., & Bernhard, U. (2012). Development and validation of a customized competency-based questionnaire: Linking social, emotional, and cognitive competencies to business unit profitability. *Cross Cultural Management, 19*(1), 90–103. https://doi.org/10.1108/13527601211195646

Sanghi, S. (2012). Introduction to competency. In *The handbook of competency mapping: Understanding, designing and implementing competency models in organizations* (pp. 3–19). SAGE Publications India Pvt Ltd. https://doi.org/10.4135/9788132108481.n1

Spencer, L. M., & Spencer, S. M. (1993). *Competence at work: Models for superior performance*. Wiley.

Succar, B., Sher, W., & Williams, A. (2013). An integrated approach to BIM competency assessment, acquisition and application. *Automation in Construction, 35*, 174–189. https://doi.org/10.1016/j.autcon.2013.05.016

Woodruffe, C. (1993). What is meant by a competency? *Leadership & Organization Development Journal, 14*(1), 29–36. https://doi.org/10.1108/eb053651

5 Construction Project Managers' Competencies

5.1 Introduction

Smart technologies have demonstrated their potential in aiding construction projects. However, to fully harness the productivity benefits, their integration with efficient and streamlined processes is essential, alongside collaborative efforts by technologically competent employees (Leviäkangas et al., 2017). In the context of the 4IR, smart technologies not only affect industrial processes but also influence entire countries, societies, products, services, business models, work environments and the knowledge and skillsets required for individuals to be competent (Brynjolfsson & McAfee, 2014; Leviäkangas et al., 2017; Marnewick & Marnewick, 2021; J. M. Müller et al., 2018; Pereira & Romero, 2017; Schwab, 2016; World Economic Forum, 2018).

The competency of PMs plays a crucial role in leading project success, influencing the efficient application of technologies and the effective control of project processes. Hence, extensive research has been conducted to study the competency of PMs (Ahmed & Anantatmula, 2017; Davis, 2011; Edum-Fotwe & McCaffer, 2000; Loufrani-Fedida & Missonier, 2015; Tabassi et al., 2016; Trivellas & Drimoussis, 2013; F. Zhang et al., 2013).

PM competencies are typically characterised by a triad of critical factors: knowledge, skills and personal attributes. The knowledge base for a PM encompasses a range of facets including, but not limited to, schedule management, cost management and quality management. The skillsets that determine a proficient PM encompass both hard and soft skills. Hard skills correspond to the abilities directly relevant to project management, such as cost analysis and schedule estimation. Conversely, soft skills are tied closely to one's personal traits and often include leadership, communication and decision-making abilities. Furthermore, the personal attributes that influence PMs' competencies include motives, attitude and traits.

In light of the accelerating pace of digital transformation, the required knowledge and skillsets for PMs have witnessed significant evolution. These changes mirror shifts in project design, construction, and management (Ngo & Hwang, 2022). Moreover, employers' expectations of soft skills and technical skills for future-ready PMs have escalated in the 4IR era (Low et al., 2021).

Professional organisations such as the PMI have also engaged in comprehensive studies on PM competencies. PMI has developed the PMCD Framework aimed at

DOI: 10.1201/9781003462231-5

fostering both hard- and soft-skill development among PMs (Zuo et al., 2018). This emphasis on identifying and enhancing common PM competencies paves the way for further examination of the evolution of technological competencies, particularly in the context of the new norms established by the 4IR.

5.2 Construction Project Management Knowledge Areas

CPM is the application of knowledge, skills, tools and technologies to direct and coordinate human and material resources throughout the lifecycle of a construction project to achieve the pre-determined project objectives (Biff, 2018; Project Management Institute, 2017a). Project success is typically defined by achieving the 'iron triangle' of time, cost and quality objectives of the project (Bronte-stewart, 2015; J. S. Chou et al., 2013; R. Müller & Turner, 2007; Zuo et al., 2018). Effective project management has been found to play a critical role in driving project success (Ceran & Dorman, 1995; Crawford, 2000; Edum-Fotwe & McCaffer, 2000; Zuo et al., 2018).

Project management was recognised as a distinct discipline in the 1950s and became a profession in the 1960s as the IPMA and PMI were established in 1965 and 1969 respectively (Biff, 2018). Both IPMA and PMI aim to develop and promote project management as a profession, establish and provide guidelines and standards for project management and offer certifications for project management professionals (Crawford, 2000). Specifically, IPMA has developed the guide for Individual Competence Baseline and PMI has developed the PMBOK as standards relating to project management competence. As PMBOK is the most widely recognised and accepted project management standard (Crawford, 2000), this study focuses on the PMBOK knowledge areas as the knowledge requirements of PMs. It is also of interest to note that although the PMBOK knowledge areas are not specific to the construction industry, the PMBOK knowledge areas are commonly applied in the context of CPM and hence will be detailed in this section.

Project management has been categorised into 10 knowledge areas in PMBOK, which define the knowledge requirements including the project management processes, inputs, outputs, tools and techniques of each identified area of project management (Project Management Institute, 2017a). The 10 identified knowledge areas of project management are as follows: (i) project integration management; (ii) project scope management, (iii) project schedule management, (iv) project cost management, (v) project quality management, (vi) project resource management, (vii) project communications management, (viii) project risk management, (ix) project procurement management and (x) project stakeholder management.

5.2.1 *Project Integration Management*

Project integration management is concerned with defining, coordinating and integrating the project activities and project management processes to ensure that project objectives are achieved (Demirkesen & Ozorhon, 2017; Project Management Institute, 2017a). Project integration management is a key responsibility of the PM and should not be delegated as PMs are responsible for the management of the project to achieve the pre-determined objectives (J. S. Chou et al., 2013; Project Management Institute, 2017a). Furthermore, project integration management is one of the most critical elements of project management, encompassing all aspects of a project, and has been found to contribute to project success considerably (Demirkesen & Ozorhon, 2017).

According to Project Management Institute (2017a), project integration management consists of seven main processes: (i) develop project charter, (ii) develop project management plan, (iii) direct and manage project work, (iv) manage project knowledge, (v) monitor and control project work, (vi) perform integrated change control and (vii) close project or phase. The tools and techniques associated with project integration management include expert judgement, data gathering, interpersonal and team skills, meetings, project management information system, knowledge management, information management, data analysis, decision making and change control (Project Management Institute, 2017a). These tools and techniques are applied throughout the project lifecycle, as project integration management involves the integration of all the other project management areas (J.-S. Chou & Yang, 2012).

5.2.2 Project Scope Management

Project scope management refers to the processes to define and control what is and is not included in a project to complete the project successfully (Project Management Institute, 2017a). Project scope management is critical especially in the early stage in order to avoid major changes that may negatively affect project performance, as project scope impacts the management of other key project management areas (Fageha & Aibinu, 2013; Khan, 2006). Furthermore, project scope management is critical for enhancing stakeholder satisfaction and successful implementation of a project as stakeholders' concerns and expectations are managed (Fageha & Aibinu, 2013). The completion of the project scope will be measured against the project management plan to determine if the project objectives have been achieved.

According to Project Management Institute (2017a), project scope management includes six main processes: (i) plan scope management, (ii) collect requirements, (iii) define scope, (iv) create WBS, (v) validate scope and (vi) control scope. The tools and techniques for project scope management include expert judgement, data gathering, data analysis, decision making, meetings, data representation, interpersonal and team skills, context diagram, prototypes, product analysis, decomposition and inspection (Project Management Institute, 2017a).

5.2.3 Project Schedule Management

Project schedule management includes the processes to ensure that the project is completed on time (Project Management Institute, 2017a). Time is one of the key criteria of project success, as traditionally defined in the 'iron triangle' of project success (Bronte-stewart, 2015; J. S. Chou et al., 2013). Project schedule management provides a plan that details when the project will deliver the expected deliverables, and serves as a communication tool among stakeholders and as a baseline for progress reporting (Project Management Institute, 2017a).

According to Project Management Institute (2017a), there are six main processes for project schedule management: (i) plan schedule management, (ii) define activities, (iii) sequence activities, (iv) estimate activity durations, (v) develop schedule and (vi) control schedule. The tools and techniques involved in project schedule management include decomposition, rolling wave planning, activity sequencing techniques, estimating techniques, schedule network analysis, critical path method, resource optimisation and schedule compression (Project Management Institute, 2017a).

5.2.4   Project Cost Management

Project cost management refers to the processes to plan, estimate, budget, finance, manage and control costs of the resources to ensure that the project is completed within budget (Project Management Institute, 2017a). Project cost is one of the key criteria to determine project success (Alashwal & Chew, 2017; Bronte-stewart, 2015; J. S. Chou et al., 2013).

According to Project Management Institute (2017a), project cost management include four main processes: (i) plan cost management, (ii) estimate costs, (iii) determine budget and (iv) control costs. The tools and techniques involved in project cost management are similar to those in project schedule management in terms of estimation, with other cost related tools and techniques including funding limit reconciliation, earned value management, to-complete performance index and reserve analysis (Project Management Institute, 2017a).

5.2.5   Project Quality Management

Project quality management includes the processes to plan, manage and control project quality requirements to meet stakeholders' objectives, project deliverables and continuous improvements (Project Management Institute, 2017a). Quality is the other key criterion within the 'iron triangle' of project success (Bronte-stewart, 2015; J. S. Chou et al., 2013). Furthermore, project quality planning has been found to impact all measures of project success (Biff, 2018; Shenhar et al., 2002).

According to Project Management Institute (2017a), project quality management has three main processes: (i) plan quality management, (ii) manage quality and (iii) control quality. The tools and techniques associated with project quality management include cost of quality, data representation, problem solving and quality improvement methods (Project Management Institute, 2017a).

5.2.6   Project Resource Management

Project resource management refers to the processes to identify, acquire and manage the resources required for successful completion of the project, ensuring that the right resources are available to the manager and project team at the right time and right place (Project Management Institute, 2017a). Resources include physical and human resources. Physical resources include equipment, materials, facilities and infrastructure; human resources include team personnel (Project Management Institute, 2017a). Physical resource management is concerned with the allocation of physical resources required for the project team for successful completion of the project efficiently and effectively (Project Management Institute, 2017a). Human resource management includes the acquiring, managing, motivating, empowering and developing the project team members who work collectively as a team to achieve the project objectives (Project Management Institute, 2017a). Project human resource management practices have been found to be affect project performance significantly (Popaitoon & Siengthai, 2014).

According to Project Management Institute (2017a), there are six main processes for project quality management: (i) plan resource management, (ii) estimate activity resources, (iii) acquire resources, (iv) develop team, (v) manage team and (vi) control resources. The tools and techniques for project resource management are similar to those for project schedule management in terms of estimation of activity resource requirements, and further include data representation, team development, individual and team assessments and problem solving (Project Management Institute, 2017a).

5.2.7 *Project Communications Management*

Project communications management refers to the processes required to ensure effective information exchange among stakeholders in the project (Project Management Institute, 2017a). This includes the development of a strategy to ensure that communication is effective for stakeholders and executing the activities necessary to implement the communication strategy such that stakeholders can receive the appropriate information in the appropriate amount in the right format at the right time and right place through the right channels (Project Management Institute, 2017a; Yap et al., 2018). Effective project communications have been found to significantly impact project performance (Badir et al., 2012; Yap et al., 2018).

According to Project Management Institute (2017a), project communications management encompasses three main processes: (i) plan communications management, (ii) manage communications and (iii) monitor communications. The tools and techniques associated with project communication management include communication requirements analysis, communication technologies, communication methods, communication skills and project reporting (Project Management Institute, 2017a). Similar to project integration management, these tools and techniques are used throughout the project lifecycle as communication among stakeholders is essential.

5.2.8 *Project Risk Management*

Project risk management includes the processes to plan, identify, analyse, manage and monitor risks of a project (Project Management Institute, 2017a). Individual project risk is an uncertain event that will impact at least one project objective and can be positive or negative risks (Hwang et al., 2014; Project Management Institute, 2017a). Project risk management aims to increase the probability or impact of positive risks and to decrease the probability or impact of negative risks to improve the chances of project success (Project Management Institute, 2017a). Effective project risk management has been found to improve project performance (Hwang et al., 2014).

According to Project Management Institute (2017a), there are seven main processes under project risk management: (i) plan risk management, (ii) identify risks, (iii) perform qualitative risk analysis, (iv) perform quantitative risk analysis, (v) plan risk responses, (vi) implement risk responses and (vii) monitor risks. The tools and techniques associated with project risk management include prompt list, risk categorisation and analysis, data representation and strategies development (Project Management Institute, 2017a).

5.2.9   Project Procurement Management

Project procurement management is concerned with the processes necessary to purchase or acquire products and services needed required from outside the project team, including the development and administration of agreements (Project Management Institute, 2017a). Effective procurement management is essential to achieve the project objectives (De Araújo et al., 2017).

According to Project Management Institute (2017a), there are three main processes under project procurement management: (i) plan procurement management, (ii) conduct procurements and (iii) control procurements. The tools and techniques under project procurement management include source selection analysis, advertising, bidder conferences, negotiation, claims administration and procurement performance review (Project Management Institute, 2017a).

5.2.10   Project Stakeholder Management

Project stakeholder management refers to the processes required to identify, analyse, engage and manage people or organisations that could impact or be impacted by the project (Project Management Institute, 2017a). Stakeholders can influence a project in many ways and can determine project success and failure (Chinyio & Akintoye, 2008; Karlsen, 2002; Yuan et al., 2010). Hence, appropriate stakeholder management strategies to effectively engage stakeholders in project decisions and execution should be developed (Project Management Institute, 2017a). The managing of multiple stakeholders and maintaining an acceptable balance between their interests is crucial to successful project delivery (Karlsen, 2002). Effective stakeholder management was also found to be critical for project success (R. J. Yang & Shen, 2015).

According to Project Management Institute (2017a), there are four main processes under project stakeholder management: (i) identify stakeholders, (ii) plan stakeholder engagement, (iii) manage stakeholder engagement and (iv) monitor stakeholder engagement. The tools and techniques associated with project stakeholder management are similar to those associated with project communication management as it requires the management of the stakeholders who could impact or be impacted by the project.

5.3   Changes in Construction Project Management Knowledge and Skills

Industrial revolutions and their associated technologies not only transform project requirements and client demands; they also change the way construction projects are designed, constructed and managed (Costa et al., 2015). Changes in construction and construction project management processes can be expected alongside with the technical changes arising from advances in technologies (Froese, 2010). This is further evident in previous evolutions within construction projects.

For example, the introduction of CAD in 1957 led to a shift from manual drafting to digitalised drawing, changing the way projects are designed including the work processes and skills required (Taylor, 2007). In the construction phase, new construction technologies have been introduced as technologies advanced. According to CII, construction technology refers to the collection of innovative tools and software used during the construction phase of a project that enables advancement in field construction methods. Technology advancement brought about mechanisation of construction industry, such as the introduction of tower cranes and excavators, presenting opportunities for high-rise buildings and changing the work processes and skills required by workers (Ngowi et al., 2005).

Communication processes have also transformed from physical communications processes such as surface mails to computer-supported communications such as electronic mails (Froese, 2010). This transformation changes project communication among project stakeholders as a high degree of integration and collaboration across project tasks and project teams is required (Froese, 2010). With the advancement in digital technologies and the need for increased collaboration and integration among stakeholders, BIM was developed to overcome the major limitations of CAD and fragmentation of the construction value chain (Ghaffarianhoseini et al., 2017; Utiome et al., 2014). Although many benefits of BIM were found throughout the literature, one of the key challenges was the need for a change in work processes (Ayinla & Adamu, 2018; Gu & London, 2010; Howard et al., 2017), highlighting the need for changes in knowledge and skills to manage collaborative work processes.

Projects also grow in complexity as project requirements change and project scale increases (J.-S. Chou & Yang, 2012; Edum-Fotwe & McCaffer, 2000). Hence, the role of PMs has evolved and extended beyond the management of the technical aspects of the projects to other non-engineering roles, such as the management of relationships (Bourne & Walker, 2004; Edum-Fotwe & McCaffer, 2000; Pant & Baroudi, 2008; Zuo et al., 2018). It was observed that the PMBOK guide emphasises the technical aspects of construction project management (Bourne & Walker, 2004; Carvalho et al., 2015; Winter et al., 2006). However, effective construction project management requires the combination of both hard and soft skills (Carvalho et al., 2015; Ceran & Dorman, 1995; Pant & Baroudi, 2008; Zuo et al., 2018). Accordingly, general and soft skills such as leadership, decision-making skills and problem solving skills have become increasingly important (Carvalho et al., 2015; Edum-Fotwe & McCaffer, 2000; Zuo et al., 2018). Correspondingly, interpersonal skills have been included in the recent revisions of the PMBOK guide, highlighting the increasing importance of soft skills in PMs (Zuo et al., 2018).

As projects evolve, the effectiveness of the existing PMBOK on project success needs to be studied. Zwikael (2009) studied the relative importance of the nine PMBOK knowledge areas according to the fourth edition of the PMBOK guide. Project stakeholder management was not included as a PMBOK knowledge area until the fifth edition of the PMBOK guide introduced in 2013. The study found that project time, risk, scope, resource, integration, quality and communications management are statistically important to project success according to their relative importance (Zwikael, 2009). Specifically, the ranking of each knowledge area in according to their relative importance within the construction and engineering industry was found to be project integration, cost, resource, risk, communications, quality, time, procurement and scope management (Zwikael, 2009). In another study, Chou and Yang (2012) found that only project communication and procurement management had significant impacts on construction project performance, owner satisfaction and project success. In contrast, it was found that effective use of PMBOK techniques, tools and skills contribute to construction project success although the relative importance of the different knowledge areas varies across different countries (J. S. Chou et al., 2013).

It was further highlighted in Fernández-Solís et al. (2015) that PMBOK has produced reliable results in manufacturing companies and processes improving workflow, consistency and performance. However, the results from PMBOK are not as consistent when applied in the construction industry due to project complexity and poor information management and transfer among stakeholders (Fernández-Solís et al., 2015). The PMBOK guide has also been criticised to lack coverage in issues such as technology management, information management and sustainability issues (Goel et al., 2019; Zwikael, 2009). Moreover, project management theories may be obsolete and should be updated according to the new and advanced methods that have been developed (Koskela & Howell, 2002).

Accordingly, global trends have changed project requirements and construction processes. Due to resource scarcity and climate change, the demand for green buildings has been increasing (Hwang & Ng, 2013). The increased demand and need for green buildings have led to the development of standards and certifications such as LEED standards in the United States, Green Mark Scheme in Singapore and BREEAM in the United Kingdom (Shan & Hwang, 2018). Due to long project duration, a green building should be treated as a process instead of a product, and is commonly mentioned together with sustainable construction (Hwang & Tan, 2012; Wu & Low, 2010). However, green building construction projects present new requirements when compared with conventional construction projects and require new construction practices (Hwang & Tan, 2012). Other studies have also identified the need to integrate sustainability into project management (Banihashemi et al., 2017; Carvalho & Rabechini, 2017; Silvius, 2017). Project sustainability management was proposed by Carvalho and Rabechini (2017) as a link between sustainability and project management performances. Although the use of sustainability practices was found to be low, the study found positive and significant impacts of the proposed sustainability management construct on project success dimensions and social and environmental performance of the projects (Carvalho & Rabechini, 2017). Apart from project sustainability management, project information management has been proposed by Froese (2010) in view of the increasing project complexity and the need for integration and collaboration between project stakeholders brought about by the emerging information technologies.

Correspondingly, in the era of 4IR, it is expected that construction processes will change as they become increasingly digitalised, automated and integrated, and with more smart buildings being constructed. Hence, it is foreseeable that CPM must adapt to manage the new construction processes for smart buildings so that PMs can continuously deliver successful construction projects.

5.4 Skills Required to Manage the Key Smart Technologies in Projects

There is an increasing emphasis on soft skills of PMs to manage projects effectively. This is further reinforced in the management of key smart technologies in projects as these technologies require integrated and collaborative project approaches to be effective (Akanmu & Anumba, 2015; De Soto et al., 2019). Moreover, the present society has been characterised by many as a knowledge society, that is 'a society in which ideas and knowledge function as commodities' (Voogt & Roblin, 2012). The knowledge society also requires individuals to have a particular understanding of information, as opposed to the notion of information exchange (Voogt & Roblin, 2012).

This is evident in the changing job profiles of the 4IR from the execution of specific work tasks to interdisciplinary work tasks, which require changing competences of specific skills and knowledge to problem-solving skills and interdisciplinary knowledge (Dombrowski & Wagner, 2014). The WEF has also recognised that the skills required in most of the jobs will be changed with the proliferation of smart technologies and change in job profiles and that new job roles will be created (World Economic Forum, 2018). This is further supported by other studies such as Fantini et al. (2020) and Card and Nelson (2019). In order to be remain competent in the knowledge society, the term '21st century skills' has been coined to represent skills that are transversal, multidimensional and associated with higher-order skills and behaviours to cope with complex problems and unpredictable situations (Voogt & Roblin, 2012; Westera, 2001).

One of the essential skills required that is widely agreed by researchers and practitioners is IT or technological skills (Alvarenga et al., 2019; Chen et al., 2019; El-Sabaa, 2001; Gann & Senker, 1998; Hwang & Ng, 2013; Odusami, 2002). This is reasonable as PMs must know how to effectively use, manage, evaluate and produce information using the different technologies to deliver successful projects (Voogt & Roblin, 2012).

Voogt and Roblin (2012) compared the frameworks of 21st-century skills and found that the core skills required include (i) communication skills, (ii) collaboration skills, (iii) ICT-related competences, (iv) social and cultural awareness, (v) creativity, (vi) critical thinking, (vii) problem-solving skills and (viii) capacity to develop quality products. Van Laar et al. (2017) reviewed the literature on the skills needed in the digital environment and proposed a framework that consists of core and contextual 21st-century digital skills. The core 21st-century digital skills were identified as (i) technical skills, (ii) information management, (iii) communication skills, (iv) collaboration skills, (v) creativity, (vi) critical thinking and (vii) problem-solving skills, while the contextual 21st-century digital skills were identified as (i) ethical awareness, (ii) cultural awareness, (iii) flexibility, (iv) self-direction and (v) lifelong learning. Van Deursen and Mossberger (2018) used a typology of skills required of individuals to manage digital technologies effectively based on the media industry which includes the following: (i) operational skills, (ii) formal skills, (iii) information skills, (iv) communication skills, (v) content creation and (vi) strategic skills. Technical, operational and formal skills refer to the skills required to use the technology to accomplish tasks; information management and skills refer to the skills required to use the technology to efficiently search, select, process, evaluate and organise information to accomplish tasks; content creation refers to the skills required to create information that is required to accomplish tasks and strategic skills refer to skills that are required to use technology as a means to achieve goals and improve one's position in society (Van Deursen & Mossberger, 2018; Van Laar et al., 2017). WEF (2018) identified the top 10 trending skills in 2022 as follows: (i) analytical thinking and innovation, (ii) active learning and learning strategies, (iii) creativity, originality and initiative, (iv) technology design and programming, (v) critical thinking and analysis, (vi) complex problem solving, (vii) leadership and social influence, (viii) EI, (ix) reasoning, problem solving and ideation and (x) systems analysis and evaluation. It was highlighted that there are increasing demands on the various forms of technology competencies, along with soft skills, EI, leadership and social influence skills (World Economic Forum, 2018).

These studies were based on the general digital environment and do not focus on CPM. For CPM, on top of technical and project management skills, soft skills of PMs are increasingly emphasised in the face of automation, digitalisation and increasing project complexity. On top of these 21st-century skills and top trending skills identified, common soft skills required of PMs include (i) active listening, (ii) composure, (iii) conceptual thinking, (iv) conflict management, (v) decision making, (vi) delegation, (vii) motivation skills, (viii) negotiation, (ix) planning and organisation skills, (x) stress management and (xi) teamwork and team building (Alvarenga et al., 2019; Chen et al., 2019; Creasy & Anantatmula, 2013; Dainty et al., 2004; Edum-Fotwe & McCaffer, 2000; El-Sabaa, 2001; Fisher, 2011; Gann & Senker, 1998; Hwang & Ng, 2013; Odusami, 2002; Succar et al., 2013; Udo & Koppensteiner, 2004; F. Zhang et al., 2013; L. Zhang & Fan, 2013a; Zuo et al., 2018). Table 5.1 summarises the skills that are required of PMs to be competent in managing projects with smart technologies.

Table 5.1 Skills Required to Manage Projects with Smart Technologies

Skills	Description	1	2	3	4	5	6	7	8	9	10	11	12	13	14	15	16	17	18	19
Technical and Operational IT Skills	Skills required to use technology to accomplish job tasks	✓	✓	✓	✓	✓	✓				✓	✓	✓	✓	✓	✓				
Project Management	Skills required to manage the project to achieve project objectives (as highlighted in PMBOK)					✓	✓			✓	✓		✓	✓	✓	✓	✓	✓		
Information Management	Skills required to use technology to search, select, evaluate and organise information to accomplish job tasks	✓	✓	✓		✓	✓		✓								✓	✓		
Planning and Organising	Skills required to plan, organise and coordinate work tasks to achieve project objectives					✓	✓			✓			✓	✓	✓	✓	✓			
Communication	Skills required for information exchange, in an effective and accurate manner, and the ability to actively listen, persuade and understand what others mean	✓	✓		✓	✓	✓	✓	✓	✓	✓	✓	✓	✓	✓		✓	✓	✓	✓
Social, Cultural and Organisational Awareness	Skills to show cultural understanding and respect other cultures, ability to read others' emotions and to work with organisation systems and culture	✓	✓		✓	✓	✓			✓	✓	✓	✓	✓			✓	✓	✓	✓
Creativity	Skills required to generate new ideas or treat familiar ideas in a different way and transform such ideas into a novel way of doing things	✓	✓		✓			✓				✓								
Problem Solving	Skills required to conceive, analyse and reason project issues and problems to be able to resolve them and make appropriate decisions to achieve the project objectives. This includes critical thinking skills, analytical thinking skills and conceptual thinking skills	✓	✓		✓	✓	✓	✓	✓	✓	✓	✓	✓	✓	✓		✓	✓	✓	✓

(*Continued*)

Table 5.1 (Continued)

Skills	Description	1	2	3	4	5	6	7	8	9	10	11	12	13	14	15	16	17	18	19
Ethical Awareness	Behave in a socially responsible way, demonstrating awareness and knowledge of legal and ethical aspects when using technologies	✓																	✓	✓
Flexibility	Skills to adapt one's thinking, attitude or behaviour to changing situations				✓	✓	✓	✓			✓							✓	✓	
Strategic Planning Skills	Skills that are required to use the technology as a means to achieve project objectives			✓		✓	✓							✓	✓	✓	✓			✓
Active Learning	Skills required to learn new knowledge effectively to apply into the project	✓			✓															
Leadership	Skills required to lead groups to achieve project objectives by providing direction, vision, alignment of project stakeholders towards project success through communication, the motivation of project team members to achieve project objectives and mentoring of project team to develop team members				✓	✓	✓	✓	✓	✓	✓	✓	✓	✓	✓	✓	✓			
Social Influence	Skills to influence, persuade, convince people to support their agenda and complete project tasks				✓	✓	✓	✓	✓	✓	✓						✓			
Composure	Ability to keep emotions under control including stress management and when faced with hostility					✓	✓	✓	✓	✓	✓			✓				✓	✓	
Conflict Management	Ability to negotiate and resolve disagreements in projects and resolve conflicts in a satisfactory manner					✓	✓	✓			✓			✓	✓	✓	✓		✓	
Decision Making	Ability to identify key issues and problems and pick the optimal solution with available information and alternatives in a timely manner					✓				✓			✓	✓	✓	✓				✓

Competency	Description	1	2	3	4	5	6	7	8	9	10	11	12	13	14	15	16	17	18	19
Delegation	Ability to allocate project tasks to team members based on their strengths and weaknesses, and to provide opportunities for development					✓	✓		✓	✓		✓	✓		✓	✓		✓	✓	✓
Motivation	Ability to motivate the project team to accomplish project tasks to fulfil project objectives						✓		✓	✓	✓	✓			✓			✓		✓
Negotiation	Ability to negotiate win–win agreements					✓	✓			✓					✓	✓		✓	✓	✓
Teamwork	Skills to work cooperatively with the project team and the project stakeholders	✓	✓			✓	✓		✓	✓		✓			✓	✓	✓	✓	✓	✓
Team Building	Ability to build effective relationships with project team members and stakeholders, to enable the project team to work together cooperatively to achieve project objectives					✓	✓				✓				✓		✓	✓		✓
Initiative	Taking proactive actions to avert problems, to improve project performance and prevent problems					✓			✓									✓		✓

[1] (Voogt & Roblin, 2012) [2] (Van Laar et al., 2017) [3] (van Deursen & Mossberger, 2018) [4] (WEF, 2018) [5] (Alvarenga et al., 2019) [6] (Chen et al., 2019) [7] (Creasy & Anantatmula, 2013) [8] (Dainty et al., 2004) [9] (Edum-Fotwe & McCaffer, 2000) [10] (El-Sabaa, 2001) [11] (Fisher, 2011) [12] (Gann & Senker, 1998) [13] (Hwang & Ng, 2013) [14] (Odusami, 2002) [15] (Succar et al., 2013) [16] (Udo & Koppensteiner, 2004) [17] (Zhang et al., 2013) [18] (Zhang & Fan, 2013) [19] (Zuo et al., 2018).

5.5 Project Management Institute Project Manager Competency Development Framework

It has been widely recognised that PMs play a critical role in ensuring project success (Ahadzie et al., 2008; Davis, 2011; Krahn & Hartment, 2006; Project Management Institute, 2017a, 2017b; F. Zhang et al., 2013). Hence, many studies aim to study the critical competencies of PMs that contribute to project success (Alvarenga et al., 2019; Loufrani-Fedida & Missonier, 2015). In addition, Crawford (2005) found that knowledge in PMBOK areas do not relate to improved perceived performance. Instead, perceptions of workplace performance may be influenced by other factors such as personality and behavioural characteristics of the individual (Crawford, 2005). This is also supported by Dainty et al. (2004), who found that key behavioural competencies underlie superior project management performance. Hence, it is essential to develop the competencies of PMs to improve project management performance and gradually the probability of successful projects.

Professional project management associations such as PMI and IPMA aim to further the project management profession and have established frameworks to develop the competency of PMs in both hard and soft skills (Zuo et al., 2018). Soft skills are increasingly important as project complexity increases and previous studies have found that the lack of competencies in soft skills is one of the main causes of project failure (Le-Hoai et al., 2008; Ling et al., 2009; Zuo et al., 2018).

Hard skills are skills that are directly related to construction project management (Edum-Fotwe & McCaffer, 2000; Zuo et al., 2018). Examples of hard skills include forecasting techniques and estimating skills (J. S. Chou et al., 2013; J.-S. Chou & Yang, 2012; Edum-Fotwe & McCaffer, 2000). Specific tools and techniques have been elaborated in each PMBOK process (Project Management Institute, 2017a). On the other hand, soft skills are skills that are transferable across different industries (Edum-Fotwe & McCaffer, 2000). Examples of soft skills include leadership, interpersonal and decision-making and problem-solving skills (Carvalho et al., 2015; Ceran & Dorman, 1995; Edum-Fotwe & McCaffer, 2000).

The PMCD framework builds on the PMBOK guide and specifies the generic skills that are required of PMs. The PMCD framework assesses PMs' competencies in three dimensions: knowledge competence, performance competence and personal competence (Project Management Institute, 2017b). These are demonstrated through the PMs' behaviours (Project Management Institute, 2017b; Zuo et al., 2018). Knowledge competence refers to what the PM knows about the application of the PMBOK processes, tools and technologies for project activities; performance competence refers to how the PM applies project management knowledge from the PMBOK guide to meet the project objectives; and personal competence refers to the behaviours of the PM, their attitudes and core personality characteristics when performing the project management activities (Project Management Institute, 2017b). According to the PMCD Framework, PMs' six critical personal competences were identified as follows: (i) communication, (ii) leadership, (iii) management; (iv) cognitive ability, (v) effectiveness and (vi) professionalism (Project Management Institute, 2017b).

The PMI Talent Triangle was also introduced in the PMCD framework, which identified technical project management, strategic and business management skills and leadership as additional factors that are critical to the successful completion of the projects (Project Management Institute, 2017b). Technical project management skills are defined as the skills to effectively apply project management knowledge to deliver successful projects and are generally described in the PMBOK guide; strategic and business management skills involve the ability to see the high-level overview of the organisation and effectively negotiate and implement decisions and actions that support strategic alignment and innovation of the project to the organisation's mission and vision; and leadership skills involve the ability to guide, motivate and direct a team to achieve the project objectives (Project Management Institute, 2017a).

To develop competencies of PMs, the PMCD framework has proposed a competency development process with four steps:

(i)   Review requirements: to define the assessment criteria by reviewing organisation and individual requirements to identify existing gaps, define goals, scope and assessment baseline.
(ii) Assess competencies: the PM's competency is assessed using the assessment criteria from the previous step. This allows areas of strengths and weaknesses of PMs to be identified to determine how the competence can be further developed.
(iii) Prepare competency development plan: based on the results of the competency assessment, a competency development plan should be defined, which prescribes or recommends activities to be undertaken.
(iv) Implement competency development plan: the competency development plan is executed. This should include evaluation of the completed activities and achievement of the defined goals to monitor and track the development of PMs' competencies.

5.6   Common Construction Project Managers' Competencies

5.6.1   Leadership Competencies

Although it has been criticised by researchers that the studies of PMs' competencies have generated extensive lists of competencies that are required of PMs, core competencies that distinguish superior PMs from average PMs lie in a relatively small set of competencies (Alvarenga et al., 2019; Dulewicz & Higgs, 2005; Loufrani-Fedida & Missonier, 2015). One key competency required of PMs is leadership competency.

Leadership competencies are crucial for PMs in leading the project team in achieving project objectives (Alvarenga et al., 2019; Podgórska & Pichlak, 2019; Project Management Institute, 2017a; Sang et al., 2018). The lack of leadership competencies was highlighted to be one of the main reasons for project failure (Ahmed & Anantatmula, 2017; Krahn & Hartment, 2006).

Ahmed and Anantatmula (2017) identified 'clarity in communication', 'define roles and responsibilities', 'communicate expectations', 'employ consistent processes' and 'establish trust' as variables of PMs' leadership competencies and found that all the variables have a significant positive relationship with project performance. However, the variables identified in Ahmed and Anantatmula (2017) are competences and do not represent the underlying intents of the PMs in their behaviours. Alvarenga et al. (2019) identified 'leadership', 'decision making', 'initiative', 'commitment', 'achievement orientation' and 'management' as factors that affect leadership of PMs and found that leadership of PMs has the highest correlation with other competencies, including self-awareness, interpersonal, communication, technical, productivity and managerial competencies. This finding demonstrates the importance of leadership competencies of PMs in driving project success. Dainty et al. (2004) developed a competency-based performance model based on the McBer Dictionary and found that team leadership predicted superior CPM performance with a high confidence level. This was due to the high level of relative autonomy given to the PMs and the team-oriented nature of construction projects, which requires the PM to be able to lead, influence and motivate the team members to achieve the project objectives (Dainty et al., 2004).

On top of cognitive intelligence which affects technical and managerial skills, effective leaders should be emotionally and socially intelligent in order to motivate team members in achieving project objectives and manage team relationships (Ahmed et al., 2013; Alvarenga et al., 2019; Boyatzis et al., 2012; Podgórska & Pichlak, 2019; Riggio & Reichard, 2008; L.-R. Yang et al., 2011). Leadership styles were also found to affect project success (Dulewicz & Higgs, 2005; R. Müller & Turner, 2007; Nixon et al., 2012; L.-R. Yang et al., 2011). Accordingly, leadership styles can be seen as the verbal and non-verbal communication of one's leadership, and should be adapted according to the purpose and complexity of each project type (Charteris-Black, 2006; De Vries et al., 2010; Podgórska & Pichlak, 2019). Communication has frequently been cited as a critical skill for effective leadership (Kerns, 2016; L.-R. Yang et al., 2011; Zuo et al., 2018). However, for communication to be effective, one must understand the perceptions, experiences and feelings of the other party, in order to deliver the message as it is intended (Kerns, 2016). Hence, EI is another common competency required of PMs.

5.6.2   Emotional Intelligence Competencies

As the role of PMs has evolved beyond the traditional engineering role and includes the management of human relationships, EI competencies are critical in managing the project team in order to achieve the project objectives and was found to play a significant role in PMs' performance (Ahmed et al., 2013; Davis, 2011; Edum-Fotwe & McCaffer, 2000; Fisher, 2011; Sang et al., 2018; F. Zhang et al., 2013). This is in alignment with Boyatzis et al. (2012), who highlighted that 'all leadership interactions are, in part, emotional activities'. Walter et al. (2011) also suggested that EI is essential to leadership. This supports the finding of Goleman (2005), who argued that EI operationalises, initiates or makes effective technical knowledge and skills. In contrast, Lindebaum and Jordan (2012) found that the PM's EI does not have a statistically significant relationship with cognitive task-related performances. However, the study supports the argument that the PM's level of EI has an association with most relational performance dimensions (Lindebaum & Jordan, 2012). (Walter et al., 2011)

Using the McBer Dictionary, Dainty et al. (2004) found that on top of team leadership, composure can predict superior CPM performance with high confidence level. Composure refers to 'the ability to keep emotions under control and restrain negative actions when faced with opposition or working under conditions of stress' and is related to EI (Cheng et al., 2005; Spencer & Spencer, 1993). This is supported by Davis (2011), who found that tools measuring adaptation, stress tolerance, optimism, flexibility, impulse control and coping demonstrated stronger and more consistent relationships with PMs' interpersonal competencies that drive performance. Furthermore, Zhang and Fan (2013) modified Goleman's EI model and found a statistically significant positive correlation between PMs' total EI and project performance. In particular, self-awareness, self-control, empathy, organisational awareness, cultural understanding and communication were found to significantly affect project performance (L. Zhang & Fan, 2013b).

Zhang et al. (2013) found that the ability to work with and lead others plays a significant role in ensuring project success. Fisher (2011) further found that understanding behavioural characteristics of others, influencing others, leading others, authentizotic behaviour, conflict management and cultural awareness were critical factors that underlie effective performance in PMs. Zhang et al. (2018) argued that PMs with higher EI prefer to develop harmonious relationships and tend to create a more open communication environment for the project team, promoting collaboration satisfaction of the project team. Furthermore, statistically significant relationships were found between PM's total EI and PM's ability to 'organise and coordinate' and 'motivate and handle conflict' (Lindebaum & Jordan, 2012). These studies show that EI plays a critical role in a PM's performance. (L. Zhang et al., 2018)

5.6.3   Digital Intelligence

Marnewick and Marnewick (2021) proposed the concept of digital intelligence for PMs. The authors recognised that a new type of intelligence is required to manage the fundamental change in the way we live, work and relate to one another in the 4IR. Through the study, digital intelligence was conceptualised to include eight main components; (i) digital identity, (ii) digital use, (iii) digital safety, (iv) digital security, (v) digital emotional intelligence, (vi) digital communication, (vii) digital literacy and (viii) digital rights (Marnewick & Marnewick, 2021). The study found that the top factors affecting PMs' digital intelligence were the ability to use technology to effectively communicate and collaborate, the ability to empathise with one's own and others' feelings online, the ability to exercise self-control to manage both online and offline life, the ability to apply AI to guide decision-making processes, the ability to identify and manage cyber risks, the ability to develop oneself as a competent change maker and the ability to build and manage a healthy identify as a digital citizen with integrity (Marnewick & Marnewick, 2021). While this study identified digital intelligence as a key competency required of PMs, the study was not conducted in the context of the construction industry, which presents a specific set of challenges in project management.

5.7   Summary

The 10 PMBOK knowledge areas and the PMI PMCD framework were introduced in this chapter, together with the evolution of CPM along with the advancements in technologies. Leadership competencies, EI and the emerging digital intelligence were also introduced in this chapter. With the expected transformation of construction projects in the era of 4IR and the increasing emphasis on smart buildings, it is reasonable to assume that CPM knowledge and skills that are required to manage projects in the 4IR will be transformed, which was further studied in this research.

References

Ahadzie, D. K., Proverbs, D. G., & Olomolaiye, P. (2008). Towards developing competency-based measures for construction project managers: Should contextual behaviours be distinguished from task behaviours? *International Journal of Project Management, 26*, 631–645. https://doi.org/10.1016/j.ijproman.2007.09.011

Ahmed, R., & Anantatmula, V. S. (2017). Empirical study of project managers leadership competence and project performance. *Engineering Management Journal, 29*(3), 189–205. https://doi.org/10.1080/10429247.2017.1343005

Ahmed, R., Azmi, N., Masood, M. T., Tahir, M., & Ahmad, M. S. (2013). What does project leadership really do ? *International Journal of Scientific and Engineering Research, 4*(1), 1–8.

Akanmu, A., & Anumba, C. J. (2015). Cyber-physical systems integration of building information models and the physical construction. *Engineering, Construction and Architectural Management, 22*(5), 516–535. https://doi.org/10.1108/ECAM-07-2014-0097

Alashwal, A. M., & Chew, M. Y. (2017). Simulation techniques for cost management and performance in construction projects in Malaysia. *Built Environment Project and Asset Management, 7*(5), 534–545. https://doi.org/10.1108/BEPAM-11-2016-0058

Alvarenga, J. C., Branco, R. R., Guedes, A. L. A., Soares, C. A. P., & E Silva, W. D. S. (2019). The project manager core competencies to project success. *International Journal of Managing Projects in Business.* https://doi.org/10.1108/IJMPB-12-2018-0274

Ayinla, K. O., & Adamu, Z. (2018). Bridging the digital divide gap in BIM technology adoption. *Engineering, Construction and Architectural Management, 25*(10), 1398–1416. https://doi.org/10.1108/ECAM-05-2017-0091

Badir, Y. F., Büchel, B., & Tucci, C. L. (2012). A conceptual framework of the impact of NPD project team and leader empowerment on communication and performance: An alliance case context. *International Journal of Project Management, 30*(8), 914–926. https://doi.org/10.1016/j.ijproman.2012.01.013

Banihashemi, S., Hosseini, M. R., Golizadeh, H., & Sankaran, S. (2017). Critical success factors (CSFs) for integration of sustainability into construction project management practices in developing countries. *International Journal of Project Management, 35*(6), 1103–1119. https://doi.org/10.1016/j.ijproman.2017.01.014

Biff, B. (2018). Project quality management practice & theory. *American Journal of Management, 18*(3), 10–17. https://doi.org/10.33423/ajm.v18i3.69

Bourne, L., & Walker, D. H. T. (2004). Advancing project management in learning organizations. *The Learning Organization, 11*(3), 226–243. https://doi.org/10.1108/09696470410532996

Boyatzis, R. E., Good, D., & Massa, R. (2012). Emotional, social, and cognitive intelligence and personality as predictors of sales leadership performance. *Journal of Leadership & Organizational Studies, 19*(2), 191–201. https://doi.org/10.1177/1548051811435793

Bronte-Stewart, M. (2015). Beyond the iron triangle: Evaluating aspects of success and failure using a project status model. *Computing & Information Systems, 19*(2), 21–37.

Brynjolfsson, E., & McAfee, A. (2014). *The second machine age: Work, progress, and prosperity in a time of brilliant technologies.* Norton & Company.

Card, D., & Nelson, C. (2019). How automation and digital disruption are shaping the workforce of the future. *Strategic HR Review, 18*(6), 242–245. https://doi.org/10.1108/SHR-08-2019-0067

Carvalho, M. M., Patah, L. A., & de Souza Bido, D. (2015). Project management and its effects on project success: Cross-country and cross-industry comparisons. *International Journal of Project Management, 33*(7), 1509–1522. https://doi.org/10.1016/j.ijproman.2015.04.004

Carvalho, M. M., & Rabechini, R. (2017). Can project sustainability management impact project success? An empirical study applying a contingent approach. *International Journal of Project Management, 35*(6), 1120–1132. https://doi.org/10.1016/j.ijproman.2017.02.018

Ceran, T., & Dorman, A. A. (1995). The complete project manager. *Journal of Architectural Engineering, 1*(2), 67–72. https://doi.org/10.1061/(ASCE)1076-0431(1995)1:2(67)

Charteris-Black, J. (2006). The communication of leadership: The design of leadership style. In *The communication of leadership: The design of leadership style.* Routledge Taylor & Francis Group. https://doi.org/10.4324/9780203968291

Chen, T., Fu, M., Liu, R., Xu, X., Zhou, S., & Liu, B. (2019). How do project management competencies change within the project management career model in large Chinese construction companies? *International Journal of Project Management, 37*(3), 485–500. https://doi.org/10.1016/j.ijproman.2018.12.002

Cheng, M. I., Dainty, A. R. J., & Moore, D. R. (2005). Towards a multidimensional competency-based managerial performance framework: A hybrid approach. *Journal of Managerial Psychology, 20*(5), 380–396. https://doi.org/10.1108/02683940510602941

Chinyio, E. A., & Akintoye, A. (2008). Practical approaches for engaging stakeholders: Findings from the UK. *Construction Management and Economics, 26*(6), 591–599. https://doi.org/10.1080/01446190802078310

Chou, J. S., Irawan, N., & Pham, A. D. (2013). Project management knowledge of construction professionals: Cross-country study of effects on project success. *Journal of Construction Engineering and Management, 139*(11). https://doi.org/10.1061/(ASCE)CO.1943-7862.0000766

Chou, J.-S., & Yang, J.-G. (2012). Project management knowledge and effects on construction project outcomes: An empirical study. *Project Management Journal, 43*(5), 47–67. https://doi.org/10.1002/pmj.21293

Costa, A. A., Lopes, P. M., Antunes, A., Cabral, I., Grilo, A., & Rodrigues, F. M. (2015). 3I Buildings: Intelligent, interactive and immersive buildings. *Procedia Engineering, 123,* 7–14. https://doi.org/10.1016/j.proeng.2015.10.051

Crawford, L. (2000). Profiling the competent project manager. *PMI Research Conference 2000* (pp. 3–15). https://www.pmi.org/learning/library/profiling-competent-project-manager-8524

Crawford, L. (2005). Senior management perceptions of project management competence. *International Journal of Project Management, 23*(1), 7–16. https://doi.org/10.1016/j.ijproman.2004.06.005

Creasy, T., & Anantatmula, V. S. (2013). From every direction – how personality traits and dimensions of project managers can conceptually affect project success. *Project Management Journal, 44*(6), 36–51. https://doi.org/10.1002/pmj.21372

Dainty, A. R. J., Cheng, M. I., & Moore, D. R. (2004). A competency-based performance model for construction project managers. *Construction Management and Economics, 22*(8), 877–886. https://doi.org/10.1080/0144619042000202726

Davis, S. A. (2011). Investigating the impact of project managers' emotional intelligence on their interpersonal competence. *Project Management Journal, 42*(4), 37–57. https://doi.org/10.1002/pmj.20247

De Araújo, M. C. B., Alencar, L. H., & De Miranda Mota, C. M. (2017). Project procurement management: A structured literature review. *International Journal of Project Management, 35*(3), 353–377. https://doi.org/10.1016/j.ijproman.2017.01.008

Demirkesen, S., & Ozorhon, B. (2017). Impact of integration management on construction project management performance. *International Journal of Project Management, 35*(8), 1639–1654. https://doi.org/10.1016/j.ijproman.2017.09.008

De Soto, G. B., Agustí-Juan, I., Joss, S., & Hunhevicz, J. (2019). Implications of Construction 4.0 to the workforce and organizational structures. *International Journal of Construction Management*, 1–13. https://doi.org/10.1080/15623599.2019.1616414

De Vries, R. E., Bakker-Pieper, A., & Oostenveld, W. (2010). Leadership = communication? The relations of leaders' communication styles with leadership styles, knowledge sharing and leadership outcomes. *Journal of Business and Psychology, 25*(3), 367–380. https://doi.org/10.1007/S10869-009-9140-2

Dombrowski, U., & Wagner, T. (2014). Mental strain as field of action in the 4th industrial revolution. *Procedia CIRP, 17*, 100–105. https://doi.org/10.1016/j.procir.2014.01.077

Dulewicz, V., & Higgs, M. (2005). Assessing leadership styles and organisational context. *Journal of Managerial Psychology, 20*(2), 105–123. https://doi.org/10.1108/02683940510579759

Edum-Fotwe, F. T., & McCaffer, R. (2000). Developing project management competency: Perspectives from the construction industry. *International Journal of Project Management, 18*, 111–124.

El-Sabaa, S. (2001). The skills and career path of an effective project manager. *International Journal of Project Management, 19*(1), 1–7. https://doi.org/10.1016/S0263-7863(99)00034-4

Fageha, M. K., & Aibinu, A. A. (2013). Managing project scope definition to improve stakeholders' participation and enhance project outcome. *Procedia – Social and Behavioral Sciences, 74*, 154–164. https://doi.org/10.1016/j.sbspro.2013.03.038

Fantini, P., Pinzone, M., & Taisch, M. (2020). Placing the operator at the centre of Industry 4.0 design: Modelling and assessing human activities within cyber-physical systems. *Computers & Industrial Engineering, 139*, 105058. https://doi.org/10.1016/j.cie.2018.01.025

Fernández-Solís, J. L., Rybkowski, Z. K., Xiao, C., Lü, X., & Chae, L. S. (2015). General contractor's project of projects – a meta-project: Understanding the new paradigm and its implications through the lens of entropy. *Architectural Engineering and Design Management, 11*(3), 213–242. https://doi.org/10.1080/17452007.2014.892470

Fisher, E. (2011). What practitioners consider to be the skills and behaviours of an effective people project manager. *International Journal of Project Management, 29*(8), 994–1002. https://doi.org/10.1016/j.ijproman.2010.09.002

Froese, T. M. (2010). The impact of emerging information technology on project management for construction. *Automation in Construction, 19*(5), 531–538. https://doi.org/10.1016/j.autcon.2009.11.004

Gann, D., & Senker, P. (1998). Construction skills training for the next millennium. *Construction Management and Economics, 16*(5), 569–580. https://doi.org/10.1080/014461998372105

Ghaffarianhoseini, A., Tookey, J., Ghaffarianhoseini, A., Naismith, N., Azhar, S., Efimova, O., & Raahemifar, K. (2017). Building Information Modelling (BIM) uptake: Clear benefits, understanding its implementation, risks and challenges. *Renewable and Sustainable Energy Reviews, 75*, 1046–1053. https://doi.org/10.1016/j.rser.2016.11.083

Goel, A., Ganesh, L. S., & Kaur, A. (2019). Sustainability integration in the management of construction projects: A morphological analysis of over two decades' research literature. *Journal of Cleaner Production, 236*, 117676. https://doi.org/10.1016/j.jclepro.2019.117676

Goleman, D. (2005). *Emotional intelligence: Why it can matter more than IQ*. Bantam Books Inc.

Gu, N., & London, K. (2010). Understanding and facilitating BIM adoption in the AEC industry. *Automation in Construction*, *19*, 988–999. https://doi.org/10.1016/j.autcon.2010.09.002

Howard, R., Restrepo, L., & Chang, C.-Y. (2017). Addressing individual perceptions: An application of the unified theory of acceptance and use of technology to building information modelling. *International Journal of Project Management*, *35*(2), 107–120. https://doi.org/10.1016/j.ijproman.2016.10.012

Hwang, B.-G., & Ng, W. J. (2013). Project management knowledge and skills for green construction: Overcoming challenges. *International Journal of Project Management*, *31*(2), 272–284. https://doi.org/10.1016/j.ijproman.2012.05.004

Hwang, B.-G., & Tan, J. S. (2012). Green building project management: Obstacles and solutions for sustainable development. *Sustainable Development*, *20*(5), 335–349. https://doi.org/10.1002/sd.492

Hwang, B.-G., Zhao, X., & Toh, L. P. (2014). Risk management in small construction projects in Singapore: Status, barriers and impact. *International Journal of Project Management*, *32*(1), 116–124. https://doi.org/10.1016/j.ijproman.2013.01.007

Karlsen, J. T. (2002). Project stakeholder management. *EMJ – Engineering Management Journal*, *14*(4), 19–24. https://doi.org/10.1080/10429247.2002.11415180

Kerns, C. (2016). High-impact communicating: A key leadership practice. *Journal of Applied Business and Economics*, *18*(5), 11–22.

Khan, A. (2006). Project scope management. *Cost Engineering*, *48*(6), 12–16. https://doi.org/10.1201/b12717-6

Koskela, L., & Howell, G. (2002). The underlying theory of project management is obsolete. *Proceedings of the PMI Research Conference* (pp. 293–302). https://www.pmi.org/learning/library/underlying-theory-project-management-obsolete-8971

Krahn, J., & Hartment, F. (2006). Effective project leadership: Project manager skills and competencies. *PMI Research Conference: New Directions in Project Management*. https://www.pmi.org/learning/library/leadership-project-manager-skills-competencies-8115

Le-Hoai, L., Lee, Y. D., & Lee, J. Y. (2008). Delay and cost overruns in Vietnam large construction projects: A comparison with other selected countries. *KSCE Journal of Civil Engineering*, *12*(6), 367–377. https://doi.org/10.1007/s12205-008-0367-7

Leviäkangas, P., Paik, S. M., & Moon, S. (2017). Keeping up with the pace of digitization: The case of the Australian construction industry. *Technology in Society*, *50*, 33–43. https://doi.org/10.1016/j.techsoc.2017.04.003

Lindebaum, D., & Jordan, P. J. (2012). Relevant but exaggerated: The effects of emotional intelligence on project manager performance in construction. *Construction Management and Economics*, *30*(7), 575–583. https://doi.org/10.1080/01446193.2011.593184

Ling, F. Y. Y., Pham, V. M. C., & Hoang, T. P. (2009). Strengths, weaknesses, opportunities, and threats for architectural, engineering, and construction firms: Case study of Vietnam. *Journal of Construction Engineering and Management*, *135*(10), 1105–1113. https://doi.org/10.1061/(ASCE)CO.1943-7862.0000069

Loufrani-Fedida, S., & Missonier, S. (2015). The project manager cannot be a hero anymore! Understanding critical competencies in project-based organizations from a multilevel approach. *International Journal of Project Management*, *33*, 1220–1235. https://doi.org/10.1016/j.ijproman.2015.02.010

Low, S. P., Gao, S., & Ng, E. W. L. (2021). Future-ready project and facility management graduates in Singapore for Industry 4.0: Transforming mindsets and competencies. *Engineering, Construction and Architectural Management*, *28*(1), 270–290. http://doi.org/10.1108/ECAM-08-2018-0322

Marnewick, C., & Marnewick, A. (2021). Digital intelligence: A must-have for project managers. *Project Leadership and Society*, *2*, 100026. https://doi.org/10.1016/j.plas.2021.100026

Müller, J. M., Kiel, D., & Voigt, K. I. (2018). What drives the implementation of Industry 4.0? The role of opportunities and challenges in the context of sustainability. *Sustainability*, *10*(1), 247–270. https://doi.org/10.3390/su10010247

Müller, R., & Turner, J. R. (2007). Matching the project manager's leadership style to project type. *International Journal of Project Management, 25*(1), 21–32. https://doi.org/10.1016/j.ijproman.2006.04.003

Ngo, J., & Hwang, B. -G. (2022). Critical project management knowledge and skills for managing projects with smart technologies. *Journal of Management in Engineering, 38*(6), 05022013. https://doi.org/10.1061/(ASCE)ME.1943-5479.0001095

Ngowi, A. B., Pienaar, E., Talukhaba, A., & Mbachu, J. (2005). The globalisation of the construction industry – A review. *Building and Environment, 40*(1), 135–141. https://doi.org/10.1016/j.buildenv.2004.05.008

Nixon, P., Harrington, M., & Parker, D. (2012). Leadership performance is significant to project success or failure: A critical analysis. *International Journal of Productivity and Performance Management, 61*(2), 204–216. https://doi.org/10.1108/17410401211194699

Odusami, K. T. (2002). Perceptions of construction professionals concerning important skills of effective project leaders. *ASCE Journal of Management in Engineering, 18*(2), 61–67. https://doi.org/10.1061/(ASCE)0742-597X(2002)18:2(61)

Pant, I., & Baroudi, B. (2008). Project management education: The human skills imperative. *International Journal of Project Management, 26*(2), 124–128. https://doi.org/10.1016/j.ijproman.2007.05.010

Pereira, A. C., & Romero, F. (2017). A review of the meanings and the implications of the Industry 4.0 concept. *Procedia Manufacturing, 13*, 1206–1214. https://doi.org/10.1016/j.promfg.2017.09.032

Podgórska, M., & Pichlak, M. (2019). Analysis of project managers' leadership competencies: Project success relation: What are the competencies of polish project leaders? *International Journal of Managing Projects in Business*. https://doi.org/10.1108/IJMPB-08-2018-0149

Popaitoon, S., & Siengthai, S. (2014). The moderating effect of human resource management practices on the relationship between knowledge absorptive capacity and project performance in project-oriented companies. *International Journal of Project Management, 32*(6), 908–920. https://doi.org/10.1016/j.ijproman.2013.12.002

Project Management Institute. (2017a). *A guide to the project management body of knowledge*. Project Management Institute.

Project Management Institute. (2017b). *Project manager competency development framework*. Project Management Institute.

Riggio, R. E., & Reichard, R. J. (2008). The emotional and social intelligences of effective leadership: An emotional and social skill approach. *Journal of Managerial Psychology, 23*(2), 169–185. https://doi.org/10.1108/02683940810850808

Sang, P., Liu, J., Zhang, L., Zheng, L., Yao, H., & Wang, Y. (2018). Effects of project manager competency on green construction performance: The Chinese context. *Sustainability (Switzerland), 10*(10). https://doi.org/10.3390/su10103406

Schwab, K. (2016). *The fourth industrial revolution*. PENGUIN GROUP.

Shan, M., & Hwang, B. (2018). Green building rating systems: Global reviews of practices and research efforts. *Sustainable Cities and Society, 39*, 172–180. https://doi.org/10.1016/j.scs.2018.02.034

Shenhar, A. J., Tishler, A., Dvir, D., Lipovetsky, S., & Lechler, T. (2002). Refining the search for project success factors: A multivariate, typological approach. *R&D Management, 32*(2), 111–126. https://doi.org/10.1111/1467-9310.00244

Silvius, G. (2017). Sustainability as a new school of thought in project management. *Journal of Cleaner Production, 166*, 1479–1493. https://doi.org/10.1016/j.jclepro.2017.08.121

Spencer, L. M., & Spencer, S. M. (1993). *Competence at work: Models for superior performance*. Wiley.

Succar, B., Sher, W., & Williams, A. (2013). An integrated approach to BIM competency assessment, acquisition and application. *Automation in Construction, 35*, 174–189. https://doi.org/10.1016/j.autcon.2013.05.016

Tabassi, A. A., Roufechaei, K. M., Ramli, M., Bakar, A. H. A., Ismail, R., & Pakir, A. H. K. (2016). Leadership competences of sustainable construction project managers. *Journal of Cleaner Production, 124*, 339–349. https://doi.org/10.1016/j.jclepro.2016.02.076

Taylor, J. E. (2007). Antecedents of successful three-dimensional computer-aided design implementation in design and construction networks. *Journal of Construction Engineering and Management, 133*(12), 993–1002. https://doi.org/10.1061/(ASCE)0733-9364(2007)133: 12(993)

Trivellas, P., & Drimoussis, C. (2013). Investigating leadership styles, behavioural and managerial competency profiles of successful project managers in Greece. *Procedia – Social and Behavioral Sciences, 73*, 692–700. https://doi.org/10.1016/j.sbspro.2013.02.107

Udo, N., & Koppensteiner, S. (2004). Core competencies of a successful project manager skill. *PMI Global Congress 2004.* https://www.pmi.org/learning/library/core-competencies-successful-skill-manager-8426

Utiome, E., Drogemuller, R., & Docherty, M. (2014). Reducing building information fragmentation: A BIM-specifications approach. *Proceedings of the CIB 2014 International Conference on Construction in a Changing World* (pp. 1–12). https://www.researchgate.net/publication/295402655_Reducing_building_information_fragmentation_a_BIM-specifications_approach

Van Deursen, A. J. A. M., & Mossberger, K. (2018). Any thing for anyone? A new digital divide in Internet-of-Things skills. *Policy & Internet, 10*(2), 122–140. https://doi.org/10.1002/poi3.171

Van Laar, E., Van Deursen, A. J. A. M., Van Dijk, J. A. G. M., & De Haan, J. (2017). The relation between 21st-century skills and digital skills: A systematic literature review. *Computers in Human Behavior, 72*, 577–588. https://doi.org/10.1016/j.chb.2017.03.010

Voogt, J., & Roblin, N. P. (2012). A comparative analysis of international frameworks for 21st century competences: Implications for national curriculum policies. *Journal of Curriculum Studies, 44*(3), 299–321. https://doi.org/10.1080/00220272.2012.668938

Walter, F., Cole, M. S., & Humphrey, R. H. (2011). Emotional intelligence: Sine qua non of leadership or folderol? *Source: Academy of Management Perspectives, 25*(1), 45–59. www.jstor.org/stable/23045035

Westera, W. (2001). Competences in education: A confusion of tongues. *Journal of Curriculum Studies, 33*(1), 75–88. https://doi.org/10.1080/00220270120625

Winter, M., Smith, C., Morris, P., & Cicmil, S. (2006). Directions for future research in project management: The main findings of a UK government-funded research network. *International Journal of Project Management, 24*(8), 638–649. https://doi.org/10.1016/j.ijproman.2006.08.009

World Economic Forum. (2018). *The future of jobs* (p. 147). World Economic Forum.

Wu, P., & Low, S. P. (2010). Project management and green buildings: Lessons from the rating systems. *Journal of Professional Issues in Engineering Education and Practice, 136*(2), 64–70. https://doi.org/10.1061/(ASCE)EI.1943-5541.0000006

Yang, L.-R., Huang, C.-F., & Wu, K.-S. (2011). The association among project manager's leadership style, teamwork and project success. *International Journal of Project Management, 29*(3), 258–267. https://doi.org/10.1016/j.ijproman.2010.03.006

Yang, R. J., & Shen, G. Q. P. (2015). Framework for stakeholder management in construction projects. *Journal of Management in Engineering, 31*(4), 04014064. https://doi.org/10.1061/(ASCE)ME.1943-5479.0000285

Yap, J. B. H., Abdul-Rahman, H., & Wang, C. (2018). Preventive mitigation of overruns with project communication management and continuous learning: PLS-SEM approach. *Journal of Construction Engineering and Management, 144*(5), 04018025. https://doi.org/10.1061/(ASCE)CO.1943-7862.0001456

Yuan, J., Skibniewski, M. J., Li, Q., & Zheng, L. (2010). Performance objectives selection model in public-private partnership projects based on the perspective of stakeholders. *Journal of Management in Engineering, 26*(2), 89–104. https://doi.org/10.1061/(ASCE)ME.1943-5479.0000011

Zhang, F., Zuo, J., & Zillante, G. (2013). Identification and evaluation of the key social competencies for Chinese construction project managers. *International Journal of Project Management, 31*(5), 748–759. https://doi.org/10.1016/j.ijproman.2012.10.011

Zhang, L., Cao, T., & Wang, Y. (2018). The mediation role of leadership styles in integrated project collaboration: An emotional intelligence perspective. *International Journal of Project Management, 36*(2), 317–330. https://doi.org/10.1016/j.ijproman.2017.08.014

Zhang, L., & Fan, W. (2013a). Improving performance of construction projects: A project manager's emotional intelligence approach. *Engineering, Construction and Architectural Management, 20*(2), 195–207. https://doi.org/10.1108/09699981311303044

Zhang, L., & Fan, W. (2013b). Improving performance of construction projects: A project manager's emotional intelligence approach. *Engineering, Construction and Architectural Management, 20*(2), 195–207. https://doi.org/10.1108/09699981311303044

Zuo, J., Zhao, X., Nguyen, Q. B. M., Ma, T., & Gao, S. (2018). Soft skills of construction project management professionals and project success factors: A structural equation model. *Engineering, Construction and Architectural Management, 25*(3), 425–442. https://doi.org/10.1108/ECAM-01-2016-0016

Zwikael, O. (2009). The relative importance of the PMBOK® guide's nine knowledge areas during project planning. *Project Management Journal, 40*(4), 94–103. https://doi.org/10.1002/pmj.20116

6 Theories of Technology Adoption and Acceptance

6.1 Introduction

The exploration of human behaviours and the acceptance level of technology has been the focus of academic studies, elucidating how individual attributes affect the efficiency and effectiveness of technology adoption. Among the theoretical models that explain and predict an individual's adoption of new technologies in the workplace, the TRA stands out as one of the most widely accepted foundations for subsequent technology-acceptance-related theories. Despite its wide-ranging acceptance, the TRA often encounters criticism due to its broad scope and restricted applications in specific contexts.

The TPB, an extension of the TRA, introduced an additional construct – perceived behavioural control, aimed at predicting one's behavior. Subsequently, Davis (1986) proposed the TAM to specifically elucidate and forecast an individual's acceptance of computer-based ISs within a professional environment. The TAM postulates that an individual's ATT can be influenced by PU and PEU. To augment its explanatory power, later research has modified and expanded the TAM by incorporating additional impact factors that influence individuals' PU and PEU.

Another critical theory in the realm of technology acceptance is the IDT. The IDT explicates the dispersion of technologies at multiple levels, encompassing global, organisational and individual perspectives. Nevertheless, it exhibits less predictive power when it comes to individual technology adoption.

Additional technology acceptance theories such as the SCT, MPCU, MM, UTAUT, PIIT and TRI offer further insights into the determinants, particularly personal attributes, that influence an individual's technology acceptance. These theories collectively contribute to a comprehensive understanding of the various factors that drive the acceptance and utilisation of technology in modern professional settings.

6.2 Technology Acceptance Model

The TAM is one of the most widely cited models to explain and predict the adoption of new technologies of an individual in the workplace (Lai, 2017; Taherdoost, 2018; Venkatesh, 2000). This section provides the theoretical underpinnings of

DOI: 10.1201/9781003462231-6

the TAM and the variations of the TAM that have been proposed to improve the adaptability, explanatory and prediction power of TAM (Momani & Jamous, 2017; Taherdoost, 2018).

6.2.1 Theory of Reasoned Action

TRA, developed by Fishbein and Ajzen (1975), was one of the earliest theories developed to predict and explain any human behaviour. TRA adopts the constructs *attitude towards behaviour* and *subjective norm* as determinants of human BI, which subsequently determines actual behaviour. *Attitude towards behaviour* is defined as 'an individual's positive or negative feelings (evaluative affect) about performing the target behaviour' and subjective norm is defined as 'an individual's perception that most people who are important to him think he should or should not perform the behaviour in question' (Fishbein & Ajzen, 1975). Figure 6.1 shows the constructs and their relationship with actual behaviour according to TRA.

Although TRA forms the foundation of many theories, TRA is limited in some aspects. TRA explains and predicts the target behaviour within a specified target, action, context and time frame, and hence the psychological variables of the model must be defined and measured specifically at the corresponding behavioural criterion to be explained (Davis, 1986). TRA is also based on a relatively static construct of attitude towards behaviour, which may change depending on the context of use and environment (Momani & Jamous, 2017; Sheppard et al., 1988). Furthermore, TRA does not consider other variables that influence BI such as fear, mood, values, habits or previous experiences (Momani & Jamous, 2017; Taherdoost, 2018). Lastly, as TRA is developed to explain human behaviours in general, modifications need to be made for TRA to be effective in explaining specific human behaviours.

6.2.2 Theory of Planned Behavior

TPB is an extension of TRA developed by Ajzen in 1985 with an added construct *perceived behavioural control*, which was theorised as an additional determinant of BI. PBC is defined as 'an individual's perception of the ease or difficulty of performing the behaviour of interest which in turn depends on the individual's self-efficacy

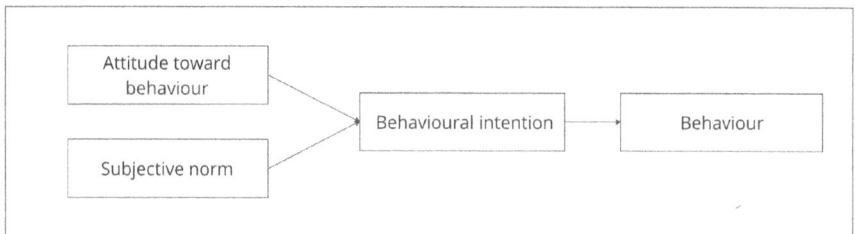

Figure 6.1 TRA

or the judgement of how well one can execute courses of action required to deal with prospective situations' (Ajzen, 1991). In this sense, PBC is grounded in the self-efficacy theory that has been proposed by Bandura (1986). Further, PBC assumes that individuals' actions are not under volitional control (Taherdoost, 2018). As TPB is essentially an extension of TRA, TPB shares similar limitations as TRA (Momani & Jamous, 2017). The TPB model is shown in Figure 6.2.

6.2.3   *Original Technology Acceptance Model*

Both TRA and TPB were developed to explain and predict human behaviour in general. In contrast, Davis (1986) developed the TAM to explain and predict specifically an individual's acceptance of computer-based ISs within the setting of the workplace. The TAM is developed within the field of IS using TRA as the theoretical foundation (Davis, 1986). The construct *subjective norm* has been excluded in the TAM due to the uncertainty in measuring the individual's specific beliefs which influence the determination of *subjective norm* (Davis, 1986). BI was also excluded in the TAM as the user acceptance testing context differs from the prediction of behaviour in general, in which the measurements of the user's motivation to use the system are taken directly after demonstration. This is in contrast with TRA's measurement of BI, where the time period between the BI and actual behaviour is affected by time (Davis, 1986).

The construct attitude towards behaviour was explained and determined by the constructs PU and PEU. PU is defined as 'the degree to which an individual believes that using a particular system would enhance his job performance' and PEU is defined as 'the degree to which an individual believes that using a particular system would be free of physical and mental effort' (Davis, 1986, p. 26). In TAM, the design features of the IS influence the PU and PEU of the individual, and PEU affects PU. Both PU and PEU are the cognitive response of the individual towards

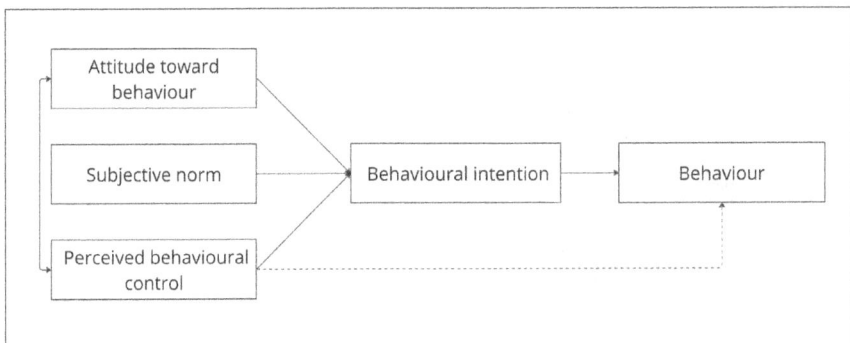

Figure 6.2 TPB

the IS, and is hypothesised to affect the individual's *attitude towards using* the IS, forming the individual's motivation of using the IS. Subsequently, the user's *attitude towards using* the IS will be a major determinant of his actual system use. The original TAM is shown in Figure 6.3.

As the original TAM proposed in Davis (1986) measures the user's motivation to use the IS immediately after the demonstration, the TAM was first modified in Davis et al. (1989) to include BI as a link between *attitude towards using the IS* and the actual system use.

The modified TAM was further refined in Venkatesh and Davis (1996), who excluded the construct *attitude towards using* in the TAM. It was found that users' BI is a better predictor of actual system use than *attitude towards using*, and that PU and PEU were found to directly influence BI, and hence *attitude towards using* was omitted to explain BI parsimoniously (Lai, 2017; Venkatesh et al., 2003; Venkatesh & Davis, 1996). The finalised version of the TAM is shown in Figure 6.4.

TAM was found to consistently explain about 40 per cent of variance in an individual's BI to use the IS (Venkatesh & Bala, 2008), and is one of the most widely used models to explain and predict technology adoption of individuals due to its simplicity, robustness and explanatory power (King & He, 2006; Lai, 2017; Sharma & Mishra, 2014; Venkatesh, 2000). However, TAM does not consider the influences of social factors on technology adoption. Hence, the application of TAM beyond the context of the workplace may be limited (Taherdoost, 2018). Furthermore, TAM is an extension of TRA and carries similar limitations from the model (Momani & Jamous, 2017). The TAM also assumes a linear relationship and does not consider factors that may enhance adoption such as integration, flexibility and completeness of information (Legris et al., 2003; Momani & Jamous, 2017). Finally, the parsimonious nature of TAM is also its key limitation, where the understanding of the determinants that contribute to PU and PEU are not sufficiently provided (Lee et al., 2003; Venkatesh, 2000).

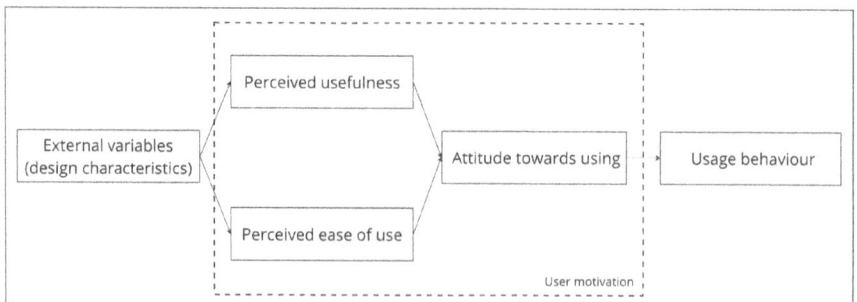

Figure 6.3 TAM

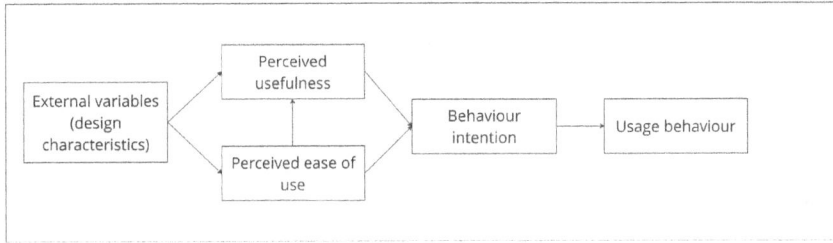

Figure 6.4 Finalised TAM

6.2.4   Extensions of Technology Acceptance Model

To improve the explanatory power of TAM, the TAM has been extended to include antecedents of PU and PEU to explain the constructs from the views of social influences and cognitive instrumental processes (Momani & Jamous, 2017; Venkatesh & Davis, 2000). The antecedents of PU were developed in TAM2, which included subjective norm, voluntariness, image, job relevance, output quality and results demonstrability (Venkatesh & Davis, 2000), while the antecedents of PEU were developed in TAM3, which included external control, computer self-efficacy, computer anxiety, computer playfulness, perceived enjoyment and objective usability (Venkatesh & Bala, 2008).

Previous studies on TAM have shown that PU is a stronger determinant of BI (Davis, 1986; Davis et al., 1989; Venkatesh & Davis, 2000). Experience was also found to influence PEU, which in turn influences PU (Venkatesh & Davis, 1996, 2000). Accordingly, TAM2 extended the TAM to include additional determinants of PU and considers the influence of users' experience over time with the target system on the additional determinants.

Subjective norm, voluntariness and image were included to reflect the social influences on an individual's intention to adopt a new IS. Voluntariness is defined as 'the extent to which potential adopters perceive the adoption decision to be non-mandatory' (Moore & Benbasat, 1991; Venkatesh & Davis, 2000). Voluntariness is posited as a moderating variable in TAM2 as previous studies have found that *subjective norm* only had significant effect on BI in mandatory settings (Hartwick & Barki, 1994; Venkatesh & Davis, 2000). *Image* is defined as 'the degree to which use of an innovation is perceived to enhance one's . . . status in one's social system' (Moore & Benbasat, 1991; Venkatesh & Davis, 2000), and is hypothesised to be positively influenced by subjective norm as 'performing the behaviour that important members of the individual's social group believe he should perform will tend to elevate his standing within the group' (Venkatesh & Davis, 2000). It was further hypothesised that the direct positive effect of subjective norm on BI (for mandatory systems) and on PU (for both mandatory and non-mandatory systems) will attenuate with increased experience (Venkatesh & Davis, 2000).

Job relevance, output quality, results demonstrability and PEU were the four cognitive instrumental determinants of PU in TAM2. Job relevance is defined as 'an individual's perception regarding the degree to which the target system is applicable to his job' (Venkatesh & Davis, 2000); output quality is defined as 'perception of how well the system performs the job-related tasks' (Venkatesh & Davis, 2000) and results demonstrability is defined as 'the tangibility of the results of using the innovation' (Moore & Benbasat, 1991; Venkatesh & Davis, 2000). It was hypothesised that there are no specific temporal shifts in the strength of the effects of cognitive instrumental determinants on PU and BI.

TAM2 was found to explain up to 60 per cent of the variance of BI and that *subjective norm* had direct effects on BI for mandatory but not voluntary contexts (Venkatesh & Davis, 2000). Further, job relevance and output quality were found to have interactive effects on determining PU, where an individual's judgements on a system's usefulness is influenced by one's cognitive matching of his job and the expected consequences of system use (Venkatesh & Davis, 2000). This finding is consistent with other models such as the Task Technology Fit (Goodhue & Thompson, 1995; Venkatesh & Davis, 2000).

TAM2 was further extended into TAM3 by combining TAM2 (Venkatesh & Davis, 2000) and the model of determinants of PEU (Venkatesh, 2000) by Venkatesh and Bala in 2008 as a complete integrated model to understand an individual's technology acceptance and use. In TAM3, PEU has been theorised to be associated with an individual's self-efficacy and procedural knowledge, whose perceptions are anchored by external control, computer self-efficacy, computer anxiety and computer playfulness, and further adjusted with perceived enjoyment and objective usability. The determinants of PEU represent several traits and emotions of individuals towards computers and computer usage, and are theorised not to be affected by social influence or cognitive instrumental processes, and hence will not influence PU (Venkatesh & Bala, 2008). Beyond TAM2 and the determinants of PEU, experience is hypothesised to moderate the relationships between PEU and PU, *computer anxiety* and PEU, and PEU and BI (Venkatesh & Bala, 2008). Accordingly, the influence of PEU on PU was found to increase with experience; the influence of computer anxiety on PEU was found to decrease with experience; and the influence of PEU on BI was found to decrease with experience.

Perception of external control is defined as 'the degree to which an individual believes that organisational and technical resources exist to support the use of the system' (Venkatesh et al., 2003; Venkatesh & Bala, 2008); computer self-efficacy is defined as 'the degree to which an individual believes that he has the ability to perform a specific task using the computer' (Venkatesh & Bala, 2008); computer anxiety is defined as 'the degree of an individual's apprehension, or even fear, when he is faced with the possibility of using computers' (Venkatesh, 2000; Venkatesh & Bala, 2008); computer playfulness is defined as 'the degree of cognitive spontaneity in microcomputer interactions' (Venkatesh & Bala, 2008); perceived enjoyment is defined as 'the extent to which the activity of using a specific system is perceived to be enjoyable in its own right, aside from any performance consequences resulting from system use' (Venkatesh, 2000; Venkatesh & Bala, 2008); and objective usability is defined as 'a comparison of systems based on the actual level of effort required to complete specific tasks' (Venkatesh, 2000; Venkatesh & Bala, 2008). TAM3 is shown in Figure 6.5.

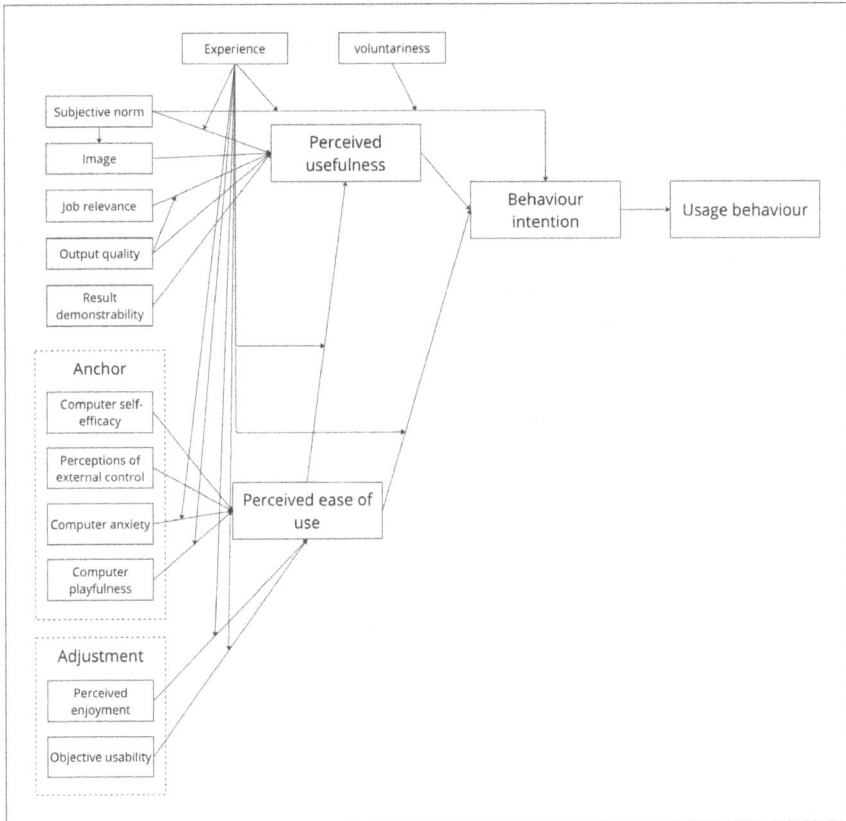

Figure 6.5 TAM3

TAM3 overcomes the key limitation of TAM, whose parsimonious nature provides insufficient actionable guidance for practitioners in developing IS that will be accepted and used by individuals, and provides comprehensive understanding of an individual's decision to accept and adopt an IT (Lee et al., 2003; Venkatesh & Bala, 2008). At the same time, the comprehensiveness of TAM3 results in a lack of parsimony.

6.3 Innovation Diffusion Theory

The IDT was developed by Rogers in 1962 to study the diffusion of any innovation. Innovation is used in IDT as a synonym for technology. Diffusion is defined as 'the process by which an innovation is communicated through certain channels over time among the members of a social system' (Rogers, 1983). Accordingly, there are four elements that influences the diffusion of innovations: (i) characteristics of the innovation, (ii) communication channels, (iii) time and (iv) social system.

The characteristics of the innovation perceived by individuals influence the diffusion of the innovation. These characteristics include relative advantage, compatibility, complexity, trialability and observability. Relative advantage refers to 'the degree to which an innovation is perceived as better than the idea it supersedes'; compatibility refers to 'the degree to which an innovation is perceived as being consistent with the existing values, past experiences, and needs of potential adopters'; complexity refers to 'the degree to which an innovation is perceived as difficult to understand and use'; trialability refers to 'the degree to which an innovation may be experimented with on a limited basis'; and observability refers to 'the degree to which the results of an innovation are visible to others' (Rogers, 1983). Innovations that are perceived as having greater relative advantage, compatibility, trialability, observability and less complexity will be adopted more rapidly, which has been found to significantly explain the rate of adoption (Rogers, 1983).

Communication channel refers to 'the means by which messages can get from one individual to another' and social system refers to 'a set of interrelated units that are engaged in joint problem solving to accomplish a common goal' (Rogers, 1983). Time is an important element in the diffusion process as it is involved in the innovation-decision process in which an individual executes from the first knowledge of an innovation to the decision to adopt or reject the innovation, the relative time of adoption of an innovation and the innovation's rate of adoption within a social system.

The innovation-decision process refers to 'the process through which an individual passes from first knowledge of an innovation to forming an attitude towards the innovation, to a decision to adopt or reject, to implementation of the new idea, and to confirmations of this decision' (Rogers, 1983). There are five main steps in the process: (i) knowledge; (ii) persuasion; (iii) decision; (iv) implementation and (v) confirmation, which typically occurs in a time-ordered sequence.

Knowledge takes place when 'an individual is exposed to the innovation's existence, and gains some understanding of how it functions'; persuasion takes place when 'an individual forms an attitude towards the innovation'; decision takes place when 'an individual engages in activities that lead to a choice to adopt or reject the innovation'; implementation takes place when 'an individual puts an innovation into use'; and confirmation takes place when 'an individual seeks reinforcement of an innovation decision that has already been made, but he may reverse this previous decision if exposed to conflicting messages about the innovation' (Rogers, 1983). Depending on an individual's relative time of adoption of an innovation, the individual is categorised into one of the five adopter categories: innovators, early adopters, early majority, late majority and laggards. Innovators are 'active information seekers about new ideas' and are the first to adopt a new idea (Rogers, 1983).

IDT was developed to explain the diffusion of innovations as opposed to predicting an individual's decision to adopt an innovation. Hence, although it has been applied to discuss technology adoption at global, organisational and individual levels, it is reasonable that IDT has less explanatory power and is less practical for the prediction of an individual's adoption of IT as compared to other models (Taherdoost, 2018).

6.4 Social Cognitive Theory

SCT was developed by Bandura in 1986 as a theory of human behaviour. SCT posits that behaviour, personal factors and environmental influences constantly interact and influence one another, reciprocally determining each other, also referred to as a 'triadic reciprocality' relationship (Compeau & Higgins, 1995). Two key features of SCT are the outcome expectations and self-efficacy. *Outcome expectations* refer to the expected consequences of the behaviour; and self-efficacy as defined by Bandura (1986) is 'an individual's judgement of their capabilities to organise and execute courses of action required to attain designated types of performances'. The key aspect to the construct of self-efficacy is one's judgement of what one can do with the skills one possesses, as opposed to the skills that one possesses (Bandura, 1986). Self-efficacy influences an individual's expectations of the outcomes of the behaviour, both of which may be influenced by one's previous experiences related to performing the certain behaviour (Compeau & Higgins, 1995; Momani & Jamous, 2017).

The SCT has been extended to the context of computer utilisation in Compeau and Higgins (1995), which will be detailed in this section. In Compeau and Higgins' (1995) SCT, computer self-efficacy refers to 'a judgement of one's capability to use a computer'. The constructs that represented the behavioural, personal and environmental elements include outcome expectations, computer self-efficacy, affect and anxiety. Outcome expectations were further categorised into outcome expectations–performance and outcome expectations–personal. Outcome expectations–performance relates to the expected consequences of performing the behaviour on job performance while outcome expectations–personal relates to the expected consequences of performing the behaviour on the individual's esteem and sense of accomplishment (Compeau & Higgins, 1995). Affect refers to 'an individual's liking for a particular behaviour, in this case, computer usage' and anxiety refers to 'an individual's anxious feelings towards performing a particular behaviour, in this case, the use of computers' (Compeau & Higgins, 1995). Accordingly, individuals are more likely to engage in behaviour that they expected to be rewarded for; higher computer self-efficacy leads to higher affect and lower anxiety of an individual towards computer use, which leads to higher probability of computer usage (Compeau & Higgins, 1995).

SCT explained 32 per cent of the variance in usage in Compeau and Higgins (1995). Although the original SCT proposed by Bandura posits a 'triadic reciprocality' structure, it was noted that a majority of SCT studies including Compeau and Higgins' (1995) model uses behaviour as the dependent variable (Carillo, 2010). The bi-directional interactions between the factors were also not considered in the SCT model for computer usage (Carillo, 2010).

6.5 Model of PC Utilisation

MPCU was developed by adapting Triandis' (1977) TIB for the context of IS and to predict computer usage (Thompson et al., 1991). In TIB, BI was argued to be influenced by social factors, expected consequences, and affect towards use, while behaviour is influenced by one's habits and facilitating conditions (Thompson et al., 1991). Social factors include subjective norms, roles and values, and refers to 'the individual's internationalisation of the reference groups' subjective culture and specific interpersonal agreements that the individual has made with others in specific social situations'; affect towards use was defined as 'the feelings of joy, elation, or pleasure, or depression, disgust, displeasure, or hate associated by an individual with a particular act'; and facilitating conditions refer to 'the objective factors in the environment that observers agree make an act easy to accomplish' (Thompson et al., 1991; Venkatesh et al., 2003).

Thompson et al. (1991) adapted TIB to predict computer usage in voluntary settings. In MPCU, the core constructs include social factors, affect towards use, long-term consequences, complexity, job-fit and facilitating conditions. Long-term consequences refer to 'outcomes that have a pay-off in the future' and job-fit refers to 'the extent to which an individual believes that using a technology can enhance the performance of his job' (Thompson et al., 1991). Facilitating conditions is concerned with the availability and provision of technical support for the computer (Thompson et al., 1991; Venkatesh et al., 2003). The MPCU is concerned with the actual use of computers and not the intention to use computers, and hence the construct BI was excluded from the model (Thompson et al., 1991). Habit was also excluded in the MPCU due to its tautological relationship with current use (Thompson et al., 1991). Figure 6.6 shows the MPCU model.

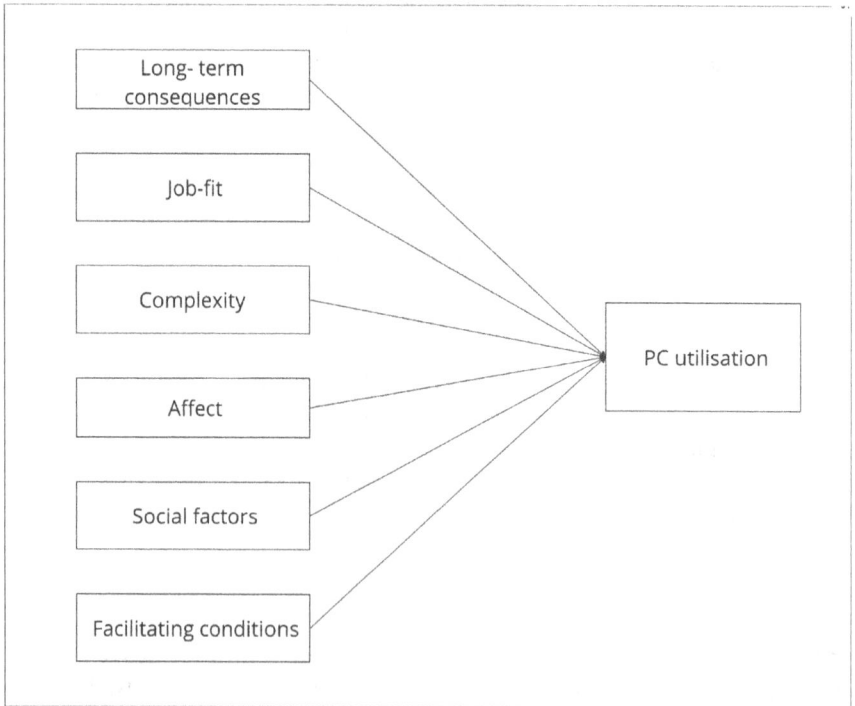

Figure 6.6 MPCU

Thompson et al. (1991) found that affect and facilitating conditions do not have significant effects on PC usage. Although the insignificant relationship between affect and usage was inconsistent with TIB and TAM, it was argued that computers may not evoke strong emotions among users, and that the MPCU utilises a different theoretical structure with the TAM (Thompson et al., 1991). In particular, attitudes include three components – affective (feelings), behavioural and cognitive (beliefs) (Thompson et al., 1991). MPCU evaluated attitudes as separate constructs (affect and complexity), and posited a direct relationship between affect and usage as opposed to TAM's indirect relationship through BI, which may have resulted in the differences in the findings (Thompson et al., 1991). It was also highlighted that the cognitive components of attitudes measured by complexity were consistent with previous studies including the TAM (Thompson et al., 1991). The insignificant relationship between *facilitating conditions* and *usage* was also inconsistent with previous studies, which could be due to the evaluation of only one aspect of facilitating conditions (i.e. availability of technical support) (Thompson et al., 1991). As the MPCU adapts the TIB, MPCU shares the same limitation of a lack of parsimony compared to TRA and TPB (Taherdoost, 2018).

6.6   Motivation Model

Davis et al. (1992) adapted the general motivation theory to the IS context to understand the use and adoption of new technologies in the workplace. In MM, motivation to perform an activity is categorised into extrinsic motivation and intrinsic motivation. Extrinsic motivation refers to 'the perception that users will want to perform an activity because it is perceived to be instrumental in achieving valued outcomes that are distinct from the activity itself, such as improved job performance' while intrinsic motivation refers to 'the perception that users will want to perform an activity for no apparent reinforcement other than the process of performing the activity per se' (Davis et al., 1992). In reference to TAM, PU is an example of extrinsic motivation and enjoyment on intention to use computers in the workplace is an example of intrinsic motivation in MM (Davis et al., 1992).

It was found that an individual's intention to use computers in the workplace is influenced mainly by PU while the degree of enjoyment is a secondary determinant of computer usage in the workplace (Davis et al., 1992). This finding is consistent with that of TAM, where it was found that PU is a stronger determinant of BI (Davis, 1986). It was also found that enjoyment strongly affects intention when the IS has a higher PU and weakly affects intention when PU is lower (Davis et al., 1992).

6.7   Unified Theory of Acceptance and Use of Technology

By studying the similarities and differences among eight common technology adoption theories and models, Venkatesh et al. (2003) developed the UTAUT to synthesise the determinants of an individual's decision to accept and use new technologies. This was to overcome the following issues: (i) each model having different sets of acceptance determinants, some of which may have the same concepts but are different constructs; and (ii) no existing model encompassing the contributions of all theories, leading to the ignorance of the contributions of other models after selecting one favoured model (Venkatesh et al., 2003).

The UTAUT has four main constructs, performance expectancy, effort expectancy, social influence, which influences use behaviour through BI, and facilitating conditions, which was posited to have a direct influence on use behaviour. Four moderators were included: gender, age, experience and voluntariness of use. Gender was posited to moderate the effects of performance expectancy, effort expectancy and social influence; age moderates the effects of all constructs; experience moderates the effects of effort expectancy, social influence and facilitating conditions; and voluntariness of use moderates the effects of social influence.

Performance expectancy is defined as 'the degree to which an individual believes that using the system will help him to attain gains in job performance' (Venkatesh et al., 2003). Performance expectancy was derived from five constructs: PU (from TAM), extrinsic motivation (from MM), job-fit (from MPCU), relative advantage (from IDT) and outcome expectations (from SCT). Effort expectancy is defined as 'the degree of ease associated with the use of the system', and is derived from PEU (from TAM), complexity (from MPCU) and ease of use (from IDT) (Venkatesh et al., 2003). Social influence is defined as 'the degree to which an individual perceives that important others believe he should use the new system' and is derived from subjective norm (from TRA, TAM2, TPB), social factors (from MPCU) and image (from IDT) (Venkatesh et al., 2003). Facilitating conditions is defined as 'the degree to which an individual believes that organisational and technical infrastructures exist to support use of the system' (Venkatesh et al., 2003). Facilitating conditions encompass the constructs of PBC (from TPB), facilitating conditions (from MPCU) and compatibility (from IDT).

In particular, the constructs of computer self-efficacy and anxiety (from SCT, TAM3) were argued to behave similarly and will be fully mediated by PEU (Venkatesh, 2000; Venkatesh et al., 2003). Furthermore, attitude towards behaviour (from TRA, TPB), intrinsic motivation (from MM) and affect (from MPCU and SCT) were argued to have no significant influence on BI as these constructs may be operationalised through performance and effort expectancies (Venkatesh et al., 2003).

UTAUT provides a comprehensive model for understanding and predicting individuals' acceptance and use of new technologies, and was found to account for 70 per cent variance of an individual's usage intention, as opposed to an average of 40 per cent by the previous models (Venkatesh et al., 2003). As such, UTAUT is becoming an increasingly used model in IS research, representing a shift from TAM (Williams et al., 2011). However, the model is complex, has a lack of parsimony in its approach and does not explain individual behaviour (Van Raaij & Schepers, 2008). Furthermore, Williams et al. (2011) found that only a small number of articles out of the 450 articles analysed used the UTAUT in their studies but cited UTAUT to build their theoretical framework.

6.8   Personal Innovativeness in IT

The PIIT was developed by Agarwal and Prasad in 1998 to further explain why an individual accepts and adopts a new IT. PIIT is defined as 'the willingness of an individual to try out any new information technology' (Agarwal & Prasad, 1998). Agarwal and Prasad (1998) conceptualised PIIT as a trait; that is PIIT is a relatively stable descriptor of individual that is generally not influenced by environmental or situational factors. PIIT was theorised as a moderator between the antecedents and consequences of the perceptions of a new technology, which utilises the constructs of *relative advantage, ease of use* and *compatibility* (from IDT) (Agarwal & Prasad, 1998).

Accordingly, as a moderator of the antecedents towards the perceptions of the IT, with the same communication channels, individuals with higher PIIT form more positive perceptions towards the IT compared with individuals with lower PIIT (Agarwal & Prasad, 1998). On the other hand, as a moderator of the consequences of the perceptions of the IT, PIIT reflects an individual's risk-taking behaviour, as innovation is associated with greater risk and uncertainty (Agarwal & Prasad, 1998; Rogers, 1983). This is consistent with Rogers (1983), who argues that innovators and early adopters of new technologies can cope with higher levels of uncertainties. Hence, at the same level of usage intention, an individual with higher PI requires less positive perception towards the IT compared with an individual with lower PI (Agarwal & Prasad, 1998).

Although PIIT was theorised to moderate the influences of antecedents and consequences of perceptions on new IT, the study found that PIIT only showed moderating effects between compatibility and usage intention as using the technology necessitates significant changes in work behaviour, which gives rise to risk (Agarwal & Prasad, 1998).

In contrast with Agarwal and Prasad (1998), PIIT was found to be a direct determinant of PU and PEU instead of a moderator in Lewis et al. (2003). Similarly, Yi et al. (2006) compared the effects of PIIT as a moderator and as a direct determinant of perceptions of IT (PU, PEU and compatibility) and found PIIT to be a significant direct determinant of PU, PEU and compatibility and had weak moderating effects on these characteristics. PIIT showed significant positive effect on PU, PEU and compatibility, and had significant effect on BI (Yi et al., 2006). Yi et al. (2006) found argued that individuals with higher PIIT tend to develop more favourable perceptions towards the technology, leading to higher adoption rate, more frequent use and stronger future *usage intention*, and is consistent with the individual characteristics of early adopters and innovators conceptualised in IDT. (Lewis et al., 2003)

6.9   Technology Readiness Index

An alternative model to explain and understand an individual's decision to adopt and use new technologies was developed by Parasuraman (2000) in the field of business management. The model focuses on the assessment of an individual's readiness to interact with technology with the goal of maximising marketing effectiveness, as opposed to system development in other theories such as the TAM (Parasuraman, 2000). The TRI measures one's TR, defined as 'people's propensity to embrace and use new technologies for accomplishing goals in home life and at work' (Parasuraman, 2000). This is similar to one's decision to accept and adopt the technology to achieve gains in performances.

The TRI is based on the concept of technology paradoxes identified in Mick and Fournier (1998). Specifically, it was argued that technologies may trigger both positive and negative feelings, which may co-exist, and eight technology paradoxes were identified: (i) control/chaos, (ii) freedom/enslavement, (iii) new/obsolete, (iv) competence/incompetence, (v) efficiency/inefficiency, (vi) fulfils/creates needs, (vii) assimilation/isolation and (viii) engaging/disengaging (Mick & Fournier, 1998). This is consistent with other technology adoption theories such as TAM3 and SCT, which identified *anxiety* and *affect* as determinants that directly or indirectly affect BI.

In the TRI, the constructs *optimism, innovativeness, discomfort* and *insecurity towards technology* were measured to determine an individual's TR (Parasuraman, 2000). Optimism is defined as 'a positive view of technology and a belief that it offers people increased control, flexibility, and efficiency in their lives' and is a driver of TR; innovativeness is defined as 'a tendency to be a technology pioneer and thought leader' and is a driver of TR; discomfort is defined as 'a perceived lack of control over technology and a feeling of being overwhelmed by it' and is an inhibitor of TR; and insecurity is defined as a 'distrust of technology and skepticism about its ability to work properly' and is an inhibitor of TR (Parasuraman, 2000). Optimism is similar to one's PU, outcome expectancy, affect, computer playfulness and enjoyment towards technology; innovativeness is similar to PIIT; discomfort and insecurity is similar to PBC and anxiety towards the use of technology.

6.10   Summary

This chapter detailed the common technology adoption and acceptance theories, including the TAM, its variations and theoretical underpinnings, IDT, SCT, MPCU, MM, UTAUT, PIIT and TRI, setting the foundation to understand what factors influence an individual's decision to adopt and use new technologies, as shown in Table 6.1. Although the different models posited various constructs and relationships among the determinants and an individual's intention to use a new technology and the actual use of the technology, core concepts that are consistent across the models include how one perceives the technology to be useful to improve one's performance, how easy one perceives the technology is to be used and the feelings one has towards the technology. The constructs from the technology adoption and acceptance models have also been frequently applied to study the acceptance of new emerging technologies, such as for Internet of Things (Chen et al., 2020; Liu et al., 2018), augmented reality (Oyman et al., 2022), robotics (Go et al., 2020; Saari et al., 2022), autonomous vehicles (Koul & Eydgahi, 2018; Müller, 2019; Nastjuk et al., 2020), laser scanning (Sepasgozar et al., 2017) and distributed ledger technology (Shrestha et al., 2021). Hence, this research will also adopt the constructs to develop the TCF.

Table 6.1 Summary of Theories of Technology Adoption and Acceptance

Theory	Findings	Advantages	Limitations
TAM	• Explains about 40% of variance in BI • PU and PEU directly influences BI • PU was found to be the primary determinant and PEU as the secondary determinant for BI • PEU also influences PU	• Parsimonious model that is robust with relatively high explanatory power	• Based on relatively static constructs and assumes a linear relationship which does not consider factors that may enhance adoption • Parsimonious nature does not sufficiently provide understanding of determinants that contribute to PU and PEU
TAM2	• Explains up to 60% of BI • Job relevance and output quality have interactive effects on determining PU • Subjective norm has direct effects on BI for mandatory contexts only	• Provides comprehensive understanding of individuals' decision to accept and adopt IT	• Lack of parsimony
TAM3	• PEU associated with individuals' self-efficacy and procedural knowledge with anchors and can be further adjusted and moderated with experience	• Provides comprehensive understanding of individuals' decision to accept and adopt IT	• Lack of parsimony
IDT	• Diffusion of innovation influenced by the relative advantage, compatibility, complexity, trialability and observability of the innovation	• Explains diffusion of innovations on global, organisational and individual levels	• Less practical for the prediction of an individual's adoption of IT
SCT	• Explains about 32% of computer usage • Individuals more likely to engage in behavior they are expected to be rewarded for and higher computer self-efficacy leads to higher affect and lower anxiety, leading to higher probability of computer usage	• Takes into consideration the behavioural, personal and environmental elements	• Does not consider bi-directional interactions between the factors of behavior, personal factors and environmental influences
MPCU	• Long-term consequences, job fit, complexity and social factors found to affect PC utilisation	• Measures actual use of the computers	• Lack of parsimony

(*Continued*)

Table 6.1 (Continued)

Theory	Findings	Advantages	Limitations
MM	• Primary influence of an individual's intention to use computers is PU while degree of enjoyment is a secondary determinant	• Same as TAM	• Same as TAM
UTAUT	• Explains up to 70% of variance of an individual's usage intention • Performance expectancy, effort expectancy, social influence found to affect BI while facilitating conditions have direct impact on use behavior, with age, gender, experience and voluntariness of use as moderators	• Comprehensive model for understanding and predicting individual's acceptance and use of new technologies	• Lack of parsimony
PIIT	• PIIT was found to be a significant determinant of PU, PEU and compatibility, and has strong effect on BI	• PIIT is posited as a trait, and is a relatively stable descriptor of an individual that is generally not influenced by environmental or situational factors	• Did not assess on all characteristics based on IDT
TRI	• Optimism, innovativeness, discomfort and insecurity towards technology can determine one's TR	• Considers that technologies trigger both positive and negative feelings, which can co-exist • Assesses individual's TR	• Does not predict individual adoption

References

Agarwal, R., & Prasad, J. (1998). A conceptual and operational definition of personal innovativeness in the domain of information technology. *Information Systems Research, 9*(2), 204–215. https://doi.org/10.1287/isre.9.2.204

Ajzen, I. (1991). The theory of planned behavior. *Organizational Behavior and Human Decision Processes, 50*, 179–211.

Bandura, A. (1986). *Social foundations of thought and action: Social cognitive theory.* Prentice Hall.

Carillo, K. D. (2010). Social cognitive theory in is research – Literature review, criticism, and research agenda. *Communications in Computer and Information Science, 54*, 20–31. https://doi.org/10.1007/978-3-642-12035-0_4

Chen, J. H., Ha, N. T. T., Tai, H. W., & Chang, C. A. (2020). The willingness to adopt the Internet of Things (IoT) conception in Taiwan's construction industry. *Journal of Civil Engineering and Management*, *26*(6), 534–550. https://doi.org/10.3846/jcem.2020.12639

Compeau, D. R., & Higgins, C. A. (1995). Computer self-efficacy: Development of a measure and initial test. *MIS Quarterly*, *19*(2), 189. https://doi.org/10.2307/249688

Davis, F. D. (1986). *A technology acceptance model for empirically testing new end-user information systems* [Massachusetts Institute of Technology]. https://dspace.mit.edu/handle/1721.1/15192

Davis, F. D., Bagozzi, R. P., & Warshaw, P. R. (1989). User acceptance of computer technology: A comparison of two theoretical models. *Management Science*, *35*(8), 982–1003. https://doi.org/10.1287/mnsc.35.8.982

Davis, F. D., Bagozzi, R. P., & Warshaw, P. R. (1992). Extrinsic and intrinsic motivation to use computers in the workplace1. *Journal of Applied Social Psychology*, *22*(14), 1111–1132. https://doi.org/10.1111/j.1559-1816.1992.tb00945.x

Fishbein, M., & Ajzen, I. (1975). *Belief, attitude, intention, and behavior: An introduction to theory and research*. Addison-Wesley Pub. Co.

Go, H., Kang, M., & Suh, S. B. C. (2020). Machine learning of robots in tourism and hospitality: Interactive technology acceptance model (iTAM) – cutting edge. *Tourism Review*, *75*(4), 625–636. https://doi.org/10.1108/TR-02-2019-0062/FULL/PDF

Goodhue, D. L., & Thompson, R. L. (1995). Task-technology fit and individual performance. *MIS Quarterly: Management Information Systems*, *19*(2), 213–233. https://doi.org/10.2307/249689

Hartwick, J., & Barki, H. (1994). Explaining the role of user participation in information system use. *Management Science*, *40*(4), 440–465. https://doi.org/10.1287/mnsc.40.4.440

King, W. R., & He, J. (2006). A meta-analysis of the technology acceptance model. *Information & Management*, *43*(6), 740–755. https://doi.org/10.1016/j.im.2006.05.003

Koul, S., & Eydgahi, A. (2018). Utilizing Technology Acceptance Model (TAM) for driverless car technology. *Journal of Technology Management & Innovation*, *13*(4), 37–46. https://doi.org/10.4067/S0718-27242018000400037

Lai, P. (2017). The literature review of technology adoption models and theories for the novelty technology. *Journal of Information Systems and Technology Management*, *14*(1). https://doi.org/10.4301/s1807-17752017000100002

Lee, Y., Kozar, K. A., Larsen, K. R. T., Lee, Y., Kozar, K. A., Lee, Y., Kozar, K. A., & Larsen, K. R. T. (2003). The technology acceptance model: Past, present, and future. *Communications of the Association for Information Systems*, *12*, 752–780. https://doi.org/10.17705/1CAIS.01250

Legris, P., Ingham, J., & Collerette, P. (2003). Why do people use information technology? A critical review of the technology acceptance model. *Information & Management*, *40*(3), 191–204. https://doi.org/10.1016/S0378-7206(01)00143-4

Lewis, W., Agarwal, R., & Sambamurthy, V. (2003). Sources of influence on beliefs about information technology use: An empirical study of knowledge workers. *MIS Quarterly: Management Information Systems*, *27*(4), 657–678. https://doi.org/10.2307/30036552

Liu, D., Lu, W., & Niu, Y. (2018). Extended technology-acceptance model to make smart construction systems successful. *Journal of Construction Engineering and Management*, *144*(6), 04018035. https://doi.org/10.1061/(ASCE)CO.1943-7862.0001487

Mick, D. G., & Fournier, S. (1998). Paradoxes of technology: Consumer cognizance, emotions, and coping strategies. *Journal of Consumer Research*, *25*(2), 123–143. https://doi.org/10.1086/209531

Momani, A. M., & Jamous, M. M. (2017). The evolution of technology acceptance theories. *International Journal of Contemporary Computer Research (IJCCR)*, *1*(1), 51–58. https://doi.org/10.1002/anie.201003816

Moore, G. C., & Benbasat, I. (1991). Development of an instrument to measure the perceptions of adopting an information technology innovation. *Information Systems Research*, *2*(3), 192–222. https://doi.org/10.1287/isre.2.3.192

Müller, J. M. (2019). Comparing technology acceptance for autonomous vehicles, battery electric vehicles, and car sharing – A study across Europe, China, and North America. *Sustainability*, *11*(16), 4333. https://doi.org/10.3390/SU11164333

Nastjuk, I., Herrenkind, B., Marrone, M., Brendel, A. B., & Kolbe, L. M. (2020). What drives the acceptance of autonomous driving? An investigation of acceptance factors from an end-user's perspective. *Technological Forecasting and Social Change*, *161*, 120319. https://doi.org/10.1016/J.TECHFORE.2020.120319

Oyman, M., Bal, D., & Ozer, S. (2022). Extending the technology acceptance model to explain how perceived augmented reality affects consumers' perceptions. *Computers in Human Behavior*, *128*, 107127. https://doi.org/10.1016/J.CHB.2021.107127

Parasuraman, A. (2000). Technology Readiness Index (TRI): A multiple-item scale to measure readiness to embrace new technologies. *Journal of Service Research*, *2*(4), 307–320. https://doi.org/10.1177/109467050024001

Rogers, E. M. (1983). *Diffusion of innovations*. Macmillan Publishing Co., Inc.

Saari, U. A., Tossavainen, A., Kaipainen, K., & Mäkinen, S. J. (2022). Exploring factors influencing the acceptance of social robots among early adopters and mass market representatives. *Robotics and Autonomous Systems*, *151*, 104033. https://doi.org/10.1016/J.ROBOT.2022.104033

Sepasgozar, S. M. E., Shirowzhan, S., & Wang, C. (2017). A scanner technology acceptance model for construction projects. *Procedia Engineering*, *180*, 1237–1246. https://doi.org/10.1016/J.PROENG.2017.04.285

Sharma, R., & Mishra, R. (2014). A review of evolution of theories and models of technology adoption. *Indore Management Journal*, *6*(2), 17–29.

Sheppard, B. H., Hartwick, J., & Warshaw, P. R. (1988). The theory of reasoned action: A meta-analysis of past research with recommendations for modifications and future research. *Journal of Consumer Research*, *15*(3), 325. https://doi.org/10.1086/209170

Shrestha, A. K., Vassileva, J., Joshi, S., & Just, J. (2021). Augmenting the technology acceptance model with trust model for the initial adoption of a blockchain-based system. *PeerJ Computer Science*, *7*, 1–38. https://doi.org/10.7717/PEERJ-CS.502/SUPP-7

Taherdoost, H. (2018). A review of technology acceptance and adoption models and theories. *Procedia Manufacturing*, *22*, 960–967. https://doi.org/10.1016/j.promfg.2018.03.137

Thompson, R. L., Higgins, C. A., & Howell, J. M. (1991). Personal computing: Toward a conceptual model of utilization. *MIS Quarterly*, *15*(1), 125. https://doi.org/10.2307/249443

Triandis, H. C. (1977). *Interpersonal behavior*. Cole Publishing Company.

Van Raaij, E. M., & Schepers, J. J. L. (2008). The acceptance and use of a virtual learning environment in China. *Computers & Education*, *50*(3), 838–852. https://doi.org/10.1016/j.compedu.2006.09.001

Venkatesh, V. (2000). Determinants of perceived ease of use: Integrating control, intrinsic motivation, and emotion into the technology acceptance model. *Information Systems Research*, *11*(4), 342–365. https://doi.org/10.1287/isre.11.4.342.11872

Venkatesh, V., & Bala, H. (2008). Technology Acceptance Model 3 and a research agenda on interventions. *Decision Sciences*, *39*(2), 273–315. https://doi.org/10.1111/j.1540-5915.2008.00192.x

Venkatesh, V., & Davis, F. D. (1996). A model of the antecedents of perceived ease of use: Development and test. *Decision Sciences*, *27*(3), 451–481. https://doi.org/10.1111/j.1540-5915.1996.tb00860.x

Venkatesh, V., & Davis, F. D. (2000). Theoretical extension of the Technology Acceptance Model: Four longitudinal field studies. *Management Science*, *46*(2), 186–204. https://doi.org/10.1287/mnsc.46.2.186.11926

Venkatesh, V., Morris, M. G., Davis, G. B., & Davis, F. D. (2003). User acceptance of information technology: Toward a unified view. *MIS Quarterly*, *27*(3), 425–478. https://doi.org/10.2307/30036540

Williams, M. D., Rana, N. P., Dwivedi, Y. K., & Lal, B. (2011). Is UTAUT really used or just cited for the sake of it? A systematic review of citations of utaut's originating article.

19th European Conference on Information Systems, ECIS 2011. https://aisel.aisnet.org/ecis2011/231

Yi, M. Y., Fiedler, K. D., & Park, J. S. (2006). Understanding the role of individual innovativeness in the acceptance of IT-based innovations: Comparative analyses of models and measures. *Decision Sciences, 37*(3), 393–426. https://doi.org/10.1111/j.1540-5414.2006.00132.x

7 Proposed Conceptual Model

Technological Competency

7.1 Introduction

Drawing upon the application of smart technologies in CPM, theories of competency and theories of technology acceptance and adoption, this chapter proposes a conceptual TCF tailored specifically for PMs managing projects with smart technologies.

Rooted in competency theory, the constituents of an individual's competency comprise knowledge, skills and personal attributes. The knowledge foundation encapsulates the ten knowledge domains proposed by PMI PMBOK, in conjunction with an understanding of the smart technologies, including CPS and IoT, BD and AI, AV and robotics, AR and VR, AM, 3D imaging and DLT, outlined in Chapter 2. Beyond the foundational knowledge, PMs are expected to attain the skillsets incorporating both hard and soft skills required by the 4IR. When considering personal attributes, common personal characteristics that remain consistent despite variations in work processes are identified for the construction of the TCF (Spencer & Spencer, 1993).

In order for PMs to effectively manage projects employing smart technologies, it is crucial that they exhibit acceptance of these technologies and demonstrate a willingness to integrate them into their project workflows. Therefore, determinants derived from theories of technology acceptance and adoption are incorporated to shape a PM's attitude and traits towards technology in this chapter.

This chapter concludes by presenting the hypotheses to be examined in this book. These hypotheses elucidate the relationships between the identified determinants and the technological competency levels of PMs.

7.2 Determinants from the Competency Theory

As highlighted in Chapter 4, determinants of competency include knowledge, skills and personal attributes (attitude, traits and motives). This section will introduce the determinants adopted from the competency theory for PMs to be competent in managing projects with smart technologies.

DOI: 10.1201/9781003462231-7

7.2.1 *Knowledge*

For PMs to be competent in managing projects, it is essential for PMs to have knowledge of project management processes, tools and techniques as described in PMBOK. The project management knowledge areas have been highlighted in Chapter 5. Specifically, there are ten knowledge areas: project integration management, project scope management, project schedule management, project cost management, project quality management, project resource management, project communications management, project risk management, project procurement management and project stakeholder management.

At the same time, with the proliferation of smart technologies and the increasing number of applications into construction projects, PMs need to have knowledge of the smart technologies in terms of their uses, applicability, potential benefits and limitations. In this study, the set of smart technologies identified are detailed in Chapter 2 and 3: CPS and IoT, BD and AI, AV and robotics, AR and VR, AM, 3D imaging and DLT. Figure 7.1 shows the knowledge proposed to be significant for PMs to be technologically competent.

7.2.2 *Skills*

On top of the essential knowledge, PMs should also possess the skills required to manage projects with smart technologies. Essential project management skills have been highlighted in the PMBOK guide. Furthermore, the skills required to manage

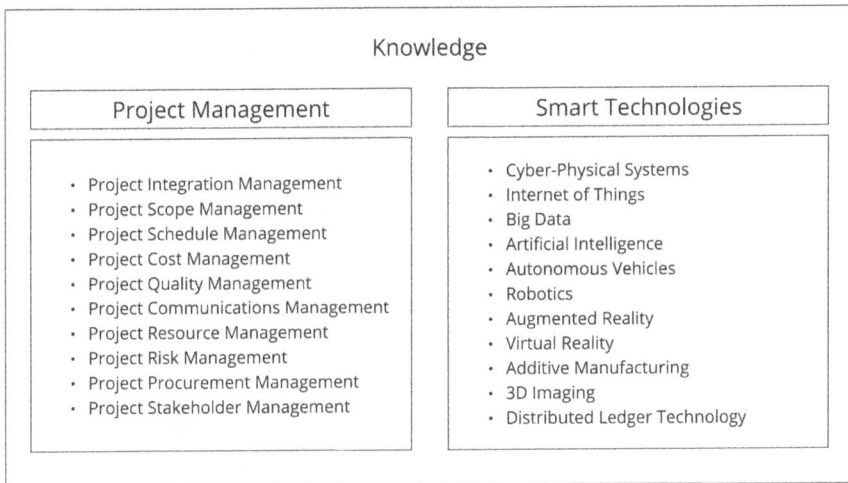

Figure 7.1 Knowledge Required of PMs to be Technologically Competent.

Skills

• Project Management Skills
• Technical and Operational Technology Skills
• Information Management Skills
• Planning and Organising Skills
• Communication Skills
• Social Awareness
• Cultural Awareness
• Organisational Awareness
• Creativity

• Problem Solving Skills
• Ethical Awareness
• Strategic Planning Skills
• Active Learning Skills
• Conflict Management Skills
• Decision Making Skills
• Delegation Skills
• Motivation Skills
• Negotiation Skills
• Team Building Skills

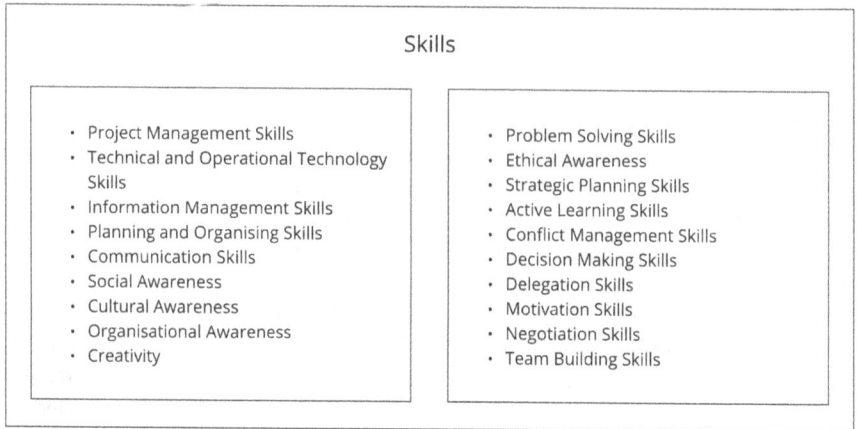

Figure 7.2 Skills Required of PMs to be Technologically Competent.

projects with smart technologies as highlighted in Chapter 5 are also essential for PMs to be competent. Figure 7.2 shows the skills that are proposed to be important for PMs to be technologically competent.

7.2.3   *Personal Attributes*

Personal attributes encompass an individual's attitudes, traits and motives. The underlying personal characteristics are provided in the McBer Dictionary and the 12 most common ones are highlighted in Spencer and Spencer (1993). As competencies are made up of knowledge, skills, attitudes, traits and motives, some of the personal characteristics may overlap with the skills required to manage projects with smart technologies. These underlying personal characteristics that consistently distinguish superior performers from average-performing managers are achievement orientation, initiative, information seeking, focus on client's needs, impact and influence, directiveness or assertiveness, teamwork and cooperation, team leadership, analytical thinking, conceptual thinking, self-control and flexibility (Spencer & Spencer, 1993). As these characteristics are argued to remain constant despite changes in the way tasks are required to be carried out (Spencer & Spencer, 1993), these are hypothesised to be relevant to PMs in managing projects with smart technologies, as shown in Figure 7.3.

7.3   Determinants from the Theories of Technology Acceptance

For PMs to be competent in managing projects with smart technologies, PMs must accept the smart technologies and be willing to adopt the smart technologies into projects. Hence, determinants of the theories of technology adoption and acceptance are adopted to form a PM's attitude and traits towards technologies in this study.

Common Personal Characteristics	
• Achievement Orientation • Initiative • Information Seeking • Focus on Client's Needs • Impact and Influence • Directiveness or Assertiveness	• Teamwork and Cooperation • Team Leadership • Analytical Thinking • Conceptual Thinking • Self-Control • Flexibility

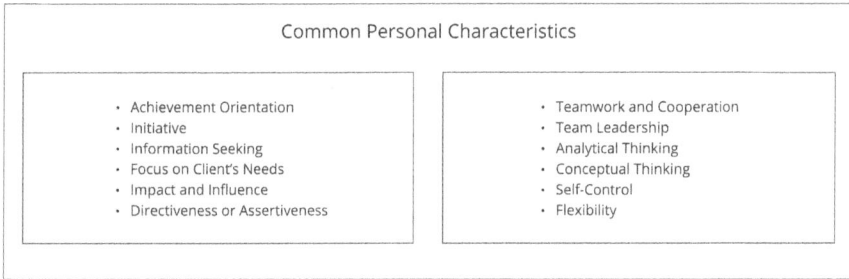

Figure 7.3 Personal Characteristics Required of PMs to be Technologically Competent.

7.3.1 Attitude Towards Technology

This study focuses on one's ATT instead of the individual's intention to use the technologies and actual use of the technologies as PMs may or may not use all the smart technologies in the projects. There are three components to 'attitude' – affective, behavioural and cognitive. The affective component of attitude refers to one's feelings towards the technologies; the behavioural component of attitude refers to how an individual tends to behave towards a given object; and the cognitive component of attitude refers to an individual's belief of technologies (Thompson et al., 1991). In particular, how an individual tends to behave towards technology can be operationalised through the affective and cognitive components of ATT. Hence, this study will focus on the affective and cognitive components of ATT.

7.3.1.1 Affective Component of Attitude Towards Technology

The affective component of ATT adopts the constructs TAFF and TANX as proposed in the SCT and as two of the determinants contributing to PEU in UTAUT. In this study, the construct TAFF refers to 'an individual's liking for technologies' and TANX refers to 'an individual's anxious feelings towards the use of technologies'. These are hypothesised to be influenced by one's TSE. The affective component of ATT is shown in Figure 7.4.

7.3.1.2 Cognitive Component of Attitude Towards Technology

The cognitive component of ATT adopts the constructs PU and PEU from the TAM, as shown in Figure 7.5. PU is hypothesised to be influenced by one's experience with using the smart technologies while PEU is hypothesised to be influenced by TAFF and TANX. Similarly, PEU is hypothesised to influence PU.

7.3.2 Traits

In the theories of technology adoption and acceptance, PIIT was theorised as an individual's trait towards technology. Hence, this study adopts the construct *PIT* as

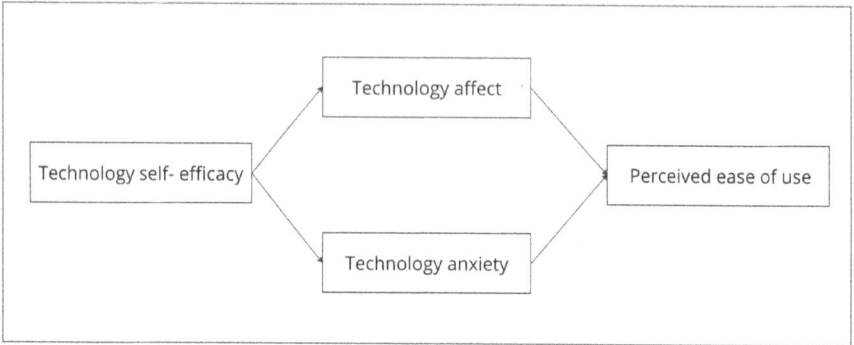

Figure 7.4 Affective Component of Attitude towards Technology.

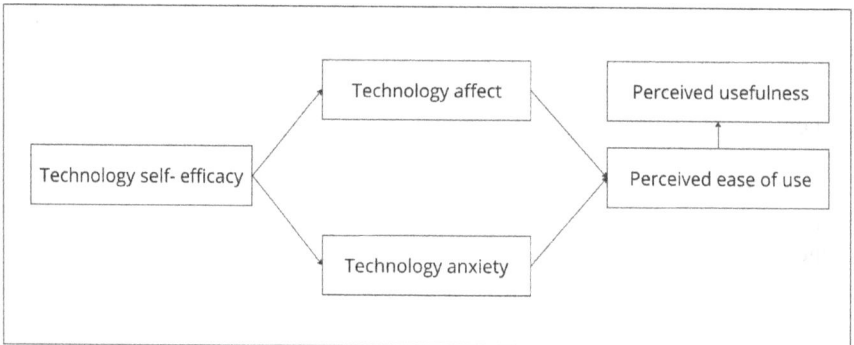

Figure 7.5 Cognitive Component of Attitude towards Technology.

a determinant of trait. Similarly, PIT is hypothesised to influence ATT directly, as shown in Figure 7.6.

7.3.3   *Motives*

Motives towards use or acceptance of technologies can be represented using the MM. The MM adopted *PU* and *perceived enjoyment* as constructs to study one's motivation to use a new technology. Similarly, in this study, extrinsic motivation will be operationalised through PU and intrinsic motivation will be operationalised through TAFF.

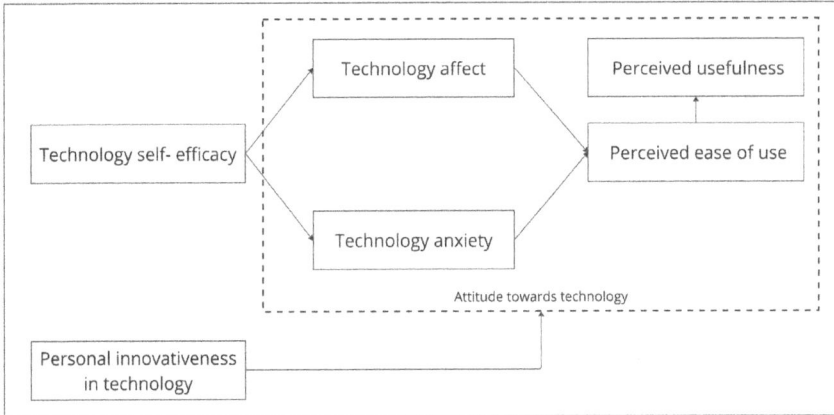

Figure 7.6 Personal Attributes in Technological Competency.

7.4 Proposed Conceptual Model for Technological Competency of Project Managers

This study proposed the following model of technological competency of PMs to manage projects with smart technologies by integrating the determinants identified from the competency theory with the determinants identified from the theories of technology adoption and acceptance. Key components of technological competency include the knowledge, skills and personal attributes that are relevant for managing projects with smart technologies by PMs. Figure 7.7 shows the proposed conceptual model for technological competency.

7.4.1 Technological Competency: Knowledge

As introduced in Section 7.2.1, the knowledge PMs should have to be technologically competent include project management knowledge as specified in PMBOK and the knowledge of the smart technologies. The knowledge of the PM is hypothesised to be a direct determinant of his technological competency in managing projects with smart technologies.

H1: Knowledge of PMs is a significant determinant of PMs' technological competency in managing projects with smart technologies.

H1.1: Project management knowledge is a significant determinant of PMs' technological competency in managing projects with smart technologies.

H1.2: Knowledge in smart technologies is a significant determinant of PMs' technological competency in managing projects with smart technologies.

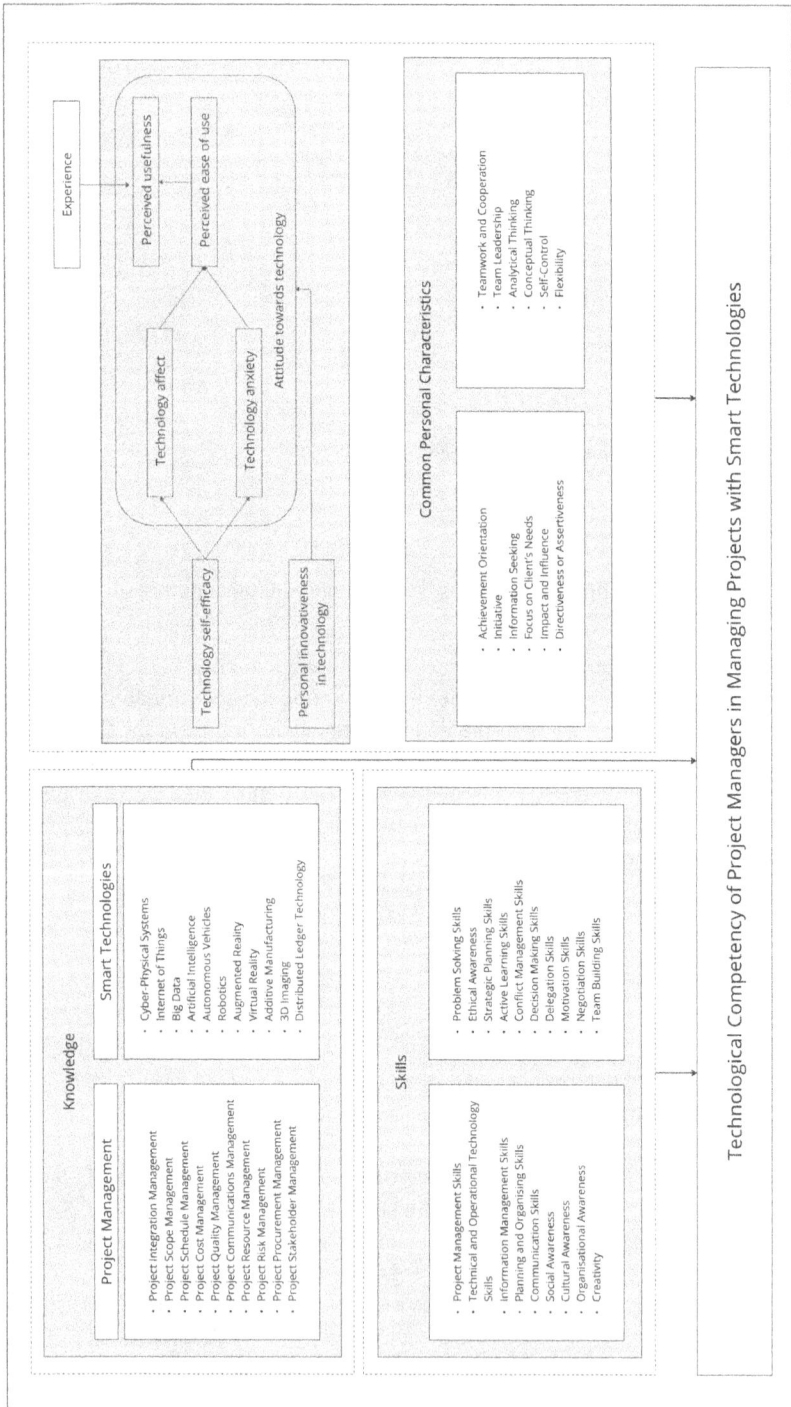

Figure 7.7 Proposed Conceptual Model for Technological Competency of PM.

7.4.2 Technological Competency: Skills

The skills that are essential for PMs to be technologically competent include project management skills as specified in PMBOK, hard skills and soft skills to manage projects and smart technologies, as highlighted in Section 5.4 and Section 7.2.2. Skills possessed by PMs are hypothesised to be a direct determinant of their technological competency in managing projects with smart technologies.

H2: Skills of PMs are a significant determinant of PMs' technological competency in managing projects with smart technologies.

7.4.3 Technological Competency: Personal Attributes

Personal attributes essential for PMs to be technologically competent include the common personal characteristics that are highlighted in Spencer and Spencer (1993), the individual's ATT and the individual's PIT. The personal characteristics highlighted in Spencer and Spencer (1993) are theorised to remain the same despite changes in work tasks as highlighted in Section 7.2.3.

The individual's ATT is determined by his TAFF, TANX, PU and PEU. TAFF and TANX are theorised to be determined by one's TSE and form the affective component of an individual's ATT, which is hypothesised to also influence the PEU of the technology. Accordingly, the affective component of ATT influences the cognitive component of ATT through the construct of PEU. The PEU in turn influences the PU of the technology. PU and PEU are theorised to form the cognitive component of an individual's ATT.

The individual's PIT is used as the construct representing one's trait towards technology. PIT is hypothesised to influence one's ATT directly. An individual with high PIT may form more positive views towards technology, and subsequently have a more open ATT.

H3: PMs' personal attributes are significant determinants of PMs' technological competency in managing projects with smart technologies.
H3.1: The common personal characteristics (from Spencer and Spencer (1993)) positively influence one's technological competency.
H3.2: An individual's TSE has a positive relationship with one's TAFF.
H3.3: An individual's TSE has a negative relationship with one's TANX.
H3.4: An individual's TAFF has a positive relationship with PEU of the technology.
H3.5: An individual's TANX has a negative relationship with PEU of the technology.
H3.6: An individual's TAFF has a positive relationship with one's ATT.
H3.7: An individual's TANX has a negative relationship with one's ATT.
H3.8: An individual's PEU of technology positively influences PU of the technology.
H3.9: An individual's PEU of technology has a positive relationship with one's ATT.
H3.10: An individual's PU of technology has a positive relationship with one's ATT.
H3.11: An individual's PIT has a positive relationship with one's ATT.
H3.12: PMs' ATT is a significant determinant of PMs' technological competency in managing projects with smart technologies.

7.5   Summary

This chapter presented a proposed conceptual model for PMs' technological competency in handling projects involving smart technologies. The TCF synthesises elements from theories of technology adoption, acceptance and competency. It hypothesises that PMs' competence in managing projects with smart technologies is contingent on their expertise in smart technologies and CPM and their adaptive personal attributes. The research hypotheses to be tested were also delineated in this chapter.

References

Spencer, L. M., & Spencer, S. M. (1993). *Competence at work: Models for superior performance*. Wiley.

Thompson, R. L., Higgins, C. A., & Howell, J. M. (1991). Personal computing: Toward a conceptual model of utilization. *MIS Quarterly*, *15*(1), 125. https://doi.org/10.2307/249443

8 Research Methodology

8.1 Introduction

This chapter introduces the research process, which consists of 3 stages and 13 major tasks to achieve the research objectives, as shown in Figure 8.1.

8.1.1 Stage 1: Understanding

Tasks 1 to 3 were conducted through a comprehensive literature review to provide an understanding of (i) the applications of the key smart technologies in projects, (ii) changes to the knowledge and skillsets of PMs to manage projects with smart technologies and (iii) the factors that influence technological competency of PMs respectively. The list of smart technologies in the 4IR was identified based on its applicability to construction projects, and the potential to bring about significant change in project management processes. Based on the findings of Tasks 1 to 3, interview questions and a survey questionnaire were developed to investigate the impact of smart technology applications on the knowledge and skillsets required of PMs to manage projects with smart technologies and the factors that influence technological competency of PMs. Case studies were conducted in Task 4 to enrich the findings of the literature review by identifying (i) the common applications of the smart technologies in projects, (ii) the challenges faced during the implementation of the smart technology applications, (iii) the changes in knowledge, skills and personal attributes required of PMs to manage projects with smart technologies and (iv) recommendations to facilitate technology adoption and improve the technological competency of PMs. This was used to support the development of the TCF. This concludes Stage 1 of the research approach.

8.1.2 Stage 2: Investigation

Stage 2 of the research approach began with pilot interviews, which were conducted with academic and industry professionals with the relevant experience in the construction industry and smart technologies to validate the relevance of the identified smart technologies in the construction industry, and to ensure the validity and reliability of the survey questionnaire in Task 5. The validated survey

DOI: 10.1201/9781003462231-8

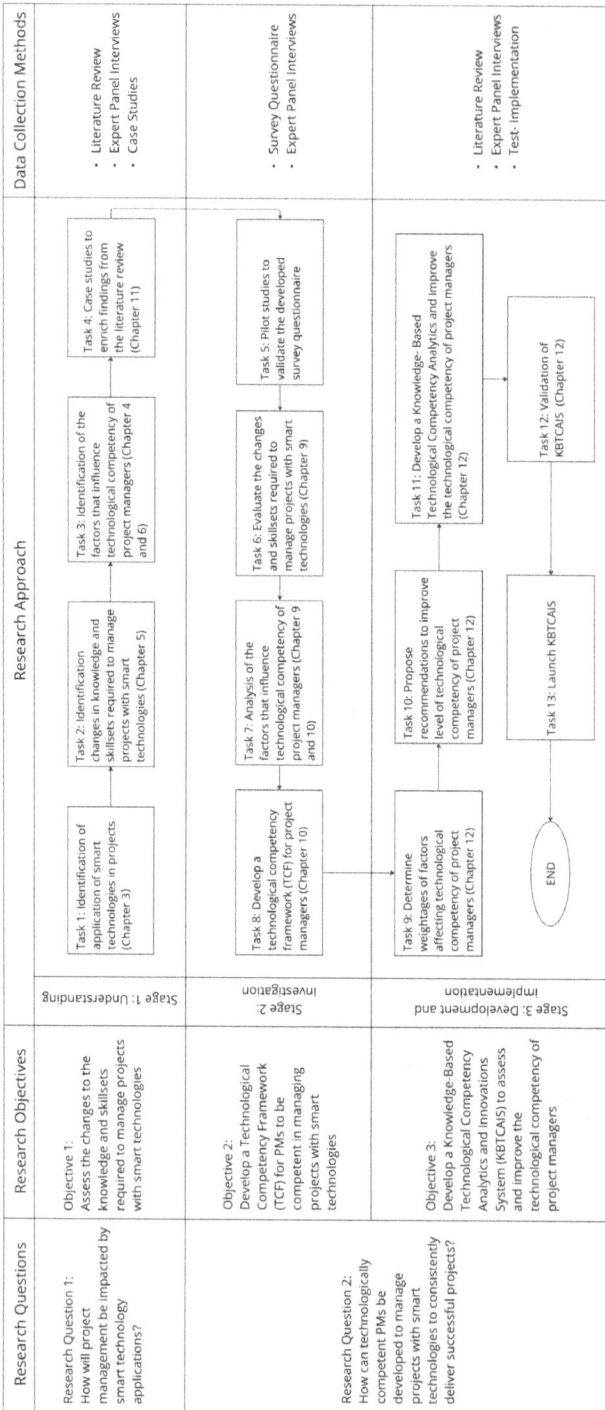

Figure 8.1 Research Approach

questionnaire was then distributed to PMs in Singapore. Based on the responses, the changes to knowledge and skillsets required of PMs to manage projects with smart technologies and the factors that influence technological competency of PMs were investigated in Tasks 6 and 7 respectively. The findings were consolidated and used to develop the TCF in Task 8. The TCF explains the components of technological competency and was used as the basis to develop the KBTCAIS. This concludes the second stage of the research approach.

8.1.3   Stage 3: Development and Implementation

Following the development of the TCF in Task 8, the weightages of the factors influencing technological competency of PMs was determined in Task 9. This was used in the knowledge base and computation of technological competency level in the proposed KBTCAIS. KBTCAIS is envisaged to be an assessment tool that can assess and provide recommendations to improve the technological competency of PMs. Hence, Task 10 identified recommendations to improve the technological competency of PMs through literature review and from the case studies conducted in Task 4. This was also used in the knowledge base in the proposed KBTCAIS. The KBTCAIS was developed in Task 11, using the developed TCF, finalised factor weightages and proposed recommendations. The KBTCAIS was validated through test-implementations and interviews in Task 12, before being launched in Task 13, concluding the research.

8.2   Data Collection Methods

This research adopted a mixed research approach including literature review, survey, interviews and case studies. The use of a combination of qualitative and quantitative data collection methods has been recommended as no single data collection method is ideal (Abowitz & Toole, 2010).

8.2.1   Literature Review

Literature review is a qualitative method of data collection and was used to gather knowledge for the understanding of the applications of smart technologies in projects, changes to knowledge and skillsets required to manage projects with smart technologies, factors affecting PMs' technological competency and the recommendations to improve the level of technological competency in PMs. This serves as a solid foundation on which the subsequent research was built on.

8.2.2   Survey

Surveys have the ability to gather large amounts of information about a particular population efficiently by drawing inferences based on data drawn from a small portion of the population and is suitable to gather broad and shallow characteristics of the population (Gravetter & Forzano, 2018; Rea & Parker, 2014;

Tan, 2017). In contrast, case studies are more appropriate to gain an in-depth understanding or interpretation of a particular situation (Creswell & Poth, 2018; Gravetter & Forzano, 2018; Tan, 2017). Survey questionnaires have been recognised as one of the most cost-effective and popular method of collecting a wide variety of information (Tan, 2017). Questionnaires are also widely used by researchers in construction management research (Chan et al., 2017). Hence, this research designed a survey questionnaire to collect the relevant data to achieve the research objectives.

One cross-section survey has been conducted in this research. The survey aimed to investigate (i) the existing implementation of smart technologies in projects, (ii) the changes in the importance of project management knowledge areas in managing projects with smart technologies, (iii) the changes in the importance of skillsets required to manage projects with smart technologies and (iv) the factors that affect PMs' attitude towards technologies. The target respondents were PMs who are responsible to manage projects. Non-probability sampling, in particular, purposive sampling and snowball sampling were used for this study. Purposive sampling refers to the selection of the sample unit from a stratified sample based on the judgement of the researcher. In particular, homogenous sampling was conducted, where the sample members were similar. In this case, the sample consisted of PMs. Snowball sampling is a technique in which existing subjects refer potential subjects who are suitable for the research study. PMs were identified through web directories such as the Society of Project Managers Singapore, Public Sector Panel of Consultants, Real Estate Developers Association Singapore and BCA Singapore Directory of Registered Contractors and Licensed Builders. The respondents were further asked to provide referrals for potential respondents.

The survey was held on an online platform and the survey link was sent to the target respondents by email. The survey questionnaire was self-administered, allowing the respondents to provide their response at their convenience and at their own pace. At the same time, respondents can have time to think about the questions before responding (Zhao, 2014). However, the response rates of online survey questionnaires are typically lower compared to personal administration of the questionnaire. To mitigate the low response rates, follow-up emails and calls were made when responses were not received by a stipulated time.

The survey questionnaire consisted of eight sections and a glossary of terms. The first section included an introduction explaining the research significance and objectives and provided the contact details. The second section asked for the profile of the respondents, such as their age, job role and working experience. The third section asked for the profile of the organisations that the respondents are affiliated to, such as the organisation domain, organisation size and years of experience in the construction industry. The fourth section gathered information on the types of projects the organisation and respondent are involved in, and whether or not the projects adopt smart technologies. The glossary of the terms used in the survey questionnaire was provided to ensure that the respondents are clear about the smart technologies that were referred to in the survey. The respondents were also asked to indicate if they have experience using the smart technologies and to rate their familiarity with the smart technologies on a 7-point Likert scale (1 = not familiar at all, 7 = very familiar). The same section also required the respondents to rate their attitude towards technologies on a 7-point Likert scale based on four statements. The fifth section required the respondent to rate the importance of the project management knowledge areas in managing conventional projects and projects with at least one smart technology, using two 7-point Likert scales (1 = not important at all, 7 = very important). Open-ended questions were presented to ask of any other knowledge areas that are deemed to be important to manage projects with smart technologies. The sixth section required the respondent to rate the importance of the skills required to manage conventional projects and projects with at least one smart technology using two 7-point Likert scale (1 = not important at all, 7 = very important). Open-ended questions were also presented to ask of any other skills are deemed to be important to manage projects with smart technologies. The seventh section requires the respondent to rate the applicability and impact of the factors affecting attitude towards technologies using two 7-point Likert scales (1 = not applicable at all, 7 = very applicable; 1 = no impact at all, 7 = very significant impact). The measurement items in the questionnaire were adapted from the questionnaires that measure the constructs as proposed in the respective models to the context of this study. Table 8.1 shows the constructs and the measurement items as adapted from the respective models.

8.2.3   Case Studies

With a given set of resources, there is greater opportunity for more in-depth observations with a fewer number of case studies (Voss et al., 2002). At the same time, if only a single case study is used, there are limitations related to generalisability of the findings and risks of misjudging of a single case (Voss et al., 2002). A small number of case studies present better validity and generalisability of the findings. Hence, this research conducted three case studies, with a developer, consultant and contractor organisation. Case studies were conducted with organisations from different domains as PMs representing different domains have different interests and may view factors affecting technological competency differently. Hence, to develop a TCF that encompasses the responsibilities of PMs in general, the opinions of PMs from different organisation domains were sought and corroborated.

While survey questionnaires may be adopted to quantitatively identify the implementation status of smart technologies in projects, with respect to changes in knowledge and skillsets required and the factors that affect one's attitude towards technologies, survey questionnaires cannot capture the reasons which contribute to the phenomenon. Hence, case studies were conducted to enrich the findings from the literature review and corroborate with the findings from the survey questionnaire. Furthermore, as case studies were conducted with experts who have the relevant experience in the construction industry and smart technologies, the challenges of implementing smart technologies in projects and recommendations to facilitate technology adoption and improve the level of technological competency in PMs were also investigated in the case studies.

Table 8.1 Constructs and Measurement Items

Construct	Code	Measurement Item	Reference
Technology Self-Efficacy	TSE	I could complete a job/task using the technology . . .	
	TSE1	If there was no one around to tell me what to do as I go	(Compeau & Higgins, 1995; Venkatesh et al., 2003; Venkatesh & Davis, 1996)
	TSE2	If I had only the user manuals for reference	(Compeau & Higgins, 1995; Venkatesh & Davis, 1996)
	TSE3	If I had seen someone else using it before trying it myself	(Compeau & Higgins, 1995; Venkatesh & Davis, 1996)
	TSE4	If I could call someone for help if I got stuck	(Compeau & Higgins, 1995; Venkatesh et al., 2003; Venkatesh & Davis, 1996)
	TSE5	If someone else had helped me get started	(Compeau & Higgins, 1995; Venkatesh & Davis, 1996)
	TSE6	If I had a lot of time to complete the job for which the technology was provided for	(Compeau & Higgins, 1995; Venkatesh et al., 2003; Venkatesh & Davis, 1996)
	TSE7	If I had just the built-in help facility for assistance	(Compeau & Higgins, 1995; Venkatesh et al., 2003; Venkatesh & Davis, 1996)
	TSE8	If someone showed me how to do it first	(Compeau & Higgins, 1995; Venkatesh & Davis, 1996)
	TSE9	If I had never used a technology like it before	(Compeau & Higgins, 1995; Venkatesh & Davis, 1996)
	TSE10	If I had used similar technology before to do the same job	(Compeau & Higgins, 1995; Venkatesh & Davis, 1996)
Technology Affect	TAFF1	I find using new smart technologies to be enjoyable	(Venkatesh, 2000; Venkatesh & Bala, 2008)
	TAFF2	The actual process of using new smart technologies is pleasant	(Venkatesh, 2000; Venkatesh & Bala, 2008)
	TAFF3	I have fun using new smart technologies	(Venkatesh, 2000; Venkatesh & Bala, 2008)
	TAFF4	New smart technologies make work more interesting	(Venkatesh et al., 2003)
	TAFF5	I like working with new smart technologies	(Venkatesh et al., 2003)
	TAFF6	I look forward to those aspects of my job that require me to use new smart technologies	(Heinssen et al., 1987)
	TAFF7	New smart technologies are necessary tools in work settings	(Heinssen et al., 1987)

(Continued)

Table 8.1 (Continued)

Construct	Code	Measurement Item	Reference
Technology Anxiety	TANX1	I feel apprehensive about using new smart technologies	(Heinssen et al., 1987; Venkatesh et al., 2003)
	TANX2	It scares me to think that I could lose a lot of information using the new smart technology by hitting the wrong key	(Heinssen et al., 1987; Venkatesh et al., 2003)
	TANX3	I hesitate to use new smart technologies for fear of making mistakes I cannot correct	(Heinssen et al., 1987; Venkatesh et al., 2003)
	TANX4	New smart technologies are intimidating to me	(Heinssen et al., 1987; Nickell & Pinto, 1986; Venkatesh et al., 2003)
	TANX5	Working with a new smart technology makes me nervous	(Venkatesh, 2000; Venkatesh & Bala, 2008)
	TANX6	New smart technologies make me feel uncomfortable	(Nickell & Pinto, 1986; Venkatesh, 2000; Venkatesh & Bala, 2008)
	TANX7	New smart technologies make me feel uneasy	(Venkatesh, 2000; Venkatesh & Bala, 2008)
	TANX8	I feel insecure about my ability to interpret an outcome from new smart technologies	(Heinssen et al., 1987)
	TANX9	I am afraid that if I begin to use new smart technologies, I will become dependent upon them and lose some of my reasoning skills	(Heinssen et al., 1987)
	TANX10	New smart technologies are difficult to understand and frustrating to work with	(Nickell & Pinto, 1986)
Perceived Usefulness	PU1	Using new smart technologies in my job would enable me to accomplish tasks more quickly	(Davis, 1989; Venkatesh et al., 2003)
	PU2	Using new smart technologies would improve my job performance	(Davis, 1989; Venkatesh, 2000; Venkatesh & Bala, 2008; Venkatesh & Davis, 2000)
	PU3	Using new smart technologies in my job would increase my productivity	(Davis, 1989; Venkatesh, 2000; Venkatesh et al., 2003; Venkatesh & Davis, 2000)
	PU4	Using new smart technologies would enhance my effectiveness on the job	(Davis, 1989; Venkatesh, 2000; Venkatesh & Bala, 2008; Venkatesh & Davis, 2000)

(Continued)

Table 8.1 (Continued)

Construct	Code	Measurement Item	Reference
	PU5	Using new smart technologies would make it easier to do my job	(Davis, 1989)
	PU6	Overall, I would find new smart technologies useful in my job	(Davis, 1989; Venkatesh et al., 2003; Venkatesh & Bala, 2008; Venkatesh & Davis, 2000)
Perceived Ease of Use	PEU1	Learning to use new smart technologies would be easy for me	(Davis, 1989; Venkatesh et al., 2003)
	PEU2	I would find it easy to get the new smart technology to do what I want it to do	(Davis, 1989; Venkatesh, 2000; Venkatesh & Bala, 2008)
	PEU3	Interacting with the new smart technologies will not require a lot of my mental effort	(Davis, 1989; Venkatesh, 2000; Venkatesh & Bala, 2008)
	PEU4	My interaction with the new smart technologies would be clear and understandable	(Davis, 1989; Venkatesh, 2000; Venkatesh et al., 2003; Venkatesh & Bala, 2008)
	PEU5	I would find the new smart technologies to be flexible to interact with	(Davis, 1989)
	PEU6	It would be easy for me to become skillful at using the new smart technologies	(Davis, 1989; Venkatesh et al., 2003)
	PEU7	Overall, I would find the new smart technologies easy to use	(Davis, 1989; Venkatesh et al., 2003)
Personal Innovativeness in Technologies	PIT1	If I heard about a new technology, I would look for ways to experiment with it	(Agarwal & Prasad, 1998)
	PIT2	Among my peers, I am usually the first to try out new technologies	(Agarwal & Prasad, 1998)
	PIT3	In general, I am hesitant to try out new technologies* (reverse scaled)	(Agarwal & Prasad, 1998)
	PIT4	I like to experiment with new technologies	(Agarwal & Prasad, 1998)

*PIT was assessed using a reversed scale, where 1 = very applicable/very significant impact; 7 = not applicable/no impact at all.

8.2.4   Interviews

Interviews were conducted with experts for the following purposes: (i) validate the developed survey questionnaire, (ii) determine the weightages of factors affecting PM's technological competency, (iii) validate the survey findings, (iv) validate the developed TCF, (v) validate the recommendations provided in the KBTCAIS and (vi) validate the developed KBTCAIS. The interviews conducted were semi-structured as semi-structured interviews provide a structure based on the research framework while providing flexibility for open-ended answers that lead to new findings (Tan, 2017). Semi-structured interviews are also more commonly used in research studies (Tan, 2017).

A first round of interviews was conducted to ensure the validity and reliability of the developed survey questionnaire. The readability, comprehensiveness and accuracy of the statements were assessed in the pilot interviews. The pilot interviews also enriched the findings from the literature review. The pilot interviews were conducted with academic and industry professionals with the relevant experience in the construction industry and smart technologies. Academic professionals were identified based on their research interests, specifically in digital technologies, training and development and project management, in the context of construction projects. Industry professionals were identified through web directories such as the Society of Project Managers Singapore.

Another round of interviews was conducted post survey to determine the weightages of the factors affecting technological competency of PMs. The interviewees were not involved in the first round of interviews, nor did they participate in the survey questionnaire. Similarly, experts include industry practitioners with relevant experience in the construction industry and smart technologies. This was conducted using the Analytical Hierarchy Process, which is discussed in the next section.

The final round of interviews was conducted to validate the findings from the survey questionnaire, developed TCF and proposed recommendations and to test-implement the KBTCAIS for validation purposes. Similarly, the interviewees were not part of the previous interviews or the survey responses. In this round of interviews, the interviewees were first presented the findings of the survey questionnaire and the developed TCF. Their opinions on the findings and developed TCF were sought. Next, the pre-assessment of the interviewees' technological competency level was conducted. In this step, the interviewees were asked to self-assess their level of proficiency in the components of technological competency. The interviewees were then directed to use the KBTCAIS. Based on the assessment results from the KBTCAIS, the interviewees were asked if the factors of technological competency presented in the KBTCAIS were comprehensive in the assessment of the level of technological competency, if the recommendations provided by the KBTCAIS are actionable and useful to improve the level of technological competency, on the user friendliness of the KBTCAIS and how likely it is for KBTCAIS to be used in their projects or organisations. Finally, the interviewees were presented the entire set of recommendations and their opinions sought on whether the recommendations were rightly categorised, actionable and useful to improve the level of technological competency based on the different categories of technological competency level.

8.3 Data Analysis Methods

To analyse the data, the following tests were used to assess the reliability and internal consistency of the responses.

8.3.1   *Data Cleaning*

The survey responses were first cleaned to remove responses that are incomplete, have invalid responses, are grossly different from all other observations (outliers) or are duplicates.

8.3.2   *Validity Tests*

The tests for validity of the survey questionnaires are critical to ensure that the questionnaire measures what it is meant to measure. Three main types of validity are content validity, construct validity and criterion-related validity.

8.3.2.1   *Content Validity*

Content validity refers to 'a test's inclusion of questions that are truly representative of the qualities it is trying to measure' (Karras, 1997). Content validity is most commonly assessed through expert interviews (Bolarinwa, 2015), which were assessed during the pilot studies prior to the distribution of the survey questionnaire in this research.

8.3.2.2   *Construct Validity*

Construct validity refers to 'the extent to which the questionnaire measures what it is meant to measure' (Bolarinwa, 2015). Construct validity is commonly assessed through the convergent and discriminant validity of the questionnaire. Accordingly, convergent validity refers to the high correlation of scores from the instrument to scores measuring the same construct from other instruments; discriminant validity refers to the low correlation between the scores from the instrument to scores from other instruments that measure theoretically different but related constructs (Onwuegbuzie et al., 2007). High correlation values represent convergent validity while low correlation values represent discriminant validity. The total item correlations were used to assess the construct validity of the survey questionnaires.

8.3.2.3   *Criterion-Related Validity*

Criterion-related validity refers to 'the extent to which the measure predicts an outcome for another measure' (Lucko & Rojas, 2010). There are three main types of criterion-related validity: concurrent, predictive and retrospective validity. Concurrent validity refers to the degree in which the scores from the instrument are related to scores from other existing instruments; predictive validity refers to the ability of the instrument to forecast future events, behaviours or outcomes; retrospective validity refers to the degree to which an instrument measures a particular behaviour in the past (Jolliffe et al., 2003). In this study, concurrent validity was used to measure the criterion-related validity of the developed KBTCAIS.

8.3.3 Reliability Tests

Apart from validity, it is also essential to measure the reliability of the survey questionnaires. Reliability refers to the degree to which the survey questionnaire produces stable and consistent results over time and between different participants (Bolarinwa, 2015; Lucko & Rojas, 2010). Reliability tests are often categorised into internal and external reliability.

8.3.3.1 Internal Consistency Reliability: Cronbach's Alpha

Internal reliability assesses the consistency of results across measures within a questionnaire, and can be measured using the Cronbach's alpha test (Lucko & Rojas, 2010). A Cronbach's alpha value of above 0.7 was considered acceptable.

8.3.3.2 External Reliability: Test–Retest Method

External reliability assesses the consistency of a measure over a period of time and across different participants. External reliability is also commonly known as the test–retest method (Lucko & Rojas, 2010). The test–retest method assesses the consistency of the questionnaire over time (Bolarinwa, 2015; Lucko & Rojas, 2010). It requires the administration of the questionnaire to the same group of people at two points of time. The test–retest method assumes that the characteristics measured do not change over time (Bolarinwa, 2015). As the focus of this study (i.e. factors affecting technological competency and technological competency assessment) may change over time, the test–retest method was not conducted.

8.3.4 Frequency Analysis

Frequency analysis was conducted on responses from Part II to IV of the survey questionnaires to gain an understanding of the profile of the respondents, organisations and projects, the experience of the respondents and organisations in using smart technologies and the general attitude towards technologies in the respondents.

8.3.5 Normality Test: Shapiro-Wilk Test

Non-parametric tests have been argued to be most suitable to analyse ordinal data (Fagerland, 2012; Kvam & Vidakovic, 2007). In contrast, others have argued that the type of data is irrelevant for the determination of the type of statistical test to be conducted (Harwell, 1988; Mircioiu & Atkinson, 2017). Hence, the Shapiro-Wilk test was used to test for the normality of the data to determine if parametric or non-parametric tests should be conducted for the subsequent data analysis (Ghasemi & Zahediasl, 2012; Salkind, 2015). Non-parametric tests will be used if the p-value is less than the significance level. A chosen alpha of 0.05 was used for the Shapiro-Wilk test with the following hypothesis:

H0: The data are normally distributed
H1: The data is not normally distributed

As the Shapiro-Wilk test results indicated that the data collected were non-normally distributed, the subsequent data analysis methods utilised non-parametric methods.

8.3.6   Mean Rank Analysis

In mean rank analysis, following the calculation of the mean of the factors, each factor was ranked according to their mean scores to determine the perceived relative importance of the factors. As the responses were not normally distributed, the mean was not used to determine the significance of the factors.

8.3.7   One-Sample Wilcoxon Signed Rank Test

To test the significance of the factors, one-sample Wilcoxon signed rank test was conducted using an alpha of 0.05. Since the questionnaire utilised 7-point Likert scales for applicability and impact for each factor, the median point of 4 (for applicability) was multiplied by the median of 4 (for impact), resulting in a threshold median of 16. If the *p*-value is less than 0.05, the factor is determined to be statistically significant.

H0: $\tilde{x} \leq 16$
H1: $\tilde{x} > 16$

8.3.8   Two-Sample Wilcoxon Signed Rank Test

The two-sample Wilcoxon signed rank test was conducted to test if the respondents perceived the importance of knowledge and skillsets in managing conventional projects and projects with smart technologies similarly using an alpha level of 0.05 (Riffenburgh, 2006). If the *p*-value is less than 0.05, the factor is deemed to be perceived differently.

H0: The importance of the factor is the same when managing conventional projects and in managing projects with smart technologies
H1: The importance of the factor is different when managing conventional projects and in managing projects with smart technologies

8.3.9   Kruskal Wallis Test

The Kruskal Wallis test was conducted to examine if three or more independent groups of respondents perceive the factor similarly using an alpha level of 0.05

(Riffenburgh, 2006). If the *p*-value is less than 0.05, the factor is deemed to be perceived differently by three or more groups of respondents.

H0: Three or more groups perceive the factor similarly
H1: Three or more groups perceive the factor differently

8.3.10 *The Mann-Whitney U Test*

While the two-sample Wilcoxon signed rank test compares two sets of data from one group of respondents, the Mann-Whitney *U* test compares two sets of data from two independent groups of respondents (Riffenburgh, 2006). If the *p*-value is less than 0.05, the factor is deemed to be perceived differently by two groups of respondents.

H0: The two groups perceive the factor similarly
H1: The two groups perceive the factor differently

8.3.11 *Exploratory Factor Analysis*

EFA aims to statistically determine underlying factor groupings that represent the correlations among a set of interrelated variables and to reduce relatively large sets of variables into more manageable set of variables to form a parsimonious model (Fabrigar & Wegener, 2012; Reio & Shuck, 2015; Zhao et al., 2015). EFA is typically used to develop a theory (Reio & Shuck, 2015). The recommended ratio of the sample size to the number of variables in the study is at least 5 (Verma, 2013; Zhao et al., 2015). EFA was conducted for the analysis of the data collected in the survey questionnaire to determine the factors affecting PMs' attitude towards technologies.

Verma (2013) suggested seven steps to conduct EFA, which was employed in this study as follows:

1. Calculate the descriptive statistics for all the variables in the study.
2. Prepare correlation matrix for all the variables in the study.
3. Conduct Kaiser-Meyer-Olkin (KMO) test to check for the adequacy of the data to run the factor analysis. A KMO value above 0.5 indicates that the sample is adequate, and values between 0.7 and 0.9 indicate that the data can be used to define specific and strong factor groups.
4. Conduct the Bartlett's test of sphericity to check that the correlation matrix is not an identity matrix. If the Bartlett's test of sphericity is significant, the correlation matrix I not an identity matrix and factor analysis can be run.
5. Conduct factor extraction by using Principal Axis Factoring to identify the underlying grouped factors. Factors with eigenvalues of >1 will be retained in the model.
6. Apply an appropriate rotation technique to obtain the final set of factor groupings. The promax rotation technique was adopted in this study. Although varimax

rotation technique is most commonly used in factor analysis, it assumes that the factors are not correlated to one another (Verma, 2013). However, in the area of social science, it is unrealistic to assume that the factors are not correlated (Zhao et al., 2015). Hence, the promax rotation technique was used in this study. The promax rotation technique was also used in previous studies that adopted EFA in the field of construction management (Chan et al., 2010; Lam et al., 2008; Zhao et al., 2015). Factor loadings of 0.40 and above are considered acceptable (Zhao et al., 2015).

7. Name the identified factor groups based on the nature of the individual variables in it.

8.3.12   *Partial Least Squares–Structural Equation Modelling*

SEM is one of the most suitable techniques to analyse the structural relationships among the variables (Eybpoosh et al., 2011; Xiong et al., 2014; Zhao et al., 2015). SEM performs both Confirmatory Factor Analysis (CFA) and path analysis simultaneously, enabling a maximally efficient fit between the data and a structural model (Lim et al., 2011; Zhao et al., 2015).

PLS-SEM was adopted as PLS-SEM does not require a large sample size or normally distributed data (Hair et al., 2012; Zhao et al., 2015). At the same time, PLS-SEM has been increasingly adopted in the construction management research field (Doloi, 2014; Le et al., 2014; Zhao et al., 2015). Moreover, this is in line with Hair et al. (2011), who recommended the use of PLS-SEM if the research is exploratory or used to extend an existing structural theory. PLS-SEM was adopted to investigate the interrelationships among the sub-components of attitude towards technologies and to further refine the measurement and structural model of PMs' attitude towards technologies. The resulting PLS-SEM model was then integrated into the developed TCF.

The PLS-SEM was assessed for its reliability and validity. According to Hair et al. (2011), the PLS-SEM should first be assessed by its measurement model, followed by the evaluation of the structural model. The measurement model used the path analysis with a bootstrapping sample of 1,000. The PLS-SEM measurement model should fulfil the following (Hair et al., 2011):

- Reflective indicator loadings should be more than 0.708.
- Cronbach's alpha should be at least 0.70.
- CR should be at least 0.70.
- AVE should be at least 0.50.
- Heterotrait–monotrait ratio should be less than 0.90.

Following the finalisation of the measurement model, the structural model needs to be evaluated. Path analysis was used to determine the relationship among the variables in the model. The bootstrapping technique was conducted to estimate the significance of the path coefficients using the recommended number of bootstrap samples of 5,000 (Hair et al., 2011). Based on the recommendations of Hair et al.

(2017), the structural model was evaluated according to the guidelines as shown in the following:

- Variance inflation factor of less than 3
- Q^2 value larger than 0 (0 depicting small predictive accuracy, 0.25 depicting medium predictive accuracy and 0.5 depicting large predictive accuracy of the PLS-SEM model)
- Q^2 predicts values of more than 0
- Compare RMSE value with the LM value of each indicator and check if PLS-SEM analysis yields higher prediction errors in terms of RMSE for all (no predictive power), majority (low predictive power), minority or the same number (medium predictive power) or none of the indicators (high predictive power)

The CTA-PLS was also adopted to check the measurement model. The I-PMA was also used to identify the priority of the components to improve PMs' attitude towards technologies. MGA according to one's experience in smart technologies was also adopted to investigate if respondents with and without experience in smart technologies perceive the direction and strength of the relationships between the constructions differently (Gudergan et al., 2008; Hair et al., 2011; Ringle & Sarstedt, 2016).

8.3.13 *Analytical Hierarchical Process*

Following the development of the TCF, AHP was conducted to determine the weightages of the factors affecting PMs' technological competency level. A two-stage SEM-AHP technique was adopted in this research as the methods may be used to overcome the limitations of the other (Komlan, 2017). PLS-SEM was used to understand the interrelationships between the sub-components affecting PMs' attitude towards technologies while AHP was adopted as a multi-criteria decision-making framework to determine the priority of the identified factors in contributing to one's technological competency level. AHP is one of the main mathematical models to support decision theory (Vargas, 2010). One of the key features of AHP is the ability to quantify intangible criteria through the use of pairwise comparisons between two alternatives (Thibadeau, 2007). Although AHP can quantify intangible criteria and support decision making, as it relies on pairwise comparisons among the criteria, the pairwise comparison matrix entails of $n(n-1)/2$ comparisons for n number of elements (Komlan, 2017). However, the large number of pairwise comparisons to be conducted may result in difficulties for the decision maker (Komlan, 2017). Hence, the integration of SEM has been increasingly used to reduce the number of comparisons to be made by decision makers in AHP (Komlan, 2017; Punniyamoorty et al., 2012).

To conduct the AHP, the hierarchy of the factors in contributing to the main goal of the decision making needs to first be established. After the hierarchy of the factors is determined, pairwise comparisons between two factors should be conducted using the Saaty comparison scale. The comparison scale utilises a scale of 1 to 9, with 1 representing equal importance between the two factors and 3, 5, 7 and 9 corresponding to the judgements where the factor in the row is 'slightly more important', 'significantly more important', 'very significantly more important' and 'absolutely more important' than the factor in the column (Saaty, 2006). The reciprocal is then calculated to form the comparison matrix. This is followed by the computation of the priority vector and the assessment of the consistency of the matrix (Saaty, 1978). The results present the priority of the factors in contributing to the goal of decision making.

8.4   Data Presentation

This section presents the profile of interviewees involved in the research. Pilot interviews were carried out with a panel of five experts who possessed substantial experience in the construction industry to validate the survey questionnaire. Following the dissemination of the approved questionnaire across the industry, a total of 97 valid responses were compiled. The respondent group was diverse, spanning a broad range of experience levels, ages, educational backgrounds and organisational types.

Case studies were undertaken with four experts, all of whom had approximately 10 years of experience managing projects that implemented smart technologies. Their insights proved invaluable in understanding the utilisation of these technologies. Nine experts, using the AHP method, determined the weightages of the identified competency factors. Finally, a group of 10 seasoned experts participated in the final validation of the research findings, TCF and KBTCAIS. The subsequent sections provide comprehensive biographical details extracted from the interviews.

8.4.1   *Profile of Pilot Interviewees*

Five experts, each boasting over a decade of experience within the construction industry and proven involvement in implementing at least one type of identified smart technology in construction projects, were invited to participate in the pilot interview. Their expertise was sought to validate the survey questionnaire. To ensure that the various perspectives of the different organisation types in the construction industry are captured, the expert panel included academics and practitioners representing developer, consultant and contractor firms. The profile of the interviewees is shown in Table 8.2.

8.4.2   *Profile of Survey Respondents*

The data was collected from July 2020 to June 2021. A total of 1,200 survey questionnaires were disseminated. The respondents were identified through web directories including the Society of Project Managers Singapore, Real Estate Developers Association Singapore and Public Sector Panel of Consultant and BCA Singapore Directory of Registered Contractors and Licensed Builders. A total of 97 valid responses were collected and further analysed in this study, equating to a response rate of 8.08 per cent. The low response rate may be attributed to the pandemic as construction organisations have been ordered to stop work and face increased crashing activities after restarting work (Ling et al., 2022). While the response rate is relatively low, low response rates in sampling frames larger than 500 with at least

Table 8.2 Profile of Interviewees for Pilot Interviews

Interviewee	Organisation Type	Years of Experience in the Construction Industry	Years of Experience in Smart Technologies	Smart Technologies Involved In
P1	Academic	17	17	CPS, IoT, big data, autonomous vehicles, robotics
P2	Academic	17	12	Big data
P3	Developer	16	10	Internet-of-Things, big data, autonomous vehicles, robotics, virtual reality
P4	Consultant	25	5	Internet-of-Things, big data, artificial intelligence, augmented reality, virtual reality
P5	Contractor	11	11	Internet-of-Things, big data, autonomous vehicles, robotics, additive manufacturing

50 valid responses were found to be not equated to nonresponse bias (Curtin et al., 2000; Fosnacht et al., 2017; Keeter et al., 2000). The respondents had various numbers of years of experience in the construction industry, age and education level and represented both females and males. Furthermore, the respondents represented organisations of different types, sizes and years of experience in the construction industry. The heterogeneity of the survey respondents allows for meaningful and valid data analysis to be conducted (Carley-Baxter et al., 2018). The profile of the survey respondents and their organisations are presented in Table 8.3.

8.4.3 Profile of Case Study Interviewees

Case studies were conducted with experts who were heavily involved in the digital transformation of their organisations and represented the different organisation types in the construction industry. Interviewees P3 to P5 were involved in both the pilot interviews and the case study interviews. As contractor firms are more heavily involved in the management of construction projects on site, three cases identified from the contractor firm provide a better understanding of the use of smart technologies in construction projects. The profile of the case study interviewees is shown in Table 8.4.

8.4.4 Profile of Interviewees for Analytical Hierarchy Process

To determine the weightages of the factors affecting technological competency of PMs, AHP was conducted with nine experts representing the different organisation types with various numbers of years of experience in the construction industry and in smart technologies. The profile of the interviewees for AHP is presented in Table 8.5.

Table 8.3 Profile of Survey Respondents

		Frequency	Percentage of Total Respondents
Respondent's Job Role	Project Manager	97	100.00
Respondent's Years of Experience in the Construction Industry	Less than 3 years	31	31.96
	3 to 10 years	20	20.62
	More than 10 years	46	47.42
Respondent's Age	25 to 34 years old	49	50.52
	35 to 44 years old	21	21.65
	45 to 54 years old	16	16.49
	55 to 64 years old	11	11.34
Respondent's Education	Diploma and below	22	22.68
	Bachelor's Degree	40	41.24
	Postgraduate Degree	35	36.08
Respondent's Experience in Smart Technology	Yes	43	44.33
	No	54	55.67
Organisation Type	Consultant	40	41.24
	Contractor	35	36.08
	Developer	22	22.68
Organisation Size	Large Enterprise (More than 200 employees)	56	57.73
	Small or Medium Enterprise (Less than 200 employees)	41	42.27
Organisation's Years of Experience	Less than 10 years	8	8.25
	10 to 20 years	13	13.40
	21 to 30 years	20	20.62
	More than 30 years	56	57.73
Organisation's Experience in Smart Technology	Yes	65	67.01
	No	32	32.99

Table 8.4 Profile of Case Study Interviewees

Interviewee	Organisation Type	Years of Experience in the Construction Industry	Years of Experience in Smart Technologies	Smart Technologies Involved In
P3	Developer	16	10	Internet-of-Things, big data, autonomous vehicles, robotics, virtual reality
P4	Consultant	25	5	Internet-of-Things, big data, artificial intelligence, augmented reality, virtual reality
P5	Contractor	11	11	Internet-of-Things, big data, autonomous vehicles, robotics, additive manufacturing
P6	Contractor	21	11	Internet-of-Things, big data, autonomous vehicles, robotics, additive manufacturing

Table 8.5 Profile of Interviewees for AHP

Interviewee	Organisation Type	Years of Experience in the Construction Industry	Years of Experience in Smart Technologies	Smart Technologies Involved In
A1	Developer	14	7	Big data, autonomous vehicles, robotics, virtual reality
A2	Developer	36	5	Big data, virtual reality
A3	Consultant	5	5	Big data, virtual reality
A4	Consultant	6	6	Big data, virtual reality
A5	Consultant	10	8	Big data, autonomous vehicles, virtual reality, additive manufacturing
A6	Consultant	7	5	Big data, autonomous vehicles, augmented reality, virtual reality, 3D imaging
A7	Contractor	25	3	Artificial intelligence, virtual reality, additive manufacturing
A8	Contractor	33	6	Autonomous vehicles, virtual reality, additive manufacturing
A9	Contractor	10	6	Internet-of-Things, autonomous vehicles, virtual reality, 3D imaging

8.4.5 Profile of Interviewees for Validation and Test Implementation

To validate the survey findings and developed TCF and KBTCAIS, interviews were conducted with 10 experts with various numbers of years in the construction industry and smart technologies representing the different organisation domains. The profile of the respondents is shown in Table 8.6.

8.5 Summary

This research adopted a mixed research approach to achieve the research objectives. One cross-sectional survey was conducted to investigate (i) the existing implementation of smart technologies in projects, (ii) the changes in the importance of project management knowledge areas in managing projects with smart technologies, (iii) the changes in the importance of skillsets required to manage projects with smart technologies and (iv) the factors that affect PMs' attitude towards technologies. Case studies were conducted to enrich the findings from the literature review, to support the development of the TCF and to identify recommendations to facilitate technology adoption and improve the technological competency of PMs. This research also adopted various statistical analysis methods to analyse the data collected from the survey to develop the TCF and KBTCAIS. Several rounds of interviews have been conducted to (i) validate the developed survey questionnaire, (ii) determine the weightages of factors affecting PMs' technological competency,

Table 8.6 Profile of Interviewees for Validation and Test Implementation

Interviewee	Organisation Type	Years of Experience in the Construction Industry	Years of Experience in Smart Technologies	Smart Technologies Involved In
V1	Developer	25	7	Big data, autonomous vehicles, robotics, virtual reality
V2	Developer	32	3	Big data, virtual reality
V3	Developer	10	3	Big data, autonomous vehicle, virtual reality
V4	Consultant	6	4	Internet-of-Things, big data, augmented reality, virtual reality
V5	Consultant	12	7	Big data, autonomous vehicles, virtual reality
V6	Consultant	17	10	Big data, autonomous vehicles, 3D imaging
V7	Contractor	22	4	Autonomous vehicles, robotics
V8	Contractor	30	2	Autonomous vehicles, virtual reality
V9	Contractor	11	6	Internet-of-Things, big data
V10	Contractor	6	3	3D imaging, autonomous vehicles

(iii) validate the survey findings, (iv) validate the developed TCF, (v) validate the recommendations provided in the KBTCAIS and (vi) validate the developed KBT-CAIS. Moreover, the profiles of the expert panel interviewees who were interviewed for pilot interviews, case studies, determination of factor weightages and validation of survey findings, along with those who were involved in the TCF and the developed KBTCAIS, are presented in this chapter. The demographic information about the survey respondents who completed the survey questionnaire is also included.

References

Abowitz, D. A., & Toole, T. M. (2010). Mixed method research: Fundamental issues of design, validity, and reliability in construction research. *Journal of Construction Engineering and Management, 136*(1), 108–116. https://doi.org/10.1061/(asce)co.1943-7862.0000026

Agarwal, R., & Prasad, J. (1998). A conceptual and operational definition of personal innovativeness in the domain of information technology. *Information Systems Research, 9*(2), 204–215. https://doi.org/10.1287/isre.9.2.204

Bolarinwa, O. A. (2015). Principles and methods of validity and reliability testing of questionnaires used in social and health science researches. *Nigerian Postgraduate Medical Journal, 22*(4), 195. https://doi.org/10.4103/1117-1936.173959

Carley-Baxter, L. R., Hill, C. A., Roe, D. J., Twiddy, S. E., Baxter, R. K., & Ruppenkamp, J. (2018). Does response rate matter? Journal editors use of survey quality measures in manuscript publication decisions. *Survey Practice*, *2*(7), 1–7. https://doi.org/10.29115/sp-2009-0033

Chan, A. P. C., Darko, A., & Ameyaw, E. E. (2017). Strategies for promoting green building technologies adoption in the construction industry-An international study. *Sustainability*, *9*(6), 969–986. https://doi.org/10.3390/su9060969

Chan, A. P. C., Lam, P. T. I., Chan, D. W. M., Cheung, E., & Ke, Y. (2010). Critical success factors for PPPs in infrastructure developments: Chinese perspective. *Journal of Construction Engineering and Management*, *136*(5), 484–494. https://doi.org/10.1061/(ASCE)CO.1943-7862.0000152

Compeau, D. R., & Higgins, C. A. (1995). Computer self-efficacy: Development of a measure and initial test. *MIS Quarterly*, *19*(2), 189. https://doi.org/10.2307/249688

Creswell, J. W., & Poth, C. N. (2018). *Qualitative inquiry and research design: Choosing among five approaches*. SAGE Publications.

Curtin, R., Presser, S., & Singer, E. (2000). The effects of response rate changes on the index of consumer sentiment. *Public Opinion Quarterly*, *64*(4), 413–428. https://doi.org/10.1086/318638

Davis, F. D. (1989). Perceived usefulness, perceived ease of use, and user acceptance of information technology. *MIS Quarterly*, *13*(3), 319–340. https://doi.org/10.2307/249008

Doloi, H. (2014). Rationalizing the implementation of web-based project management systems in construction projects using PLS-SEM. *Journal of Construction Engineering and Management*, *140*(7). https://doi.org/10.1061/(ASCE)CO.1943-7862.0000859

Eybpoosh, M., Dikmen, I., & Birgonul, M. T. (2011). Identification of risk paths in international construction projects using structural equation modeling. *Journal of Construction Engineering and Management*, *137*(12), 1164–1175. https://doi.org/10.1061/(ASCE)CO.1943-7862.0000382

Fabrigar, L. R., & Wegener, D. T. (2012). Requirements for and decisions in choosing exploratory factor common analysis. In *Understanding statistics: Exploratory factor analysis* (pp. 19–38). Oxford University Press.

Fagerland, M. W. (2012). T-tests, non-parametric tests, and large studies a paradox of statistical practice? *BMC Medical Research Methodology*, *12*(1), 78. https://doi.org/10.1186/1471-2288-12-78

Fosnacht, K., Sarraf, S., Howe, E., & Peck, L. K. (2017). How important are high response rates for college surveys? *Review of Higher Education*, *40*(2), 245–265. https://doi.org/10.1353/rhe.2017.0003

Ghasemi, A., & Zahediasl, S. (2012). Normality tests for statistical analysis: A guide for non-statisticians. *International Journal of Endocrinology and Metabolism*, *10*(2), 486–489. https://doi.org/10.5812/ijem.3505

Gravetter, F. J., & Forzano, L.-A. B. (2018). *Research methods for the behavioral sciences*. Cengage Learning.

Gudergan, S. P., Ringle, C. M., Wende, S., & Will, A. (2008). Confirmatory tetrad analysis in PLS path modeling. *Journal of Business Research*, *61*(12), 1238–1249. https://doi.org/10.1016/j.jbusres.2008.01.012

Hair, J. F., Hult, G. T. M., Ringle, C. M., & Sarstedt, M. (2017). A primer on partial least squares structural equation modeling (PLS-SEM). In *Sage*. Sage.

Hair, J. F., Ringle, C. M., & Sarstedt, M. (2011). PLS-SEM: Indeed a silver bullet. *Journal of Marketing Theory and Practice*, *19*(2), 139–152. https://doi.org/10.2753/MTP1069-6679190202

Hair, J. F., Sarstedt, M., Pieper, T. M., & Ringle, C. M. (2012). The use of partial least squares structural equation modeling in strategic management research: A review of past practices and recommendations for future applications. *Long Range Planning*, *45*(5–6), 320–340. https://doi.org/10.1016/j.lrp.2012.09.008

Harwell, M. R. (1988). Choosing between parametric and nonparametric tests. *Journal of Counseling & Development*, *67*(1), 35–38. https://doi.org/10.1002/j.1556-6676.1988.tb02007.x

Heinssen, R. K., Glass, C. R., & Knight, L. A. (1987). Assessing computer anxiety: Development and validation of the computer anxiety rating scale. *Computers in Human Behavior*, *3*(1), 49–59. https://doi.org/10.1016/0747-5632(87)90010-0

Jolliffe, D., Farrington, D. P., Hawkins, J. D., Catalano, R. F., Hill, K. G., & Kosterman, R. (2003). Predictive, concurrent, prospective and retrospective validity of self-reported delinquency. *Criminal Behaviour and Mental Health*, *13*(3), 179–197. https://doi.org/10.1002/cbm.541

Karras, D. J. (1997). Statistical methodology: II. Reliability and validity assessment in study design, part B. *Academic Emergency Medicine*, *4*(2), 144–147. https://doi.org/10.1111/j.1553-2712.1997.tb03723.x

Keeter, S., Miller, C., Kohut, A., Groves, R. M., & Presser, S. (2000). Consequences of reducing nonresponse in a national telephone survey. *Public Opinion Quarterly*, *64*(2), 125–148. https://doi.org/10.1086/317759

Komlan, G. (2017). A two-staged SEM-AHP technique for understanding and prioritizing mobile financial services perspective adoption. *European Journal of Business and Management*, *9*(30), 107–123.

Kvam, P. H., & Vidakovic, B. (2007). Nonparametric statistics with applications to science and engineering. In *Nonparametric statistics with applications to science and engineering*. John Wiley & Sons, Inc. https://doi.org/10.1002/9780470168707

Lam, E. W. M., Chan, A. P. C., & Chan, D. W. M. (2008). Determinants of successful design-build projects. *Journal of Construction Engineering and Management*, *134*(5), 333–341. https://doi.org/10.1061/(ASCE)0733-9364(2008)134:5(333)

Le, Y., Shan, M., Chan, A. P. C., & Hu, Y. (2014). Investigating the causal relationships between causes of and vulnerabilities to corruption in the Chinese public construction sector. *Journal of Construction Engineering and Management*, *140*(9), 05014007. https://doi.org/10.1061/(ASCE)CO.1943-7862.0000886

Lim, B. T. H., Ling, F. Y. Y., Ibbs, C. W., Raphael, B., & Ofori, G. (2011). Empirical analysis of the determinants of organizational flexibility in the construction business. *Journal of Construction Engineering and Management*, *137*(3), 225–237. https://doi.org/10.1061/(ASCE)CO.1943-7862.0000272

Ling, F. Y. Y., Zhang, Z., & Yew, A. Y. R. (2022). Impact of COVID-19 pandemic on demand, output, and outcomes of construction projects in Singapore. *Journal of Management in Engineering*, *38*(2), 04021097. https://doi.org/10.1061/(ASCE)ME.1943

Love, P. E. D., Holt, G. D., & Li, H. (2002). Triangulation in construction management research. *Engineering, Construction and Architectural Management*, *9*(4), 294–303. https://doi.org/10.1108/eb021224

Lucko, G., & Rojas, E. M. (2010). Research validation: Challenges and opportunities in the construction domain. *Journal of Construction Engineering and Management*, *136*(1), 127–135. https://doi.org/10.1061/(ASCE)CO.1943-7862.0000025

Mircioiu, C., & Atkinson, J. (2017). A comparison of parametric and non-parametric methods applied to a Likert scale. *Pharmacy*, *5*(4), 26. https://doi.org/10.3390/pharmacy5020026

Nickell, G. S., & Pinto, J. N. (1986). The computer attitude scale. *Computers in Human Behavior*, *2*(4), 301–306. https://doi.org/10.1016/0747-5632(86)90010-5

Onwuegbuzie, A. J., Witcher, A. E., Collins, K. M. T., Filer, J. D., Wiedmaier, C. D., & Moore, C. W. (2007). Students' perceptions of characteristics of effective college teachers: A validity study of a teaching evaluation form using a mixed-methods analysis. *American Educational Research Journal*, *44*(1), 113–160. https://doi.org/10.3102/0002831206298169

Punniyamoorty, M., Mathiyalagan, P., & Lakshmi, G. (2012). A combined application of structural equation modeling (SEM) and analytic hierarchy process (AHP) in supplier selection. *Benchmarking*, *19*(1), 70–92. https://doi.org/10.1108/14635771211218362

Rea, L. M., & Parker, R. A. (2014). *Designing and conducting survey research*. Jossey-Bass.

Reio, T. G., & Shuck, B. (2015). Exploratory factor analysis. *Advances in Developing Human Resources*, *17*(1), 12–25. https://doi.org/10.1177/1523422314559804

Riffenburgh, R. H. (2006). Tests on ranked data. In *Statistics in medicine* (pp. 281–303). Academic Press. https://doi.org/10.1016/b978-012088770-5/50056-3

Ringle, C. M., & Sarstedt, M. (2016). Gain more insight from your PLS-SEM results the importance-performance map analysis. *Industrial Management and Data Systems, 116*(9), 1865–1886. https://doi.org/10.1108/IMDS-10-2015-0449/FULL/PDF

Saaty, T. L. (1978). Modelling unstructure decision problems – the theory of analytical hierarchies. *Mathematics and Computers in Simulation, XX*, 147–158. https://doi.org/10.1016/0378-4754(78)90064-2

Saaty, T. L. (2006). There is no mathematical validity for using fuzzy number crunching in the analytic hierarchy process. *Journal of Systems Science and Systems Engineering, 15*(4), 457–464. https://doi.org/10.1007/s11518-006-5021-7

Salkind, N. (2015). Shapiro-wilk test for normality. In *Encyclopedia of measurement and statistics*. Sage Publications, Inc. https://doi.org/10.4135/9781412952644.n404

Tan, W. (2017). *Research methods: A practical guide for students and researchers*. World Scientific.

Thibadeau, B. (2007). Prioritizing project risks using AHP. *PMI Global Congress 2007* (pp. 1–9). www.pmi.org/learning/library/project-decision-making-tool-7292

Vargas, R. V. (2010). Using the Analytic Hierarchy Process (AHP) to select and prioritize projects in a portfolio. *PMI Global Congress* (pp. 1–22). www.pmi.org/learning/library/analytic-hierarchy-process-prioritize-projects-6608

Venkatesh, V. (2000). Determinants of perceived ease of use: Integrating control, intrinsic motivation, and emotion into the technology acceptance model. *Information Systems Research, 11*(4), 342–365. https://doi.org/10.1287/isre.11.4.342.11872

Venkatesh, V., & Bala, H. (2008). Technology Acceptance Model 3 and a research agenda on interventions. *Decision Sciences, 39*(2), 273–315. https://doi.org/10.1111/j.1540-5915.2008.00192.x

Venkatesh, V., & Davis, F. D. (1996). A model of the antecedents of perceived ease of use: Development and test. *Decision Sciences, 27*(3), 451–481. https://doi.org/10.1111/j.1540-5915.1996.tb00860.x

Venkatesh, V., & Davis, F. D. (2000). Theoretical extension of the Technology Acceptance Model: Four longitudinal field studies. *Management Science, 46*(2), 186–204. https://doi.org/10.1287/mnsc.46.2.186.11926

Venkatesh, V., Morris, M. G., Davis, G. B., & Davis, F. D. (2003). User acceptance of information technology: Toward a unified view. *MIS Quarterly, 27*(3), 425–478. https://doi.org/10.2307/30036540

Verma, J. P. (2013). Data analysis in management with SPSS software. In *Data analysis in management with SPSS software*. Springer India. https://doi.org/10.1007/978-81-322-0786-3

Voss, C., Tsikriktsis, N., & Frohlich, M. (2002). Case research in operations management. *International Journal of Operations and Production Management, 22*(2), 195–219. https://doi.org/10.1108/01443570210414329

Xiong, B., Skitmore, M., Xia, B., Masrom, M. A., Ye, K., & Bridge, A. (2014). Examining the influence of participant performance factors on contractor satisfaction: A structural equation model. *International Journal of Project Management, 32*(3), 482–491. https://doi.org/10.1016/j.ijproman.2013.06.003

Zhao, X. (2014). *Enterprise risk management in Chinese construction firms operating overseas*. National University of Singapore.

Zhao, X., Hwang, B.-G., Pheng Low, S., & Wu, P. (2015). Reducing hindrances to enterprise risk management implementation in construction firms. *Journal of Construction Engineering and Management, 141*(3), 04014083. https://doi.org/10.1061/(ASCE)CO.1943-7862.0000945

9 Data Analysis and Discussion

9.1 Introduction

This chapter provides a detailed exploration of the results derived from data analysis performed on the obtained survey responses. After an initial cleaning process, a sum of 97 valid responses was gathered. The obtained data was consistent and non-normally distributed. Therefore, non-parametric tests were adopted for further analysis.

In Singapore, an average of 23.51 per cent of the total projects have implemented at least one smart technology. The comparatively slow pace of technology adoption can be attributed to the lack of suitable use cases, the technology's infancy stage in the industry and cost-related concerns. Projects in building and infrastructure have witnessed a higher rate of adoption, influenced by factors such as safety monitoring and government initiatives. Moreover, VR emerged as the most familiar technology to PMs, largely attributed to the widespread use of BIM, followed by AR and BD. Meanwhile, technologies such as DLT, CPS and AM were less familiar due to technical hurdles, shortage of skilled professionals and regulatory challenges.

The data analysis results revealed that the generally low adoption rates of smart technologies may stem from integration complexities, technological limitations, human resource constraints and regulatory and cost barriers. The results of data analysis showed that practitioners in Singapore's construction industry generally believe having knowledge in these technology applications is important, where BIM has been identified as the most important one, followed by AR and VR and IoT.

This chapter also offers a comparison of the skills required to manage traditional projects versus those involving smart technologies. The one-sample Wilcoxon signed rank test concluded that all discussed skills are crucial in managing both types of projects. Conventional project management was found to hinge on skills such as project management expertise, problem solving, communication, planning, and organizing and team-building abilities. Managing projects involving smart technologies prioritised similar skills, including problem solving, information management, communication, project management and planning and organizing abilities. It should be noticed that technical and operational technology skills, information management skills, creativity, ethical awareness, strategic planning skills, active learning skills, decision-making skills and motivation skills were more important in managing projects with smart technologies compared to conventional projects.

DOI: 10.1201/9781003462231-9

In addition, this chapter analyses the factors that shape one's attitudes towards technologies. It was discovered that the PEU and TSE hold the most influence. A person's confidence in their ability to use technology considerably affects their inclination and ease of adoption of new technologies.

The analysis also revealed that respondents' demographic background has limited impacts on their perceptions towards smart technologies. Perceptions regarding the impact and significance of TSE factors varied across respondents of different age groups, while the importance of project management knowledge and skills was similarly perceived across all age groups. Additionally, respondents' educational backgrounds, years of experience in the construction industry, their exposure to smart technologies and their roles within organisations did not significantly influence the perceived importance of the knowledge and skills needed to manage projects with smart technologies. When evaluating factors that influence attitudes towards technologies, differences in perceptions were observed among respondents with varying education levels, experiences with smart technologies, organisational roles and years of working experience. The subsequent sections present detailed discussion of the data analysis results.

9.2 Data Pre-processing

The collected data was first cleaned to remove incomplete responses, invalid responses, duplicates and responses that were not from PMs, resulting in a total of 97 valid responses that were subsequently analysed. Following that, the validity and reliability of the responses were assessed to ensure that the responses allow for meaningful data analysis. The summary of the data pre-processing results is shown in Appendices A-1 to A-3 for knowledge in project management, skills and attitude towards technologies respectively.

The content validity of the survey questionnaire has been validated through the pilot interviews with the experts. Following that, the construct validity of the survey questionnaire was assessed using corrected item-total correlations (Zijlmans et al., 2019). The results indicated that the items were valid for data analysis (Pallant, 2020; Squires et al., 2011). The responses were then assessed for their consistency using Cronbach's alpha, and a value above 0.7 was preferable. The Cronbach's alpha for the entire dataset was found to be 0.962, and the results for the individual components were found to be higher than 0.8, indicating that the responses were consistent. To determine if parametric or non-parametric tests should be adopted for the subsequent data analysis, Shapiro-Wilk test was conducted using an alpha value of 0.05. The results indicated that the data is non-normally distributed and thus non-parametric tests were used for the subsequent data analysis.

9.3 Status Quo of Smart Technology Application in the Construction Industry

9.3.1 *Existing Implementation Level of Smart Technologies in Projects*

Table 9.1 shows the profile of projects organisations have been involved in in the last three years and the number of projects that have implemented at least one smart

Table 9.1 Profile of Projects

	Total Projects in the Last 3 Years	Projects with at Least One Smart Technology in the Last 3 Years	Percentage
Building	1971	490	24.86
Infrastructure	314	68	21.66
Industrial	334	61	18.26
Others	57	10	17.54
Total	2676	629	**23.51**

technology. An average of 23.51 per cent of the total projects the respondent's organisations were involved in had implemented at least one smart technology. While 67.01 per cent of the organisations and 44.33 per cent of the respondents had implemented at least one smart technology in projects, the relatively lower proportion of projects with at least one smart technology reflects the slow uptake of new technologies in the construction industry.

One reason for the relatively low level of smart technology adoption in construction projects could be the lack of suitability or justification for applying smart technologies in the specific projects. It is evident both from past studies and through interviews that smart technologies are likely to be applied in projects if there are specific use cases and expected benefits and if they can overcome targeted issues in the projects (Ercan, 2019; Pellicer et al., 2014; Prebanić & Vukomanović, 2021). Hence, not all projects require the implementation of smart technologies. At the same time, the low level of adoption of smart technologies in construction projects could also be because the use of smart technologies in the Singapore construction industry is still in its infancy, and the smart technologies are in their experimentation and testing periods or are deployed in smaller-scaled projects. Specifically, building projects were found to have the highest implementation level of smart technologies, with about one-quarter of the projects having implemented at least one smart technology. Building projects include residential, commercial office buildings and shopping malls (Construction Industry Institute, 2022). Along this note, there has been increasing number of tall buildings in the Singapore built environment. At the same time, the government has established the Construction Industry Transformation Map which focuses on the drive towards integrated project delivery, design for manufacturing and assembly, prefabricated pre-volumetric construction approaches and sustainability (Building and Construction Authority, 2017). With the prevalence of tall building projects and modular construction, it is reasonable that building projects have the highest implementation level of smart technologies among the project types, although the implementation level remains low.

This is followed by infrastructure projects, with about one-fifth of the projects having implemented at least one smart technology. Infrastructure projects consist of tunnels, railways and the like (Construction Industry Institute, 2022). The Singapore government has been expanding its land transport network with an increase in underground infrastructure (Land Transport Authority, 2020). As underground infrastructure projects present huge potential for smart technologies especially for Internet-of-Things to improve the safety performance through real-time monitoring of underground conditions (L. Wang et al., 2020; Zhou & Ding, 2017), it is reasonable that infrastructure projects have the second highest level of implementation of smart technologies.

Industrial projects were found to have the lowest level of smart technology implementation among the project types. This is reasonable as industrial projects are relatively low storey that emphasise on functionality and safety over aesthetic differentiation with a low interface with the public (Construction Industry Institute, 2022). Hence, the cost of smart technologies may not be justified with the potential benefits that may be derived through the project lifecycle.

9.3.2 Respondent's Level of Familiarity in Smart Technologies and Attitude Towards Technologies

Table 9.2 shows the respondents' level of familiarity in each of the smart technologies. It was found that VR, AR and BD were smart technologies that respondents had higher level of familiarity in while the technologies that respondents were least familiar in were DLT, CPS and AM.

It was found that the respondents were most familiar with VR. BIM enables the visualisation of the physical and functional properties of buildings and supports a variety of analyses such as clash detection and cost estimation while VR technologies integrate information systems and immersive environments, extending the capabilities of BIM by enabling immersive user experience (Kamari et al., 2020; Kim et al., 2021). With BIM being mandated in Singapore for new projects with a gross floor area larger than 5,000 square metres (Hwang et al., 2020; Jiang et al., 2021), it is expected that

Table 9.2 Respondents' Level of Familiarity in Smart Technologies

Smart Technologies	Level of Familiarity in Technology (1 = Not familiar at all, 7 = Very familiar)					
	Mean	SD	Rank	Median	Shapiro-Wilk	WSR
Cyber-Physical System	1.65	0.107	10	1	0.001*	0.000*
Internet-of-Things	2.09	0.141	7	2	0.001*	0.001*
Big Data	2.51	0.138	3	2	0.001*	0.001*
Artificial Intelligence	1.99	0.137	8	1	0.001*	0.001*
Robotics	2.31	0.126	5	2	0.001*	0.001*
Autonomous Vehicle	2.33	0.115	4	2	0.001*	0.001*
Augmented Reality	2.66	0.136	2	3	0.001*	0.001*
Virtual Reality	3.41	0.128	1	4	0.001*	0.001*
Additive Manufacturing/3D Printing	1.73	0.120	9	1	0.001*	0.001*
Laser Scanning	2.27	0.151	6	2	0.001*	0.001*
Distributed Ledger Technology/ Blockchain	1.57	0.102	11	1	0.001*	0.000*

SD, standard deviation; WSR, one-sample Wilcoxon signed rank test.
* Significant at *p*-value <0.05.

respondents are most familiar with VR technology. In addition, the relatively low level of familiarity the respondents have in VR is consistent with the finding that the capabilities of BIM have not been fully exploited (Oesterreich & Teuteberg, 2019).

Next, the respondents had a relatively low level of familiarity in AR. Similar to VR, AR can enhance the capabilities of BIM by providing information from the BIM model onto real-world physical environments and enable collaboration among stakeholders remotely (Harikrishnan et al., 2021; Noghabaei et al., 2020). Despite several use cases of AR in the construction industry, the poor collaborative environment and practices in the Singapore construction industry may result in the low adoption of AR technology (Hwang et al., 2020). In addition, technological factors such as hardware and software limitations, costs of technology and site conditions hinder the adoption of AR (Harikrishnan et al., 2021; Oesterreich & Teuteberg, 2016). Hence, this may result in a relatively low level of familiarity in AR.

The respondents had a relatively low level of familiarity in BD. Although construction projects can generate a large volume, variety and velocity of data, the Singapore construction industry was found to lack documentation leading to poor knowledge management for organisational learning (Low et al., 2016). Beyond the Singapore construction industry, poor data management practices, information transfer environment and challenges to BD adoption are other reasons that lead to low levels of BD adoption in the construction industry (Bhattacharya & Momaya, 2021; Bilal et al., 2016; Fernández-Solís et al., 2015; Gamil & Rahman, 2017; Hwang et al., 2022; Kania et al., 2020). Hence, the adoption of BD in the Singapore construction industry may be low, leading to a low level of familiarity in BD in the respondents.

On the other hand, the respondents were found to have the lowest level of familiarity in DLT. While DLT has the potential to overcome key challenges faced in the construction industry, the existing implementation status of DLT remains low, with most studies demonstrating the applicability of DLT in projects on a theoretical level (Hunhevicz & Hall, 2020). This is consistent with other studies investigating the challenges of implementing DLT in construction projects, citing challenges including lack of regulatory oversight, lack of skilled professionals and technical- and business-process-related challenges (Hunhevicz & Hall, 2020; Mohammed et al., 2021; Tezel et al., 2020). The immutable nature of DLT in addition to a lack of skilled professionals also brings about resistance towards adopting DLT with the fear of human errors and poorly written smart contracts (Tezel et al., 2020). The resulting low level of implementation of DLT in the construction industry may explain why respondents have the lowest level of familiarity in DLT.

Next, respondents were found to have a low level of familiarity in CPS. CPS can be implemented in construction projects for real-time monitoring and control, integrated data platform and communication among stakeholders. According to Greer et al. (2019), there has been a convergence of the concepts of CPS and IoT to refer to systems that collect information of the physical environment and send them to the cyber model for processing and analysis for the controlling of physical processes through actuators. Along this line, respondents may view CPS and IoT as similar technologies, resulting in a lower level of familiarity in CPS. In addition, construction sites may not have existing structures that have been constructed and sensors may be placed on moving objects such as on moving equipment, materials and workers requiring human control. This may result in a low level of familiarity in CPS.

Finally, respondents were also found to have low level of familiarity in AM. While the Singapore government has identified AM as one of the key technologies to drive construction productivity, the implementation of AM is not without barriers (Building and Construction Authority, 2016; Pan et al., 2021). Some of the key barriers include building codes and regulations, liability for 3D printed components and technology-related barriers (Ghaffar et al., 2018; Wu et al., 2018). Despite government support, the level of AM implementation in the Singapore construction industry remains low. Hence, it is reasonable that respondents have low level of familiarity in AM.

While there is a low level of familiarity in the smart technologies, the respondents displayed a slight inclination towards the use of technologies, perceiving the use of technologies in their work as good, interesting and fun and expressed a liking towards the use of technologies in their work. Table 9.3 shows the summary of the results of the respondents' attitude towards technologies. Despite the slight inclination towards the use of technologies, the adoption of smart technologies in the construction industry remains low, as reflected by the level of familiarity in the smart technologies. This may indicate the complexities of integrating the use of technologies in existing workflows, perceived limitations of existing technologies and human-resource-related, regulatory and cost-related barriers to technology adoption (Hwang et al., 2022; Oesterreich & Teuteberg, 2016, 2019). The finding shows that if the barriers to smart technology adoption are mitigated, respondents are willing to adopt smart technologies in their work. On this note, the slight inclination towards the use of technologies can be leveraged on to facilitate and encourage the adoption of smart technologies in the construction industry

Table 9.3 Respondents' Attitude Towards Technologies

Attitude Towards Technologies	Mean	SD	Median	Shapiro-Wilk	WSR
Use of Technologies in Their Work Is Bad/Good (1 = bad, 7 = good)	4.95	0.148	5	0.001*	0.001*
Use of Technologies in Their Work Makes Work Uninteresting/Interesting (1 = disinteresting, 7 = interesting)	4.73	0.140	5	0.001*	0.001*
Use of Technologies in Their Work Is Boring/Fun (1 = boring, 7= fun)	4.66	0.137	5	0.001*	0.001*
Dislike/Like the Use of Technologies in Their Work (1 = dislike, 7 = like)	4.81	0.162	5	0.001*	0.001*

SD, standard deviation; WSR, one-sample Wilcoxon signed rank test.
* Significant at *p*-value <0.05.

9.4   Technological Competency of Project Managers

9.4.1   *Respondents' Self-Assessed Level of Technological Competency*

The respondents were asked to self-assess their level of technological competency. The summary of the results is shown in Appendix A-4. It was found that the respondents generally viewed themselves to have a moderately low level of technological competency in managing projects with smart technologies, with a mean of 3.07 and standard deviation of 0.181. The moderately low level of technological competency in managing projects with smart technologies is expected as widespread use of smart technologies in the Singapore construction industry is still in its nascent stage (Hwang et al., 2020, 2022). Hence, respondents may perceive their level of technological competency to be relatively low due to their unfamiliarity with the smart technologies.

9.4.2   *Knowledge in Project Management*

9.4.2.1   *Importance in Conventional Projects and Projects with Smart Technologies*

Table 9.4 shows the summary of the data analysis results of the perceived importance of each project management knowledge area in managing conventional projects and projects with at least one smart technology. Based on the one-sample Wilcoxon signed rank test, all the project management knowledge areas were found to be important in managing both conventional projects and projects with at least one smart technology. This is expected as knowledge in project management serves as the essential domain knowledge for managing construction projects and the principles of project management remain the same regardless of the use of smart technologies in the projects. The findings also corroborate with existing studies on the characteristics of the smart technologies, and how they may be applied to projects. Furthermore, the findings were further supported by Interviewees P3, P4, P5, P6 and A8, who further highlighted that project success criteria of schedule, cost, quality and safety remain the same whether or not smart technologies are used in projects.

In conventional projects, the top three project management knowledge areas were found to be project schedule management, project cost management and project risk management. On the other hand, the top three project management knowledge areas required to manage projects with smart technologies were found to be project communication management, project cost management and project risk management.

The top project management knowledge areas for managing conventional construction projects are consistent with the triple constraints of projects of time, cost and quality and the safety performance of projects (Alzahrani & Emsley, 2013; Bronte-stewart, 2015; Chan et al., 2004). In addition, project schedule management, project cost management and project risk management were also found to be the prioritised knowledge areas in project management education programs (Nguyen et al., 2017).

Table 9.4 Summary of Data Analysis Results for Knowledge in Project Management

Project Management Knowledge Areas	For Conventional Projects					For Projects with at Least One Smart Technology					Differences Between Conventional Projects and Projects with at Least One Smart Technology	
	Mean	SD	Median	Rank	WSR	Mean	SD	Median	Rank	WSR	Two-sample Wilcoxon signed rank test	SRCC
Project Integration Management	6.02	1.031	6	6	0.000*	6.23	1.075	7	6	0.000*	0.004*	0.855*
Project Scope Management	5.73	1.150	6	8	0.001*	6.02	1.090	6	9	0.001*	0.001*	
Project Schedule Management	6.39	0.919	7	1	0.000*	6.28	1.214	7	4	0.000*	0.350	
Project Cost Management	6.34	0.978	7	2	0.000*	6.30	1.165	7	2	0.000*	0.709	
Project Quality Management	6.22	0.971	7	5	0.000*	6.27	1.141	7	5	0.000*	0.349	
Project Resource Management	5.89	1.117	6	7	0.001*	6.05	1.140	6	8	0.001*	0.069	
Project Communication Management	6.29	1.080	7	4	0.000*	6.39	1.105	7	1	0.000*	0.122	
Project Risk Management	6.33	1.028	7	3	0.000*	6.29	1.145	7	3	0.000*	0.583	
Project Procurement Management	5.61	1.160	6	10	0.001*	5.84	1.161	6	10	0.001*	0.024*	
Project Stakeholder Management	5.70	1.300	6	9	0.001*	6.08	1.179	7	7	0.001*	0.001*	

SD, standard deviation; SRCC, Spearman rank correlation coefficient; WSR, one-sample Wilcoxon signed rank test.
* Significant at p-value <0.05.

While project success criteria remain the same for projects with smart technologies, one of the key smart technology applications is to integrate the entire value chain and overcome fragmentation in the construction industry (Oesterreich & Teuteberg, 2016). This is expected to improve collaboration and communication with stakeholders through the provision of access to real-time project information (Bhattacharya & Momaya, 2021; Ngo et al., 2021). Hence, project communication management was found to be most important in managing projects with smart technologies. Specifically, project communication channels may change in order to effectively convey relevant messages to each stakeholder with the access to real-time project information when required. In addition, with the availability of data, communication with project stakeholders may shift towards data-driven information. This is further supported by BD, which can change the way data is communicated to stakeholders using data visualisation tools (Bilal et al., 2016). Furthermore, smart technologies such as AR and VR allow for simulation and improve visualisation by stakeholders (Ngo et al., 2021; Y. Wang et al., 2014). Apart from that, smart technologies can allow the project team to be formed among parties in different geographical locations, increasing the emphasis on cultural intelligence (Li et al., 2019). In order to maximise the potential of the smart technologies, it must be complemented with a corresponding shift in project communication processes (Oesterreich & Teuteberg, 2016; Prebanić & Vukomanović, 2021). At the same time, as project team members may not be familiar with the use of smart technologies in projects, it is important to communicate with the team members to address their concerns, align their objectives or gain buy-in for the use of smart technologies in projects. Hence, it is expected that project communication management becomes more important when managing projects with smart technologies.

Next, project cost management was found to be the second most important project management knowledge area when managing projects with smart technologies. Smart technologies require high upfront costs and may require a higher proportion of contingency budget due to the uncertainties and complexities in the integration of smart technologies in project processes (Hwang et al., 2022; Oesterreich & Teuteberg, 2016). Hence, project cost management may play a more important role when managing projects with smart technologies to ensure that the project is completed within budget. This finding is consistent with Hwang and Ng's (2013), who found that project cost management plays a critical role in managing green construction projects as green materials, new systems, equipment and technologies are costlier. Project cost management may also be supported by BD, where historical project data can be used for cost estimation, tender evaluations and prediction of cost overruns (Bilal et al., 2016). Traditionally, project cost estimation relies heavily on the PM's experience. The use of BD will shift away the reliance on PM's experience towards data-driven estimation (Bilal et al., 2016). Hence, PMs may need to have a better understanding of the project cost management processes to apply BD to effectively manage costs in construction projects.

The third most important project management knowledge area when managing projects with smart technologies was found to be project risk management. While health, safety and environment risks are major risks considered in a project, project risks also include contractual risks, financial risks, procurement risks, design risks and security risks (Zhao et al., 2014). On this note, as the use of smart technologies in construction projects is still in its nascent stage, the integration of smart technologies in construction projects brings about high uncertainties and requires changes in project management processes (Oesterreich & Teuteberg, 2016). This results in an increase in project risks. Apart from that, as smart technologies can be used to automate work processes, the focus of risk management will broaden to include ergonomics of human workers and human–computer interactions (Tay et al., 2017). At the same time, the use of AM enables 3D-printed components to be produced off site and on site, bringing about a different set of risks to be managed (Hunt et al., 2014; Tay et al., 2017; Wu et al., 2016). In particular, the integrity of 3D-printed components rely heavily on the material used, which has not been widely standardised or regulated (Kothman & Faber, 2016). With the use of smart technologies, the scope of project risk management will be widened, and hence, project risk management plays an increasingly important role when managing projects with smart technologies.

9.4.2.2 *Changes in Knowledge in Project Management Required*

According to the two-sample Wilcoxon signed rank test, as shown in Table 9.4, it was found that the importance of project integration management, project stakeholder management, project scope management and project procurement management significantly increases when managing projects with smart technologies compared to conventional projects. This finding is also consistent with the characteristics of smart technology applications in projects.

In particular, project integration management was expected to be more important in managing projects with smart technologies. This could be because of the changes in project management processes enabled by the end-to-end integration of the value chain (Dallasega et al., 2018; Oesterreich & Teuteberg, 2016). Specifically, how to effectively integrate the stakeholders along the value chain and project monitoring and control throughout the project lifecycle may be key areas that PMs need to be competent in. The provision of access to real-time project information to project stakeholders may transform project monitoring and control and change management processes as project stakeholders can receive real-time updates on project progress and control changes on-site remotely (Osunsanmi et al., 2020). At the same time, the availability of real-time project information enables quick decision making, impacting project change management (Akanmu & Anumba, 2015). Furthermore, as project data is collected throughout the project lifecycle and stored in a centralised platform, post-project learning could also be enhanced with more holistic data (Oesterreich & Teuteberg, 2016). Apart from that, smart technologies can also be applied to automate work processes. The automation of work processes may change the way PMs coordinate project activities as human labour may be replaced by robotics and AV, shifting the focus to human–robot interactions when planning and coordinating project activities (De Soto et al., 2019; Oesterreich & Teuteberg, 2016).

Next, project scope management was also expected to be more important when managing projects with smart technologies. In particular, smart technologies can enable simulation and visualisation of project design, improving stakeholders' understanding of the final product early, aiding in project scoping processes (Bhattacharya & Momaya, 2021; Chi et al., 2013). In addition, AM enables rapid prototyping, which can also allow stakeholders to visualise the design alternatives physically, facilitating project scoping (Wu et al., 2016). Project scope definition could also be changed with the use of BD. With historical project data, estimated project schedule and costs may be provided based on the design alternatives, which may change project scoping processes (Bilal et al., 2016). Furthermore, 3D imaging and AV can assist human workers in surveying areas which are hard to reach, enabling more holistic survey of the project site (Álvares et al., 2018). The access to real-time project information and increased transparency may also result in more frequent changes, and thus PMs need to be mindful of the project scope.

The importance of project stakeholder management was also found to increase when managing projects with smart technologies. This is expected as the management of stakeholders will change with the availability of project information and user immersion and simulation (Prebanić & Vukomanović, 2021). In particular, constant communication may be required in order to manage the expectations of the stakeholders and ensure understanding of the project progress. At the same time, as the smart technologies may be quite new to the project stakeholders, some stakeholders may still have reservations about using the technology and may need the PM to manage the expectations of the stakeholders (Arayici et al., 2012; Oesterreich & Teuteberg, 2016, 2019). Moreover, the existing status of smart technology adoption in the Singapore construction industry remains in its nascent stage. On this note, the applications of smart technologies may be limited to a small project scope. Consequently, the benefits that may be derived from the technology may be limited. In this case, it is important for the PM to manage the expectations of the stakeholders.

Finally, project procurement management was perceived to be more important when managing construction projects with smart technologies. This could be due to the broadening of supplier/vendor/contractor selection criteria arising from the need for interoperability of systems, JIT delivery, cyber-security and digital expertise in construction projects with smart technologies (Bhattacharya & Momaya, 2021; Oesterreich & Teuteberg, 2016; Zhong et al., 2017). In particular, cyber-security will play an essential role in managing construction projects with smart technologies due to the integration of the entire value chain, potential bounding of collected data to individuals, increased exposure to cyber threats and the potential for large-scale damage due to the interconnected network of devices in the project (Bilal et al., 2016; Maskuriy et al., 2019; Oesterreich & Teuteberg, 2016). In addition, digital expertise of the contracted firms plays a critical role in ensuring that the technologies may be fully utilised and properly managed for project success. Digital expertise will include the availability of skilled manpower, experience in the use of technologies and the availability of resources for the specified scope of works. Beyond that, as the use of smart technologies typically results in a common model and platform for the project, issues such as the ownership of the model and data need to be managed during the procurement stage (Maskuriy et al., 2019). Furthermore, BD can be used to support decision making and evaluation of alternatives using historical project data (Bilal et al., 2016). The use of DLT in projects also enables smart contracts, giving rise to a new set of knowledge and skills required to manage the formation, execution and monitoring of contracts within the project (Li et al., 2019). Finally, e-tendering has been increasingly adopted in the construction industry due to its potential to improve productivity of the tendering process (Al Yahya et al., 2018; Ibem & Laryea, 2017; Wimalasena & Gunatilake, 2018). However, the shift to e-tendering requires a corresponding shift in the skills and processes to complement the e-tendering process (Ibem & Laryea, 2017). This may bring about an increased focus on project procurement management when managing construction projects with smart technologies.

SRCC was also conducted to find out if there are any statistically significant relationships between the perceived importance of knowledge of project management areas in managing conventional projects and in managing projects with smart technologies. Based on an alpha value of 0.05, the results indicated a positive relationship between the rank order of the project management knowledge areas required to manage conventional projects and in managing projects with smart technologies. While project integration, scope, procurement and stakeholder management need to be emphasised in managing projects with smart technologies, the relative importance of the project management knowledge areas remain largely the same. This finding is expected as project management knowledge forms the foundations of project management and should remain the same when managing projects with smart technologies. Hence, the relative importance of project management knowledge areas remains similar. This is also consistent with the aforementioned findings.

9.4.3 Skills Required for Effective Project Management

9.4.3.1 Importance in Conventional Projects and Projects with Smart Technologies

Table 9.5 shows the summary of data analysis results for the skills required to manage conventional projects and projects with smart technologies. Based on the one-sample Wilcoxon signed rank test, all skills were found to be important in managing both conventional projects and projects with smart technologies. This finding is expected and corroborates with previous findings where project management principles have been argued to remain the same regardless of the use of smart technologies in the projects, and hence require the same underlying set of skills to manage projects (Paton & Hodgson, 2016; Vrchota et al., 2021).

The top five skills required to manage conventional projects were found to be project management skills, problem-solving skills, communication skills, planning and organising skills and team-building skills while the top five skills required to manage projects with smart technologies were found to be problem-solving skills,

Table 9.5 Summary of Data Analysis Results for Skills to Manage Projects

Skills to Manage Projects	For Conventional Projects					For Projects with at Least One Smart Technology					Differences Between Conventional Projects and Projects with at Least One Smart Technology	
	Mean	SD	Median	Rank	WSR	Mean	SD	Median	Rank	WSR	Two-sample Wilcoxon signed rank test	SRCC
Technical and Operational Technology Skills	5.02	1.145	5	16	0.001*	5.91	1.091	6	9	0.001*	0.001*	0.742*
Project Management Skills	6.29	1.000	7	1	0.000*	6.13	1.007	6	4	0.000*	0.250	
Information Management Skills	5.72	1.116	6	8	0.001*	6.18	1.173	7	2	0.000*	0.007*	
Planning and Organising Skills	6.23	0.995	7	4	0.000*	6.06	1.049	6	5	0.001*	0.173	
Communication Skills	6.24	1.116	7	3	0.000*	6.14	1.099	7	3	0.000*	0.490	
Social Awareness	5.20	1.397	5	14	0.001*	5.33	1.264	5	17	0.001*	0.389	
Cultural Awareness	5.11	1.163	5	15	0.001*	5.18	1.146	5	19	0.001*	0.601	
Organisational Awareness	5.56	1.291	6	9	0.001*	5.52	1.156	5	15	0.001*	0.686	
Creativity	4.77	1.271	5	17	0.001*	5.82	1.031	6	11	0.000*	0.001*	
Problem-Solving Skills	6.27	0.974	7	2	0.000*	6.29	0.866	7	1	0.000*	0.964	
Ethical Awareness	4.51	1.174	4	18	0.001*	5.60	1.288	6	13	0.001*	0.001*	
Strategic-Planning Skills	5.33	1.344	5	11	0.001*	6.00	1.090	6	7	0.001*	0.001*	
Active Learning Skills	4.38	1.439	4	19	0.013*	5.28	1.205	5	18	0.001*	0.001*	
Conflict Management Skills	5.74	1.261	6	7	0.001*	5.95	1.112	6	8	0.001*	0.166	
Decision-Making Skills	5.78	1.317	6	6	0.001*	6.06	1.223	7	6	0.001*	0.049*	
Delegation Skills	5.53	1.284	6	10	0.001*	5.59	1.170	6	14	0.001*	0.805	
Motivation Skills	5.26	1.341	5	13	0.001*	5.74	1.244	6	12	0.001*	0.002*	
Negotiation Skills	5.28	1.910	5	12	0.001*	5.48	1.191	5	16	0.001*	0.230	
Team-Building Skills	5.93	1.333	7	5	0.001*	5.91	1.225	6	10	0.001*	0.875	

SD, standard deviation; SRCC, Spearman rank correlation coefficient; WSR, one-sample Wilcoxon signed rank test.
* Significant at p-value <0.05.

information management skills, communication skills, project management skills and planning and organisation skills.

In conventional projects, project management skills were perceived to be most important. This finding is expected as it is the main responsibility of the PM (Alvarenga et al., 2019). Project management skills were also noted to be closely associated with problem-solving skills and planning and organising skills as these are key skills related to task-related performance of PMs, in which task performance behaviours should account for about half of the variance in the manager's performance (Ahadzie et al., 2008; Alvarenga et al., 2019). The findings were also consistent with the growing emphasis on soft skills for PMs, as communication skills and team-building skills were also two of the top skills required of PMs to manage projects (Ahadzie et al., 2008; Alvarenga et al., 2019; El-Sabaa, 2001). While soft skills have been argued to be important due to increasing project complexities, cross-functional and virtual teams and the heightened awareness on the role people play in delivering projects, domain-related skills such as project management skills still play an essential role in managing conventional projects (Alvarenga et al., 2019; Zhang & Fan, 2013).

On the other hand, a similar set of skills was found to be important in managing projects with smart technologies. This is reasonable as project objectives for projects remain the same with time, cost, quality, safety as key success criteria, albeit using a different set of tools to achieve the project objectives. With the use of smart technologies which have not been widely implemented in construction projects, there is potential for unforeseen circumstances and uncertainties during the project execution phase (Oesterreich & Teuteberg, 2016). Hence, it is reasonable that PMs may require a higher level of problem-solving skills to resolve the problems that arise. Especially with the increasing complexities in projects, the root causes and relationships among different components of the projects may be more interconnected, resulting in the need for the ability to understand the bigger picture that gives rise to the problem (Luo et al., 2017; Wyskwarski, 2021). This will also require PMs to have the ability to effectively utilise the technologies to solve the problems faced (Marnewick & Marnewick, 2021).

Next, information management skill is the second most important skill required of PMs to manage projects with smart technologies. This finding is expected as smart technology applications can collect large volume, variety and velocity of data, and also have the potential to generate useful and meaningful insights (Bilal et al., 2016; Ngo et al., 2020). However, this is largely dependent on the quality of data (Bilal et al., 2016; Hazen et al., 2014). In addition, PMs should have the skills to manage the vast amount of information made available to them and interpret the outputs from the data analysis (Marnewick & Marnewick, 2021). Furthermore, with the potential of smart technologies to collect information bound to individuals, PMs must also be able to handle personal information with discretion (Marnewick & Marnewick, 2021).

Communication skills were also found to be important for PMs in managing projects with smart technologies. While construction projects are complex by nature, the integration of smart technologies into projects further increases project complexities

(Marnewick & Marnewick, 2021). In addition, uncertainties surrounding the integration of smart technologies into construction projects need to be addressed (Oesterreich & Teuteberg, 2016). Hence, communication skills remain important when managing projects with smart technologies. Furthermore, workers may be concerned of being displaced by automation (Oesterreich & Teuteberg, 2016). PMs need to be able to address the concerns of workers and align their objectives with the project objectives. Beyond addressing the concerns of workers, PMs also need to be able to communicate the outputs from smart technologies to stakeholders effectively and have the skills to communicate with stakeholders using both face-to-face and virtual means when managing projects with smart technologies (Marnewick & Marnewick, 2021). Moreover, while automation may replace human workers in certain work processes, human workers will still be required in project teams and collaboration is expected to increase with the integration of the value chain (Oesterreich & Teuteberg, 2016; Trenerry et al., 2021). Hence, communication skills remain essential for PMs in managing projects with smart technologies.

Finally, both project management and planning and organising skills remain as the top skills required of PMs when managing projects with smart technologies. This finding is expected as project management skills remain as the foundational domain skills required of PMs to fulfil their main responsibility of managing projects. While smart technologies should be used as tools to assist PMs in managing projects, the effective integration of smart technologies into existing work processes will require extensive planning and organising skills (Oesterreich & Teuteberg, 2016). Furthermore, smart technology applications like real-time monitoring and control supported by JIT delivery can improve the productivity of the project (Zhong et al., 2017). On this note, managing JIT deliveries will require PMs to have the planning and organising skills to utilise the real-time information and schedule project activities effectively (Prasad, 1995; Qureshi et al., 2013).

9.4.3.2 *Changes in Skills Required for Effective Project Management*

The results in Table 9.5 indicate that technical and operational technology skills, information management skills, creativity, ethical awareness, strategic planning skills, active learning skills, decision-making skills and motivation skills were more important in managing projects with smart technologies compared to conventional projects. This finding corroborates with the key features of smart technology applications in projects and how these may be used to improve project performance.

It is reasonable for technical and operational technology skills, information management skills and strategic planning skills to be more important when managing projects with smart technologies. This is because PMs need to have the necessary skills to operate the technologies which will include managing the inputs and outputs for the technologies and planning for the use of technologies to achieve project objectives (Alvarenga et al., 2019; El-Sabaa, 2001). In addition, since smart technologies are not widely used in construction projects, the relatively new way of work requires PMs to think out of the box on how to integrate the smart technologies into existing work processes. This has been supported by Interviewees P6 and A5, who emphasised the growing importance of the ability to integrate smart technologies into existing work processes. Apart from that, it is important for PMs to have the ability to learn and apply the new technologies on the job to maximise the potential of smart technologies and understand the implications of the technologies on project outcomes (World Economic Forum, 2018). The uncertainty arising from new ways of working requires a higher level of decision-making skill as the uncertainties need to be managed effectively (Sniazhko, 2019). This involves the ability to recognise the risks posed by the technologies, determine the amount of risks to be managed, formulate strategies to manage the risks and coordinate with others to manage the potential risks from the technologies in a timely manner with the available information (Comfort & Wukich, 2013). While these skills will also be required in managing conventional projects, the integration of smart technologies in projects brings about uncertainties in work processes and outcomes and also requires better planning and management of information, hence requiring a higher level of skills in technical and operational technology skills, information management skills, strategic planning skills, active learning skills and decision-making skills. Next, ethical awareness is also expected to be more important as smart technologies are more intrusive, with their ability to collect information bound to individuals (Bilal et al., 2016; Oesterreich & Teuteberg, 2016). On this note, PMs have an additional responsibility to ensure that collected data is used appropriately and is in compliance with the relevant regulations (Bilal et al., 2016; Oesterreich & Teuteberg, 2016). Finally, motivation skills are also more important when managing projects with smart technologies. This is also expected as workers may have fears of being displaced by technologies; PMs need to be able to address the concerns of the workers and persuade them to develop their potential and improve project performance with the assistance of the smart technologies (Oesterreich & Teuteberg, 2016). These skills were also consistent with the top six digital-age skills required for project delivery (Project Management Institute, 2018).

SRCC was conducted using an alpha value of 0.05. The results indicated a significant positive relationship between the rank-order of the skills required to manage conventional projects and projects with smart technologies. This finding is consistent with the aforementioned, where the premise of project management remains similar when managing conventional projects and projects with smart technologies. Hence, although technical and operational technology skills, information management skills, creativity, ethical awareness, strategic planning skills, active learning skills, decision making skills and motivation skills need to be emphasised when managing projects with smart technologies, the finding reflects the view that smart technologies should be used as a tool to assist PMs in managing project work, resulting in a similar rank-order for the skills required to manage projects.

9.4.4 *Factors Affecting One's Attitude Towards Technologies*

Tables 9.6 and 9.7 show the summary of the data analysis results for the factors affecting an individual's attitude towards technologies. The one-sample Wilcoxon signed rank test for the significance of the factor used a median of 16 as the critical value. As the significance of the factors is derived by multiplying the applicability and the impact of the factor, the critical value is similarly determined by the product of the median of applicability and the median of impact of the factors. Based on the one-sample Wilcoxon signed rank test, all the factors were found to significantly affect one's attitude towards technologies.

The results indicated that the PU was most applicable in determining one's attitude towards technologies. This is consistent with the TAM, where it was found that the relationship between PU and the usage of the system was stronger compared to the PEU–usage relationship (Davis, 1989). The authors further explained that users can be primarily driven to adopt new applications for the specific functions that are critically needed despite the difficulty of use. On the other hand, new applications that are easy to use will not be adopted if they do not perform a useful function (Davis, 1989). While PU was found to be most applicable in determining one's attitude towards technologies, an individual's TSE and PEU were found to be more impactful in determining one's attitude towards technologies. Interviewees A2, A7 and A8 shared that if one is confident in his ability to use technologies, he will naturally present a liking towards the use of technologies and find it easy to pick up new technologies. This is more impactful in determining a PM's attitude towards technologies as PMs are responsible for delivering projects using the technologies specified in the contract, limiting the autonomy of PMs in deciding the smart technology to be implemented based on its usefulness. This is in alignment with previous studies which found that persons with high computer self-efficacy display higher willingness to participate in computer-related activities and perform better in solving problems faced (Yeşilyurt et al., 2016). Overall, it was found that an individual's TSE, PU and PEU of technologies were most significant in determining one's attitude towards technologies. As PMs typically do not have the autonomy to decide the smart technology to be adopted in projects, TSE was perceived to play a more significant role in determining one's attitude towards technologies as the confidence in one's ability in using and managing technologies influences one's liking towards technologies, subsequently influencing the PEU and PU of the technologies (Kulviwat et al., 2014). In addition, individuals with higher self-efficacy were found to be more likely to engage and execute the tasks in question (Makri-Botsari et al., 2004). Apart from one's TSE, individuals are more likely to adopt technologies that were perceived to be useful. This is highlighted in both research and practice. In particular, Interviewees P3 to P6 emphasised the importance of identifying specific use cases and targeted problems to be overcome by the use of technologies in projects. Doing so greatly facilitates the gaining of buy-in and individual's willingness to use the technologies. Hence, it is reasonable for PU to be significant in influencing one's attitude towards technologies. Finally, PEU is also significant in determining one's attitude towards technologies. Specifically, Interviewees P3 to P6 noted that it is essential to integrate the use of technologies into existing work processes, making it easy for workers to implement the technologies in their work tasks. This will facilitate the adoption of technologies in projects.

Under one's TSE, TSE1 and TSE9 were found to be most applicable, impactful and significant in contributing to one's attitude towards technologies. The two factors involve one's ability to complete job tasks using technologies without guidance from others. Interviewees V1 and V10 further explained that the ability to navigate the use of technologies without explicit guidance could reflect a higher level of proficiency in using technologies and users with the ability to do so will naturally have higher confidence in their ability to use technologies, resulting in a more positive attitude towards technologies.

TAFF1, TAFF2, TAFF5 and TAFF7 were found to be most applicable, impactful and significant in contributing to one's attitude towards technologies. These factors capture one's enjoyment and liking to use smart technologies, including the necessity of smart technologies in work settings. As the concept of technological competency refers to the underlying characteristics of an individual that result in superior work performance, it is reasonable that the liking to work with smart technologies and thinking that smart technologies are necessary in work settings can contribute to one's technological competency through one's attitude towards technologies. Interviewee V3 stated that while TAFF could be also influenced by the other identified factors, TAFF1, TAFF2, TAFF5 and TAFF7 could play a more concrete role in encouraging PMs to use technologies in their work.

Attitude towards technologies was also found to be greatly affected by TANX4, TANX7 and TANX8. The factors involved feelings of intimidation, uneasiness and insecurity in one's ability to interpret the outcomes from the smart technologies. Interviewees V5 and V9 were of the view that these factors capture the more general feelings associated with anxiety, and the interpretation of the outcomes from the smart technologies can be hard especially for PMs who are less technology inclined.

Next, PEU2, PEU4 and PEU6 were found to be the most applicable, impactful and significant factors contributing to one's attitude towards technologies. The three factors involved the ease of getting the technology to do what the user wants the technology to do, having clear and understandable interaction with the technology and finding it easy to be skilful in using the technology. This is reasonable as one needs to be skilled in using the technology in order to maximise the potential of the technology.

PU1, PU3 and PU5 were found to be most significant in influencing one's attitude towards technologies. While the factors for PU of technologies are all related to the job relevance of the technology, PU1, PU3 and PU5 are more related to the improvements in job productivity through improved ease in doing work and increase in job effectiveness. These are the areas in which PMs can see direct job relevance of the technologies in their job.

The factors of PIT were found to contribute to one's attitude towards technologies, especially in terms of one's willingness to experiment with new technologies. This is reasonable as willingness to experiment with new technologies can reflect one's openness to technologies, consistent with Rogers' (1995) finding in which innovators are active information seekers about new ideas.

Table 9.6 Summary of Data Analysis Results for Applicability and Impact of Factors Affecting Attitude Towards Technologies

Attitude Towards Technologies	Applicability						Impact					
	Mean	SD	Median	RW	RA	WSR	Mean	SD	Median	RW	RA	WSR
TSE1	5.70	1.355	6	1	5	0.001*	5.80	1.343	6	2	2	0.001*
TSE2	5.24	1.008	5	6	26	0.001*	5.08	1.230	5	8	29	0.001*
TSE3	5.05	1.054	5	10	31	0.001*	5.05	1.253	5	9	30	0.001*
TSE4	5.31	1.004	5	4	23	0.001*	5.29	1.172	5	4	22	0.001*
TSE5	5.20	1.057	5	8	28	0.001*	5.11	1.224	5	6	27	0.001*
TSE6	5.25	0.969	5	5	25	0.001*	5.11	1.172	5	5	26	0.001*
TSE7	5.23	1.056	5	7	27	0.001*	5.08	1.196	5	7	28	0.001*
TSE8	5.19	1.003	5	9	29	0.001*	4.88	1.431	5	10	37	0.001*
TSE9	5.65	1.581	6	2	10	0.001*	5.81	1.294	6	1	1	0.001*
TSE10	5.45	1.099	5	3	16	0.001*	5.34	1.215	6	3	20	0.001*
TAFF1	5.57	1.089	6	3	13	0.001*	5.33	1.281	5	2	21	0.001*
TAFF2	5.55	1.090	6	4	14	0.001*	5.29	1.233	5	3	23	0.001*
TAFF3	4.99	1.132	5	6	36	0.001*	4.93	1.277	5	6	34	0.001*
TAFF4	5.05	1.149	5	5	32	0.001*	4.91	1.385	5	7	35	0.001*
TAFF5	5.65	1.242	6	2	9	0.001*	5.28	1.297	5	4	24	0.001*
TAFF6	4.99	1.432	5	7	37	0.001*	4.97	1.454	5	5	32	0.001*
TAFF7	5.70	1.284	6	1	4	0.001*	5.55	1.291	6	1	10	0.001*
TANX1	4.66	1.189	4	9	43	0.001*	4.64	1.284	4	9	43	0.001*
TANX2	4.72	1.375	5	6	40	0.001*	4.74	1.333	5	8	42	0.001*
TANX3	4.72	1.420	5	7	41	0.001*	4.84	1.352	5	6	40	0.001*
TANX4	5.01	1.604	5	2	34	0.001*	4.97	1.517	5	2	33	0.001*
TANX5	4.71	1.574	5	8	42	0.001*	4.77	1.447	5	7	41	0.001*
TANX6	4.96	1.428	5	4	38	0.001*	4.88	1.570	5	5	39	0.001*
TANX7	5.00	1.507	5	3	35	0.001*	4.88	1.522	5	4	38	0.001*
TANX8	5.01	1.440	5	1	33	0.001*	4.99	1.410	5	1	31	0.001*
TANX9	4.40	1.484	5	10	44	0.001*	4.61	1.462	5	10	44	0.001*
TANX10	4.89	1.413	5	5	39	0.001*	4.89	1.574	5	3	36	0.001*
PU1	5.70	1.051	6	3	3	0.001*	5.67	1.134	6	1	9	0.001*
PU2	5.67	0.943	6	4	6	0.001*	5.52	1.081	5	6	11	0.001*
PU3	5.71	1.118	6	2	2	0.001*	5.60	1.247	6	5	7	0.001*
PU4	5.66	1.135	6	6	8	0.001*	5.60	1.143	6	3	6	0.001*
PU5	5.67	1.197	6	5	7	0.001*	5.64	1.120	6	2	5	0.001*
PU6	5.72	1.038	6	1	1	0.001*	5.59	1.116	5	5	8	0.001*
PEU1	5.37	1.424	6	6	21	0.001*	5.47	1.259	6	4	13	0.001*
PEU2	5.61	1.204	6	2	12	0.001*	5.65	1.299	6	2	4	0.001*
PEU3	5.40	1.213	5	4	19	0.001*	5.37	1.325	5	7	19	0.001*
PEU4	5.40	1.312	6	5	20	0.001*	5.52	1.267	6	3	12	0.001*
PEU5	5.36	1.378	6	7	22	0.001*	5.45	1.216	5	5	14	0.001*
PEU6	5.54	1.199	6	3	15	0.001*	5.45	1.291	6	6	15	0.001*
PEU7	5.61	1.151	6	1	11	0.001*	5.69	1.202	6	1	3	0.001*
PIT1	5.44	1.224	6	1	17	0.001*	5.42	1.273	5	1	16	0.001*
PIT2	5.09	1.275	5	4	30	0.001*	5.18	1.291	5	4	25	0.001*
PIT3	5.27	1.517	5	3	24	0.001*	5.42	1.506	5	2	17	0.001*
PIT4	5.40	1.133	5	2	18	0.001*	5.41	1.231	5	3	18	0.001*

RA, rank across components; RW, rank within component; SD, standard deviation; WSR, one-sample Wilcoxon signed rank test.
* Significant at *p*-value <0.05.

Table 9.7 Summary of Data Analysis Results for Significance of Factors Affecting Attitude Towards Technologies

Attitude Towards Technologies	Significance (Applicability × Impact)					
	Mean	SD	Median	RW	RA	WSR
TSE1	34.27	13.346	36	1	1	0.001*
TSE2	27.53	11.054	25	7	28	0.001*
TSE3	26.46	11.197	25	9	34	0.001*
TSE4	28.90	10.609	25	4	24	0.001*
TSE5	27.54	10.998	25	6	27	0.001*
TSE6	27.55	10.065	25	5	26	0.001*
TSE7	27.36	10.299	25	8	29	0.001*
TSE8	26.31	11.582	25	10	35	0.001*
TSE9	34.14	13.992	36	2	2	0.001*
TSE10	30.25	11.729	30	3	23	0.001*
TAFF1	30.80	12.221	30	3	16	0.001*
TAFF2	30.36	11.853	25	4	21	0.001*
TAFF3	25.79	11.231	25	7	39	0.001*
TAFF4	26.04	11.850	25	6	37	0.001*
TAFF5	31.07	12.151	35	2	13	0.001*
TAFF6	26.56	13.344	25	5	32	0.001*
TAFF7	33.02	13.203	36	1	7	0.001*
TANX1	22.95	11.925	16	9	43	0.001*
TANX2	23.97	12.600	25	8	42	0.001*
TANX3	24.39	12.875	20	6	40	0.001*
TANX4	27.00	14.440	25	1	30	0.001*
TANX5	24.24	13.960	20	7	41	0.001*
TANX6	26.25	14.341	25	4	36	0.001*
TANX7	26.51	14.245	25	3	33	0.001*
TANX8	26.90	13.756	25	2	31	0.001*
TANX9	21.82	12.099	21	10	44	0.001*
TANX10	25.84	14.148	25	5	38	0.001*
PU1	33.21	11.321	36	1	3	0.001*
PU2	32.05	10.731	30	6	11	0.001*
PU3	33.12	11.923	36	2	4	0.001*
PU4	32.75	11.745	35	5	10	0.001*
PU5	33.02	12.081	30	3	6	0.001*
PU6	32.97	11.572	30	4	8	0.001*
PEU1	30.43	12.840	30	5	19	0.001*
PEU2	32.85	12.739	36	2	9	0.001*
PEU3	30.37	12.874	28	6	20	0.001*
PEU4	31.04	13.151	35	4	14	0.001*
PEU5	30.26	12.733	30	7	22	0.001*
PEU6	31.38	12.675	30	3	12	0.001*
PEU7	33.03	12.121	36	1	5	0.001*
PIT1	30.86	12.798	30	1	15	0.001*
PIT2	27.75	12.518	25	4	25	0.001*
PIT3	30.52	14.453	25	2	17	0.001*
PIT4	30.48	12.510	25	3	18	0.001*

RA, rank across components; RW, rank within component; SD, standard deviation; WSR, one-sample Wilcoxon signed rank test.
* Significant at p-value <0.05.

9.4.5   *Perceived Differences by Respondents of Different Characteristics*

As this study is concerned with the factors that affect technological competency of PMs in managing projects with smart technologies, the perceived differences in factors relating to the knowledge in project management and skills to manage projects with smart technologies and attitude towards smart technologies by respondents of different ages, education, years of experience in the construction industry, experience with smart technologies, organisation domain and organisation's experience in implementing smart technologies have been analysed.

9.4.5.1   *Respondents of Different Ages*

The summary of the results for the perceived differences on the importance of the factors in influencing one's technological competency by respondent age is shown in Appendices A-5 to A-7. The respondents from the different age groups did not perceive the importance of knowledge in project management areas and skills in contributing to technological competency differently. This finding is expected as knowledge in project management and skills to manage projects are essential to one's role to manage projects regardless of age groups. This is consistent with Hoxha and McMahan (2019) who found that PM's age did not influence project success.

Apart from the knowledge and skills to manage projects, respondents from different age groups perceived the impact and significance of several TSE factors and the impact of TANX9 differently. Mann-Whitney U test has been conducted on these factors to identify the specific age groups that have perceived the factors differently. The results of the Mann-Whitney U test are shown in Table 9.8.

For TSE2, respondents between 25 and 34 years of age perceived the impact of TSE2 differently from those who were 35 to 44 years old and 55 to 64 years old; respondents between 35 and 44 years of age perceived the impact differently from those who were 45 to 54 years old; respondents who were 45 to 54 years old also perceived the impact differently from those who are 55 to 64 years old. In particular, respondents who were 45 to 54 years old perceived the impact of TSE2 on attitude towards technologies to be the least while respondents who were 55 to 64 years old perceived the impact of TSE2 on attitude towards technologies to be the greatest. TSE2 involves the ability to complete a job task using smart technologies if user manuals were available for reference.

Respondents between 45 and 54 years of age perceived the impact of TSE4 differently from those aged 35 to 44 years and 55 to 64 years. In particular, respondents who were 45 to 54 years old perceived the impact to be the least while respondents aged 55 to 64 years perceived the impact to be the greatest. TSE4 involves the ability to complete a job task using smart technologies if assistance could be called for when stuck.

Respondents between 45 and 54 years of age also perceived the impact of TSE10 differently from all other respondents. They also perceived the significance of TSE10 on attitude towards technologies differently from respondents aged 35 to 44

Table 9.8 Mann-Whitney *U* Test Results for Perceived Differences by Different Age Groups

Factor	25 to 34 (n = 49)	35 to 44 (n = 21)	45 to 54 (n = 16)	55 to 64 (n = 11)	25 to 34 vs 35 to 44	25 to 34 vs 45 to 54	25 to 34 vs 55 to 64	35 to 44 vs 45 to 54	35 to 44 vs 55 to 64	45 to 54 vs 55 to 64
	Mean	*Mean*	*Mean*	*Mean*	*Mann-Whitney U Test*					
Impact of TSE2	4.82	5.62	4.75	5.73	0.018*	0.875	0.032*	0.039*	0.805	0.032*
Impact of TSE4	5.14	5.71	4.81	5.82	0.073	0.322	0.082	0.029*	0.757	0.031*
Impact of TSE10	5.29	5.71	4.63	5.91	0.229	0.039*	0.155	0.004*	0.664	0.012*
Significance of TSE10	29.41	33.90	23.81	36.36	0.177	0.085	0.135	0.009*	0.667	0.018*
Impact of TANX9	4.29	5.33	4.50	4.82	0.006*	0.595	0.306	0.043*	0.462	0.478

* Significant at *p*-value <0.05.

years and 55 to 64 years. Specifically, respondents between 45 and 54 years of age perceived the impact and significance of TSE10 to be the least, while respondents between 55 and 64 years of age perceived the impact and significance of TSE10 to be the greatest on an individual's attitude towards technologies. TSE10 involves the ability to complete a job task using smart technologies if a similar technology has been used before to do the same job task.

It is expected that respondents aged 55 to 64 years perceived the impact and significance of TSE factors on an individual's attitude towards technologies. This is consistent with previous studies on older adults' self-efficacy in using technologies across various industries (Mariano et al., 2021; Vaportzis et al., 2017). These studies involved respondents aged 55 and above and found that older adults were influenced by the stereotype threat as they did not grow up with technologies, had lower expectations on task performance, lacked confidence in their own abilities to use technologies and found technologies to be complicated (Mariano et al., 2021; Vaportzis et al., 2017). On the other hand, respondents aged 45 to 54 years of age are categorised as under Generation X at the time of the data collection (Hernaus & Vokic, 2014). Generation X did not grow up with the Internet but learned to use it as adults (Calvo-Porral & Pesqueira-Sanchez, 2020). As they have learned to integrate technologies later on in their lives, their experience in learning new technologies may have led to a lower perception of the impact and significance of TSE factors on one's attitude towards technologies.

It was also found that respondents aged 35 to 44 years perceived the impact of TANX9 on one's attitude towards technologies differently from respondents who were aged 25 to 34 years and 45 to 54 years. In particular, respondents aged 35 to 44 years perceived the impact to be the greatest while respondents aged 25 to 34 years perceived the impact to be the least. TANX9 is related to one's fear to become dependent on smart technologies and lose some reasoning skills. Millennials refer to individuals born between 1980 and 2000 (Calvo-Porral & Pesqueira-Sanchez, 2020). Since Millennials have grown up with technologies, they are known to be technologically literate and savvy (Calvo-Porral & Pesqueira-Sanchez, 2020). While it is expected that respondents between 25 and 34 years of age perceived the impact of TSE9 on one's attitude towards technologies to be the least due to their familiarity with technologies, it was surprising that respondents between 35 and 44 years perceived the impact of TSE9 on one's attitude towards technologies to be the greatest. One possible explanation could be due to the general increase in anxiety levels in older adults (Volkom et al., 2014). Despite the increased anxiety levels in older adults, the older population may have already found ways to overcome some of the anxieties and obstacles faced when using technologies by seeking help from others (Volkom et al., 2014). The stereotype that Millennials are technologically savvy may also put pressure on respondents aged 35 to 44 years, making them more unwilling to seek help to address the anxieties.

9.4.5.2   *Respondents with Different Education Levels*

Based on the Kruskal Wallis results as shown in Appendices A-8 to A-10, respondents with different education levels did not perceive the importance of the knowledge in project management areas in contributing to technological competency differently. This finding is expected as knowledge in project management is essential to one's role to manage projects regardless of one's education level (Nijhuis, 2012). It was also found that there is little distinction on the intended learning levels of project management knowledge between bachelor's and master's degrees (Nijhuis et al., 2018). In addition, previous studies have found that project management curriculum in higher education is hard to deliver and may be best delivered through workplace experience (Nijhuis, 2012; Nijhuis et al., 2018). Furthermore, knowledge in project management can also be attained through professional certifications and may not be significantly affected by the formal education level.

Next, respondents with different education levels did not perceive the importance of the skills in contributing to technological competency differently, except for the importance of communication skills. The applicability, impact and significance of factors affecting one's attitude towards technologies were not found to be influenced by the respondent's formal education level. This finding is expected and consistent with previous technology acceptance and adoption models such as the TAM, TPB and UTAUT, which did not include formal education as a factor influencing one's decision to accept or adopt technologies.

Communication skills were perceived differently by postgraduates, as shown in Table 9.9. Specifically, postgraduates perceived the importance to be least among the education levels while respondents with an education level of diploma and below perceived communication skill to be most important in managing construction projects with smart technologies. One possible reason could be that postgraduates are more likely to manage complex construction projects, which already requires a high level of communication skill, and thus did not perceive the importance of communication skills in managing construction projects to be significantly higher (Nijhuis, 2012). Along the same line, PMs who have diploma and below are more likely to be involved in execution roles on site, requiring more frequent communication with site staff when managing projects with smart technologies (Low et al., 2021).

9.4.5.3 Respondents' Years of Experience in the Construction Industry

Based on the Kruskal Wallis test results shown in Appendices A-11 to A-13, respondents with different years of experience in the construction industry perceive the knowledge and skills required in project management and factors affecting attitude

Table 9.9 Mann-Whitney *U* Test Results for Perceived Differences by Different Education Level

Factor	Diploma and below (n = 22)	Bachelor's (n = 40)	Postgraduate (n = 35)	Diploma and below vs Bachelors	Diploma and below vs Postgraduate	Bachelor's vs Postgraduate
	Mean	Mean	Mean	Mann-Whitney U Test		
Communication Skills	6.45	6.30	5.77	0.707	0.022*	0.036*

* Significant at *p*-value <0.05.

towards technologies similarly. This finding is rather unexpected as experience in the construction industry has been found to be one of the critical factors predicting project success (Besteiro et al., 2015). In addition, experience in the construction industry was found to be one of the most effective ways to improve one's knowledge and skills in project management (Nijhuis, 2012). On the contrary, several studies also found that experience did not significantly influence project success (Hoxha & McMahan, 2014; Paton & Hodgson, 2016). In particular, having a comprehensive understanding of the domain knowledge and skills was found to be more important than number of years of experience in the industry (Paton & Hodgson, 2016). Furthermore, project management knowledge and skills may be transferable across industries as the basis of project management remains the same; the integration of smart technologies requires more technical and digital knowledge and skills to integrate the technologies into construction projects (Low et al., 2021; Paton & Hodgson, 2016). This may explain why years of experience in the construction industry did not significantly influence one's perception on the importance of knowledge and skills in managing projects and attitude towards technologies.

9.4.5.4   *Respondents' Experience in Smart Technologies*

The summary of the results of perceived differences by respondents with and without experience in smart technologies is shown in Appendices A-14 to A-16. Table 9.10 shows the summary of the factors found to be perceived differently by respondents' experience in smart technologies. Respondents with and without experience in the smart technologies did not perceive the importance of the knowledge in project management areas in contributing to technological competency differently. This finding is expected as knowledge in project management is essential for project management, and project management principles are expected to remain the same in projects with smart technologies (Marnewick & Marnewick, 2021; Paton & Hodgson, 2016).

The results indicated that respondents with and without experience in the smart technologies did not perceive the importance of the skills in contributing to technological competency differently, except for the importance of negotiation skills. Specifically, respondents with experience in smart technologies perceived the importance of negotiation skills to be more important. This is consistent with a study on the skills for future-ready graduates in the Singapore built-environment industry, which highlighted that people may not buy into your ideas despite being technically competent (Low et al., 2021). Hence, negotiation skills may be more important to persuade others to buy into the use of smart technologies in the projects.

The applicability, impact and significance of several TANX and PU of technology factors on attitude towards technologies were perceived differently by respondents with and without experience in smart technologies, as shown in Table 9.10. Respondents with experience in smart technologies perceived the applicability, impact and significance of the factors to contribute more to one's attitude towards technologies.

First, the applicability and significance of TANX1 on one's attitude towards technologies were perceived differently among respondents with and without

Table 9.10 Summary of Factors Perceived Differently by Respondents' Experience in Smart Technologies

Factor	No (n = 54)	Yes (n = 43)	No vs Yes
	Mean	Mean	Mann-Whitney
Negotiation Skills	5.30	5.72	0.046*
Applicability of TANX1	4.44	4.93	0.029*
Applicability of TANX5	4.39	5.12	0.022*
Applicability of TANX7	4.72	5.35	0.042*
Applicability of PU4	5.43	5.95	0.029*
Applicability of PU5	5.44	5.95	0.046*
Applicability of PU6	5.54	5.95	0.025*
Impact of TANX8	4.70	5.35	0.023*
Impact of PU1	5.41	6.00	0.005*
Impact of PU2	5.31	5.77	0.028*
Impact of PU3	5.33	5.93	0.028*
Impact of PU4	5.35	5.91	0.011*
Impact of PU5	5.39	5.95	0.006*
Impact of PU6	5.35	5.88	0.016*
Significance of TANX1	21.15	25.21	0.048*
Significance of TANX7	23.89	29.79	0.043*
Significance of TANX8	24.19	30.30	0.035*
Significance of PU1	30.70	36.35	0.009*
Significance of PU2	30.06	34.56	0.023*
Significance of PU3	30.89	35.93	0.033*
Significance of PU4	30.15	36.02	0.016*
Significance of PU5	30.35	36.37	0.016*
Significance of PU6	30.52	36.05	0.016*

* Significant at p-value <0.05.

experience in smart technologies. TANX1 is concerned with the feeling of apprehensiveness when using the smart technology. The applicability of TANX5 to one's attitude towards technologies was also perceived differently by respondents with and without experience in smart technologies. TANX5 is related to the feeling of nervousness when working with smart technologies. The applicability and significance of TANX7, which is concerned with the feeling of uneasiness when using smart technologies, on one's attitude towards technologies were perceived differently by respondents with and without experience in smart technologies. The results also indicated that respondents with and without experience in smart technologies perceived the impact and significance of TANX8 on one's attitude towards technologies. TANX8 is concerned with the insecurities in one's ability to interpret outcomes from the smart technologies.

In general, respondents with experience in smart technologies perceived the applicability, impact and significance of TANX factors to play a greater role in one's attitude towards technologies. One possible reason could be that TANX has been found to be an anchor in determining one's PEU of technologies, and was theorised to diminish over time (Venkatesh & Bala, 2008). While it has been found that one's TANX is likely to diminish over time with increased experience in the smart technologies, those who have experienced the use of smart technologies may be more aware of the effect of TANX on how one perceives smart technologies. Hence, it is reasonable that respondents who have experience in smart technologies perceive the applicability, impact and significance of TANX factors on one's attitude towards technologies.

The impact and significance of all PU factors on one's attitude towards technologies were perceived differently among respondents with and without experience in smart technologies. The applicability of PU4, PU5 and PU6 on one's attitude towards technologies was also perceived differently by respondents with and without experience in smart technologies. PU1 refers to the individual's thinking that the use of smart technologies can help to accomplish tasks more quickly; PU2 is concerned with the improvements in job performance through the use of smart technologies; PU3 involves the increase in productivity with the use of smart technologies; PU4 is concerned with the enhancements in job effectiveness with the use of smart technologies; PU5 refers to the improved ease in completing a job using smart technologies; and PU6 refers to the perceived overall usefulness of smart technologies. The finding is consistent with previous studies where the effect of PU on one's behavioural intention and use increases with experience (Davis, 1989; Venkatesh & Bala, 2008). With experience in smart technologies, respondents may have a better understanding of the benefits derived from the technologies, further reinforcing one's attitude towards technologies.

9.4.5.5   Organisation Domain

Based on the Kruskal Wallis results, shown in Appendices A-17 to A-19, respondents from different organisation domains did not perceive the importance of the knowledge and skills required to manage projects in contributing to technological competency differently. This finding is expected as knowledge in project management and skills to manage projects are essential domain knowledge and skills and are required by PMs representing developers, consultants or contractors to deliver projects successfully (Müller & Turner, 2007). While PMs representing the different organisation

Table 9.11 Mann-Whitney *U* Test Results for Perceived Differences by Organisation Domain

Factor	Consultant (n = 40)	Contractor (n = 35)	Developer (n = 22)	Consultant vs Contractor	Consultant vs Developer	Contractor vs Developer
	Mean	Mean	Mean	Mann-Whitney U Test		
Applicability of PIT2	4.85	5.54	4.82	0.015*	0.958	0.052
Significance of PIT2	25.25	32.11	25.36	0.019*	0.881	0.071

* Significant at *p*-value <0.05.

domains have different interests to address, the principles of project management remain the same (Müller & Turner, 2007; Vrchota et al., 2021).

Apart from that, respondents from different organisation domains generally view the factors influencing attitude towards technologies similarly, except for the applicability and significance of PIT2. The Mann-Whitney *U* test result for PIT2 is shown in Table 9.11. PIT2 involves being the first among peers to try out new technologies. Particularly, the means of the applicability of PIT2 perceived by contractors, consultants and developers are 5.54, 4.85 and 4.82 respectively. Contractors perceived the significance of PIT2 to be the greatest among respondents from the different organisation domains, followed by consultants and developers. It is reasonable for contractors to perceive PIT2 to play a greater role in one's attitude towards technologies due to the hypercompetitive nature of the construction industry (Bilal et al., 2019). While adoption of technologies is also important for consultants and developers, contractors may face greater pressure to experiment with new technologies and be the first among the firms to gain a competitive edge over others (Bilal et al., 2019).

9.4.5.6 *Organisations' Experience in Smart Technologies*

The summary of results of the perceived differences among organisations with and without experience in smart technologies is shown in Appendices A-20 to A-22. The summary of factors perceived differently according to the organisations' experience in smart technologies is shown in Table 9.12. The results indicated that respondents

Table 9.12 Mann-Whitney Results for Organisations' Experience in Smart Technologies

Factor	No (n = 32)	Yes (n = 65)	No vs Yes
	Mean	Mean	Mann-Whitney
Project Resource Management	5.69	6.23	0.032*
Project Procurement Management	5.47	6.02	0.021*
Applicability of TSE2	4.94	5.38	0.043*
Applicability of TANX1	4.22	4.88	0.016*
Impact of TSE2	4.72	5.26	0.039*
Impact of TAFF3	4.47	5.15	0.025*
Impact of TANX1	4.22	4.85	0.021*
Impact of TANX9	4.16	4.83	0.028*
Impact of PU1	5.31	5.85	0.029*
Impact of PU5	5.38	5.77	0.047*
Significance of TSE2	24.03	29.25	0.029*
Significance of TAFF3	22.31	27.51	0.024*
Significance of TANX1	18.69	25.05	0.014*
Significance of TANX9	18.41	23.51	0.035*
Significance of PU1	29.66	34.95	0.039*

* Significant at *p*-value <0.05.

from organisations with and without experience in smart technologies perceived the importance of project resource management and project stakeholder management differently. In particular, respondents from organisations with experience in smart technologies perceive project resource management and project stakeholder management to be more important than respondents from organisations without experience in smart technologies. This is expected as organisational processes will be required to change in order to effectively integrate new technologies into projects. Some of the key changes that can be expected from the use of smart technologies are the shift towards human–computer interactions, data-driven decision making and collaboration across the entire value chain (Oesterreich & Teuteberg, 2016). Hence, it is reasonable that project resource management will become more important in managing projects with smart technologies as human workers may experience fears in getting displaced by technologies and may need support to transition towards human–computer interactions (De Soto et al., 2019). At the same time, project stakeholder management is also expected to be more important when managing projects with smart technologies as collaboration plays a key role in maximising the potential of smart technologies (Oesterreich & Teuteberg, 2016).

Next, respondents from organisations with and without experience in smart technologies did not perceive the skills to manage projects with smart technologies differently. This could be because the skills required to manage projects are individual related and are equally important in managing construction projects regardless of whether the organisations have implemented smart technologies or not (Paton & Hodgson, 2016).

The results indicated that the applicability, impact and significance of several factors affecting one's attitude towards technologies were perceived differently by respondents from organisations with and without experience in smart technologies. In particular, respondents from organisations with experience in smart technologies perceived the factors to play a more important role in one's attitude towards technologies.

The respondents perceived the applicability, criticality and significance of TSE2 differently. TSE2 involves the ability to complete a job task using smart technologies if user manuals were available for reference. As organisations with experience in smart technologies may already have established several guidelines for implementing the smart technologies in projects, respondents may perceive the role of guidelines in facilitating technology adoption in organisations to be greater. The provision of user manuals may also demonstrate the relevance of the technology for the job tasks, ensuring a minimum standard of output quality while demonstrating potential results, increasing one's PU and hence the willingness to adopt the technologies (Venkatesh & Bala, 2008).

The respondents from organisations with and without experience in smart technologies also perceived the impact and criticality of TAFF3 differently. TAFF3 is concerned with having fun using smart technologies. Respondents from organisations with experience in smart technologies may have experienced how having fun can help to increase one's willingness to adopt technologies through PEU, consistent with the findings in Venkatesh (2000) and Venkatesh and Bala (2008).

Organisations may also play an important role in creating a supportive environment that makes the adoption of technologies fun, resulting in the perceived differences among respondents from organisations with and without experience in smart technologies (Venkatesh & Bala, 2008).

Respondents from organisations with experience in smart technologies also perceived the applicability of TANX1 and impact and significance of TANX1 and TANX9 on one's attitude towards technologies to be greater. TANX1 is concerned with the feeling of apprehensiveness about using smart technologies, while TANX9 is concerned with the fear of being dependent on smart technologies and losing some reasoning skills. This may be due to the adjustments in one's perception depending on the actual experience with the smart technologies (Venkatesh, 2000; Venkatesh & Bala, 2008). This may also be affected by the level of organisational support during the implementation of these technologies, hence leading to the perceived differences among respondents from organisations with and without experience in smart technologies (Venkatesh & Bala, 2008).

Finally, respondents from organisations with experience in smart technologies also perceived the impact of PU1 and PU5 and significance of PU1 on one's attitude towards technologies to be greater. PU1 is concerned with the ability to accomplish tasks more quickly with smart technologies while PU5 is concerned with the job tasks being more easily completed with smart technologies. Similarly, organisations play a key role in facilitating the conditions for effective technology adoption (Venkatesh et al., 2003). Facilitating conditions were also found to be a predictor of users' acceptance of IT, mediated through PEU and PU (Venkatesh et al., 2003). Hence, it is reasonable for respondents from organisations with experience in smart technologies to perceive the impact and significance of PU factors to play a greater role in forming one's attitude towards technologies.

9.5 Summary

This chapter explored the (i) existing implementation status of smart technologies in the Singapore construction industry, (ii) respondents' attitude towards technologies and self-assessed technological competency level, (iii) the perceived importance and changes in knowledge in project management and skills to manage projects in both conventional projects and projects with smart technologies, (iv) the applicability, impact and significance of the factors affecting one's attitude towards technologies and (v) the perceived differences in the importance of knowledge in project management, skills and factors affecting one's attitude towards technologies in contributing to one's technological competency to manage projects with smart technologies.

The findings indicate a moderately low implementation of smart technologies in the Singapore construction industry, with about a quarter of the projects implementing at least one smart technology. The respondents also displayed a slight inclination towards technologies, with an average perceived technological competency level. All identified knowledge areas in project management and skills to manage projects were found to be important in managing both conventional projects and projects with smart technologies, with several project management knowledge areas and skills to manage projects being significantly more important when managing projects with smart technologies. The identified factors were also found to be applicable, impactful and significant in determining one's attitude towards technologies, with TSE, PU and PEU perceived as more significant in determining one's attitude towards technologies. Respondents of different characteristics were also found to perceive the importance of several project management knowledge areas, skills to manage projects and factors affecting attitude towards technologies differently.

References

Ahadzie, D. K., Proverbs, D. G., & Olomolaiye, P. (2008). Towards developing competency-based measures for construction project managers: Should contextual behaviours be distinguished from task behaviours? *International Journal of Project Management, 26*, 631–645. https://doi.org/10.1016/j.ijproman.2007.09.011

Akanmu, A., & Anumba, C. J. (2015). Cyber-physical systems integration of building information models and the physical construction. *Engineering, Construction and Architectural Management, 22*(5), 516–535. https://doi.org/10.1108/ECAM-07-2014-0097

Alvarenga, J. C., Branco, R. R., Guedes, A. L. A., Soares, C. A. P., & E Silva, W. D. S. (2019). The project manager core competencies to project success. *International Journal of Managing Projects in Business.* https://doi.org/10.1108/IJMPB-12-2018-0274

Álvares, J. S., Costa, D. B., & Melo, R. R. S. de. (2018). Exploratory study of using unmanned aerial system imagery for construction site 3D mapping. *Construction Innovation, 18*(3), 301–320. https://doi.org/10.1108/CI-05-2017-0049

Al Yahya, M., Skitmore, M., Bridge, A., Nepal, M., & Cattell, D. (2018). e-Tendering readiness in construction: The posterior model. *Construction Innovation, 18*(2). https://doi.org/10.1108/CI-06-2017-0051

Alzahrani, J. I., & Emsley, M. W. (2013). The impact of contractors' attributes on construction project success: A post construction evaluation. *International Journal of Project Management, 31*(2), 313–322. https://doi.org/10.1016/J.IJPROMAN.2012.06.006

Arayici, Y., Onyenobi, T., & Egbu, C. (2012). Building Information Modelling (BIM) for Facilities Management (FM). *International Journal of 3-D Information Modeling, 1*(1), 55–73. https://doi.org/10.4018/ij3dim.2012010104

Besteiro, E. N. C., De Souza Pinto, J., & Novaski, O. (2015). Success factors in project management. *Business Management Dynamics, 4*(9), 19–34.

Bhattacharya, S., & Momaya, K. S. (2021). Actionable strategy framework for digital transformation in AECO industry. *Engineering, Construction and Architectural Management, 28*(5), 1397–1422. https://doi.org/10.1108/ECAM-07-2020-0587

Bilal, M., Oyedele, L. O., Kusimo, H. O., Owolabi, H. A., Akanbi, L. A., Ajayi, A. O., Akinade, O. O., & Davila Delgado, J. M. (2019). Investigating profitability performance of construction projects using big data: A project analytics approach. *Journal of Building Engineering, 26*, 100850. https://doi.org/10.1016/j.jobe.2019.100850

Bilal, M., Oyedele, L. O., Qadir, J., Munir, K., Ajayi, S. O., Akinade, O. O., Owolabi, H. A., Alaka, H. A., & Pasha, M. (2016). Big Data in the construction industry: A review of present status, opportunities, and future trends. *Advanced Engineering Informatics, 30*, 500–521. https://doi.org/10.1016/j.aei.2016.07.001

Bronte-Stewart, M. (2015). Beyond the iron triangle: Evaluating aspects of success and failure using a project status model. *Computing & Information Systems, 19*(2), 21–37.

Building and Construction Authority. (2016). *BCA identifies 35 key technologies for R&D to drive construction productivity in the next lap.* www.bca.gov.sg

Building and Construction Authority. (2017). *Construction industry transformation map.* www.bca.gov.sg/citm/

Calvo-Porral, C., & Pesqueira-Sanchez, R. (2020). Generational differences in technology behaviour: Comparing millennials and Generation X. *Kybernetes, 49*(11), 2755–2772. https://doi.org/10.1108/K-09-2019-0598

Chan, A. P. C., Scott, D., & Chan, A. P. L. (2004). Factors affecting the success of a construction project. *Journal of Construction Engineering and Management, 130*(1), 153–155. https://doi.org/10.1061/(asce)0733-9364(2004)130:1(153)

Chi, H.-L., Kang, S.-C., & Wang, X. (2013). Research trends and opportunities of augmented reality applications in architecture, engineering, and construction. *Automation in Construction, 33*, 116–122. https://doi.org/10.1016/j.autcon.2012.12.017

Comfort, L. K., & Wukich, C. (2013). Developing decision-making skills for uncertain conditions: The challenge of educating effective emergency managers. *Journal of Public Affairs Education, 19*(1), 53–71. https://doi.org/10.1080/15236803.2013.12001720

Construction Industry Institute. (2022). *CII – project definition rating index overview.* www.construction-institute.org/resources/knowledgebase/pdri-overview

Dallasega, P., Rauch, E., & Linder, C. (2018). Industry 4.0 as an enabler of proximity for construction supply chains: A systematic literature review. *Computers in Industry, 99*, 205–225. https://doi.org/10.1016/j.compind.2018.03.039

Davis, F. D. (1989). Perceived usefulness, perceived ease of use, and user acceptance of information technology. *MIS Quarterly, 13*(3), 319–340. https://doi.org/10.2307/249008

De Soto, G. B., Agustí-Juan, I., Joss, S., & Hunhevicz, J. (2019). Implications of Construction 4.0 to the workforce and organizational structures. *International Journal of Construction Management*, 1–13. https://doi.org/10.1080/15623599.2019.1616414

El-Sabaa, S. (2001). The skills and career path of an effective project manager. *International Journal of Project Management, 19*(1), 1–7. https://doi.org/10.1016/S0263-7863(99)00034-4

Ercan, T. (2019). Building the link between technological capacity strategies and innovation in construction companies. *Sustainable Management Practices.* https://doi.org/10.5772/INTECHOPEN.88238

Fernández-Solís, J. L., Rybkowski, Z. K., Xiao, C., Lü, X., & Chae, L. S. (2015). General contractor's project of projects – a meta-project: Understanding the new paradigm and its implications through the lens of entropy. *Architectural Engineering and Design Management, 11*(3), 213–242. https://doi.org/10.1080/17452007.2014.892470

Gamil, Y., & Rahman, I. A. (2017). Identification of causes and effects of poor communication in construction industry: A theoretical review. *Emerging Science Journal, 1*(4), 239–247. https://doi.org/10.28991/IJSE-01121

Ghaffar, S. H., Corker, J., & Fan, M. (2018). Additive manufacturing technology and its implementation in construction as an eco-innovative solution. *Automation in Construction, 93*, 1–11. https://doi.org/10.1016/J.AUTCON.2018.05.005

Greer, C., Burns, M. J., Wollman, D., & Griffor, E. (2019). *Cyber-physical systems and internet of things, special publication (NIST SP).* National Institute of Standards and Technology. https://doi.org/10.6028/NIST.SP.1900-202

Harikrishnan, A., Said Abdallah, A., Ayer, S. K., El Asmar, M., & Tang, P. (2021). Feasibility of augmented reality technology for communication in the construction industry. *Advanced Engineering Informatics, 50*, 101363. https://doi.org/10.1016/J.AEI.2021.101363

Hazen, B. T., Boone, C. A., Ezell, J. D., & Jones-Farmer, L. A. (2014). Data quality for data science, predictive analytics, and big data in supply chain management: An introduction to the problem and suggestions for research and applications. *International Journal of Production Economics, 154*, 72–80. https://doi.org/10.1016/j.ijpe.2014.04.018

Hernaus, T., & Vokic, N. P. (2014). Work design for different generational cohorts: Determining common and idiosyncratic job characteristics. *Journal of Organizational Change Management, 27*(4), 615–641. https://doi.org/10.1108/JOCM-05-2014-0104/FULL/PDF

Hoxha, L., & McMahan, C. (2014, June). Does a project manager's work experience help project success? *International Journal of Construction Project Management*, 80526.

Hoxha, L., & McMahan, C. (2019). The Influence of project manager's age on project success. *Journal of Engineering, Project, and Production Management, 9*(1), 12–19. https://doi.org/10.2478/JEPPM-2019-0003

Hunhevicz, J. J., & Hall, D. M. (2020). *Do you need a blockchain in construction? Use case categories and decision framework for DLT design options.* https://doi.org/10.1016/j.aei.2020.101094

Hunt, G., Mitzalis, F., Alhinai, T., Hooper, P. A., & Kovac, M. (2014). 3D printing with flying robots. *2014 IEEE international conference on robotics and automation (ICRA)* (pp. 4493–4499). https://doi.org/10.1109/ICRA.2014.6907515

Hwang, B.-G., & Ng, W. J. (2013). Project management knowledge and skills for green construction: Overcoming challenges. *International Journal of Project Management, 31*(2), 272–284. https://doi.org/10.1016/j.ijproman.2012.05.004

Hwang, B.-G., Ngo, J., & Her, P. W. Y. (2020). Integrated digital delivery: Implementation status and project performance in the Singapore construction industry. *Journal of Cleaner Production, 262*, 121396. https://doi.org/10.1016/j.jclepro.2020.121396

Hwang, B.-G., Ngo, J., & Teo, J. Z. K. (2022). Challenges and strategies for the adoption of smart technologies in the construction industry: The case of Singapore. *ASCE Journal of Management in Engineering, 38*(1), 05021014. https://doi.org/10.1061/(asce)me.1943-5479.0000986

Ibem, E. O., & Laryea, S. (2017). E-tendering in the South African construction industry. *International Journal of Construction Management, 17*(4), 310–328. https://doi.org/10.1080/15623599.2016.1222666

Jiang, R., Wu, C., Lei, X., Shemery, A., Hampson, K. D., & Wu, P. (2021). Government efforts and roadmaps for building information modeling implementation: Lessons from Singapore, the UK and the US. *Engineering, Construction and Architectural Management*. https://doi.org/10.1108/ECAM-08-2019-0438/FULL/PDF

Kamari, A., Paari, A., & Torvund, H. Ø. (2020). BIM-enabled virtual reality (VR) for sustainability life cycle and cost assessment. *Sustainability, 13*(1), 249. https://doi.org/10.3390/SU13010249

Kania, E., Radziszewska-Zielina, E., & Śladowski, G. (2020). Communication and information flow in polish construction projects. *Sustainability (Switzerland), 12*(21), 1–23. https://doi.org/10.3390/su12219182

Kim, J. I., Li, S., Chen, X., Keung, C., Suh, M., & Kim, T. W. (2021). Evaluation framework for BIM-based VR applications in design phase. *Journal of Computational Design and Engineering, 8*(3), 910–922. https://doi.org/10.1093/jcde/qwab022

Kothman, I., & Faber, N. (2016). How 3D printing technology changes the rules of the game: Insights from the construction sector. *Journal of Manufacturing Technology Management, 27*(7), 932–943. https://doi.org/10.1108/JMTM-01-2016-0010

Kulviwat, S., Bruner, G. C., & Neelankavil, J. P. (2014). Self-efficacy as an antecedent of cognition and affect in technology acceptance. *Journal of Consumer Marketing, 31*(3), 190–199. https://doi.org/10.1108/JCM-10-2013-0727/FULL/XML

Land Transport Authority. (2020). *LTA | Land transport master plan 2040.* www.lta.gov.sg/content/ltagov/en/who_we_are/our_work/land_transport_master_plan_2040.html

Li, J., Greenwood, D., & Kassem, M. (2019). Blockchain in the built environment and construction industry: A systematic review, conceptual models and practical use cases. *Automation in Construction, 102*, 288–307. https://doi.org/10.1016/j.autcon.2019.02.005

Low, S. P., Gao, S., & Ng, E. W. L. (2021). Future-ready project and facility management graduates in Singapore for Industry 4.0: Transforming mindsets and competencies. *Engineering, Construction and Architectural Management, 28*(1), 270–290. https://doi.org/10.1108/ECAM-08-2018-0322

Low, S. P., Gao, S., & Woo, K. F. (2016). Enhancing construction productivity through organizational learning in the Singapore construction industry. *International Journal of Construction Project Management, 8*(1), 71–89.

Luo, L., He, Q., Jaselskis, E. J., & Xie, J. (2017). Construction project complexity: Research trends and implications. *Journal of Construction Engineering and Management, 143*(7), 04017019. https://doi.org/10.1061/(asce)co.1943-7862.0001306

Makri-Botsari, E., Paraskeva, F., Koumbias, E., Dendaki, A., & Panaikas, P. (2004). Skills in computer use, self-efficacy and self-concept. *WIT Transactions on Information and Communication Technologies, 31*, 377–386. https://doi.org/10.2495/CI040161

Mariano, J., Marques, S., Ramos, M. R., Gerardo, F., Cunha, C. L. da, Girenko, A., Alexandersson, J., Stree, B., Lamanna, M., Lorenzatto, M., Mikkelsen, L. P., Bundgård-Jørgensen, U., Rêgo, S., & de Vries, H. (2021). Too old for technology? Stereotype threat and technology use by older adults. *Behaviour and Information Technology.* https://doi.org/10.1080/0144929X.2021.1882577

Marnewick, C., & Marnewick, A. (2021). Digital intelligence: A must-have for project managers. *Project Leadership and Society, 2*, 100026. https://doi.org/10.1016/j.plas.2021.100026

Maskuriy, R., Selamat, A., Ali, K. N., Maresova, P., & Krejcar, O. (2019). Industry 4.0 for the construction industry – how ready is the industry? *Applied Sciences, 9*(14), 2819. https://doi.org/10.3390/app9142819

Mohammed, A., Almousa, A., Ghaithan, A., & Hadidi, L. A. (2021). The role of blockchain in improving the processes and workflows in construction projects. *Applied Sciences, 11*(19), 8835. https://doi.org/10.3390/APP11198835

Müller, R., & Turner, J. R. (2007). Matching the project manager's leadership style to project type. *International Journal of Project Management, 25*(1), 21–32. https://doi.org/10.1016/j.ijproman.2006.04.003

Ngo, J., Hwang, B.-G., & Teo, J. Z. K. (2021). Impact of smart technologies on construction projects: Improvements in project performance. *CIB W78 2021.*

Ngo, J., Hwang, B.-G., & Zhang, C. (2020). Factor-based big data and predictive analytics capability assessment tool for the construction industry. *Automation in Construction, 110*, 103042. https://doi.org/10.1016/j.autcon.2019.103042

Nguyen, L. D., Chih, Y.-Y., & García de Soto, B. (2017). Knowledge areas delivered in project management programs: Exploratory study. *ASCE Journal of Management in Engineering, 33*(1), 04016025. https://doi.org/10.1061/(asce)me.1943-5479.0000473

Nijhuis, S. (2012). Learning for project management in a higher education curriculum. *Research and Education Conference, Limerick, Munster, Ireland,* 15. https://www.pmi.org/learning/library/learning-project-management-higher-education-curriculum-6315

Nijhuis, S., Vrijhoef, R., & Kessels, J. (2018). Tackling project management competence research. *Project Management Journal, 49*(3), 62–81. https://doi.org/10.1177/8756972818770591

Noghabaei, M., Heydarian, A., Balali, V., & Han, K. (2020). Trend analysis on adoption of virtual and augmented reality in the architecture, engineering, and construction industry. *Data, 5*(1). https://doi.org/10.3390/DATA5010026

Oesterreich, T. D., & Teuteberg, F. (2016). Understanding the implications of digitisation and automation in the context of Industry 4.0: A triangulation approach and elements of a research agenda for the construction industry. *Computers in Industry, 83*, 121–139. https://doi.org/10.1016/j.compind.2016.09.006

Oesterreich, T. D., & Teuteberg, F. (2019). Behind the scenes: Understanding the socio-technical barriers to BIM adoption through the theoretical lens of information systems research. *Technological Forecasting and Social Change, 146*, 413–431. https://doi.org/10.1016/j.techfore.2019.01.003

Osunsanmi, T. O., Aigbavboa, C. O., Emmanuel Oke, A., & Liphadzi, M. (2020). Appraisal of stakeholders' willingness to adopt Construction 4.0 technologies for construction projects. *Built Environment Project and Asset Management, 10*(4), 547–565. https://doi.org/10.1108/BEPAM-12-2018-0159

Pallant, J. (2020). *SPSS survival manual: A step by step guide to data analysis using IBM SPSS* (7th ed.). Routledge. https://doi.org/10.4324/9781003117452

Pan, Y., Zhang, Y., Zhang, D., & Song, Y. (2021). 3D printing in construction: State of the art and applications. *International Journal of Advanced Manufacturing Technology*, *115*(5–6), 1329–1348. https://doi.org/10.1007/S00170-021-07213-0/FIGURES/10

Paton, S., & Hodgson, D. (2016). Project managers on the edge: Liminality and identity in the management of technical work. *New Technology, Work and Employment*, *31*(1), 26–40. https://doi.org/10.1111/NTWE.12056

Pellicer, E., Yepes, V., Correa, C. L., & Alarcón, L. F. (2014). Model for systematic innovation in construction companies. *Journal of Construction Engineering and Management*, *140*(4). https://doi.org/10.1061/(asce)co.1943-7862.0000700

Prasad, B. (1995). JIT quality matrices for strategic planning and implementation. *International Journal of Operations and Production Management*, *15*(9), 116–142. https://doi.org/10.1108/01443579510099706/FULL/PDF

Prebanić, K. R., & Vukomanović, M. (2021). Realizing the need for digital transformation of stakeholder management: A systematic review in the construction industry. *Sustainability (Switzerland)*, *13*(22), 12690. https://doi.org/10.3390/su132212690

Project Management Institute. (2018). The project manager of the future – developing digital-age project management skills to thrive in disruptive times. In *Pulse of the profession in-depth report* (pp. 1–16). Project Management Institute.

Qureshi, M. I., Iftikhar, M., Bhatti, M. N., Shams, T., & Zaman, K. (2013). Critical elements in implementations of just-in-time management: Empirical study of cement industry in Pakistan. *SpringerPlus*, *2*(1), 1–14. https://doi.org/10.1186/2193-1801-2-645

Rogers, E. M. (1995). Lessons for guidelines from the diffusion of innovations. *The Joint Commission Journal on Quality Improvement*, *21*(7), 324–328. https://doi.org/10.1016/S1070-3241(16)30155-9

Sniazhko, S. (2019). Uncertainty in decision-making: A review of the international business literature. *Cogent Business and Management*, *6*(1). https://doi.org/10.1080/23311975.2019.1650692

Squires, J. E., Estabrooks, C. A., Newburn-Cook, C. V., & Gierl, M. (2011). Validation of the conceptual research utilization scale: An application of the standards for educational and psychological testing in healthcare. *BMC Health Services Research*, *11*, 107. https://doi.org/10.1186/1472-6963-11-107

Tay, Y. W. D., Panda, B., Paul, S. C., Noor Mohamed, N. A., Tan, M. J., & Leong, K. F. (2017). 3D printing trends in building and construction industry: A review. *Virtual and Physical Prototyping*, *12*(3), 261–276. https://doi.org/10.1080/17452759.2017.1326724

Tezel, A., Papadonikolaki, E., Yitmen, I., & Hilletofth, P. (2020). Preparing construction supply chains for blockchain technology: An investigation of its potential and future directions. *Frontiers of Engineering Management*, *7*(4), 547–563. https://doi.org/10.1007/s42524-020-0110-8

Trenerry, B., Chng, S., Wang, Y., Suhaila, Z. S., Lim, S. S., Lu, H. Y., & Oh, P. H. (2021). Preparing workplaces for digital transformation: An integrative review and framework of multi-level factors. *Frontiers in Psychology*, *12*, 822. https://doi.org/10.3389/FPSYG.2021.620766/BIBTEX

Vaportzis, E., Clausen, M. G., & Gow, A. J. (2017, October). Older adults perceptions of technology and barriers to interacting with tablet computers: A focus group study. *Frontiers in Psychology*, *8*. https://doi.org/10.3389/FPSYG.2017.01687

Venkatesh, V. (2000). Determinants of perceived ease of use: Integrating control, intrinsic motivation, and emotion into the technology acceptance model. *Information Systems Research*, *11*(4), 342–365. https://doi.org/10.1287/isre.11.4.342.11872

Venkatesh, V., & Bala, H. (2008). Technology Acceptance Model 3 and a research agenda on interventions. *Decision Sciences*, *39*(2), 273–315. https://doi.org/10.1111/j.1540-5915.2008.00192.x

Venkatesh, V., Morris, M. G., Davis, G. B., & Davis, F. D. (2003). User acceptance of information technology: Toward a unified view. *MIS Quarterly*, *27*(3), 425–478. https://doi.org/10.2307/30036540

Volkom, M. V., Stapley, J. C., & Amaturo, V. (2014). Revisiting the digital divide: Generational differences in technology use in everyday life. *North American Journal of Psychology*, *16*(3), 557–574.

Vrchota, J., Řehoř, P., Maříková, M., & Pech, M. (2021). Critical success factors of the project management in relation to Industry 4.0 for sustainability of projects. *Sustainability (Switzerland)*, *13*(1), 1–19. https://doi.org/10.3390/su13010281

Wang, L., Xu, S., Qiu, J., Wang, K., Ma, E., Li, C., & Guo, C. (2020). Automatic monitoring system in underground engineering construction: Review and prospect. *Advances in Civil Engineering*, *2020*. https://doi.org/10.1155/2020/3697253

Wang, Y., Wang, X., Truijens, M., Hou, L., & Zhou, Y. (2014). Integrating augmented reality with building information modeling: Onsite construction process controlling for liquefied natural gas industry. *Automation in Construction*, *40*, 96–105. https://doi.org/10.1016/j.autcon.2013.12.003

Wimalasena, N. N., & Gunatilake, S. (2018). The readiness of construction contractors and consultants to adopt e-tendering: The case of Sri Lanka. *Construction Innovation*, *18*(3), 350–370. https://doi.org/10.1108/CI-03-2017-0025

World Economic Forum. (2018). *The future of jobs* (p. 147). World Economic Forum.

Wu, P., Wang, J., & Wang, X. (2016). A critical review of the use of 3-D printing in the construction industry. *Automation in Construction*, *68*, 21–31. https://doi.org/10.1016/j.autcon.2016.04.005

Wu, P., Zhao, X., Baller, J. H., & Wang, X. (2018). Developing a conceptual framework to improve the implementation of 3D printing technology in the construction industry. *Architectural Science Review*, *61*(3), 133–142. https://doi.org/10.1080/00038628.2018.1450727

Wyskwarski, M. (2021). Requirements for project managers – What do job advertisements say? *Sustainability (Switzerland)*, *13*(23), 04017007. https://doi.org/10.3390/su132312999

Yeşilyurt, E., Ulaş, A. H., & Akan, D. (2016). Teacher self-efficacy, academic self-efficacy, and computer self-efficacy as predictors of attitude toward applying computer-supported education. *Computers in Human Behavior*, *64*, 591–601. https://doi.org/10.1016/J.CHB.2016.07.038

Zhang, L., & Fan, W. (2013). Improving performance of construction projects: A project manager's emotional intelligence approach. *Engineering, Construction and Architectural Management*, *20*(2), 195–207. https://doi.org/10.1108/09699981311303044

Zhao, X., Hwang, B.-G., & Phng, W. (2014). Construction project risk management in Singapore: Resources, effectiveness, impact, and understanding. *KSCE Journal of Civil Engineering*, *18*(1), 27–36. https://doi.org/10.1007/s12205-014-0045-x

Zhong, R. Y., Peng, Y., Xue, F., Fang, J., Zou, W., Luo, H., Thomas Ng, S., Lu, W., Shen, G. Q. P., & Huang, G. Q. (2017). Prefabricated construction enabled by the Internet-of-Things. *Automation in Construction*, *76*, 59–70. https://doi.org/10.1016/j.autcon.2017.01.006

Zhou, C., & Ding, L. Y. (2017). Safety barrier warning system for underground construction sites using Internet-of-Things technologies. *Automation in Construction*, *83*, 372–389. https://doi.org/10.1016/J.AUTCON.2017.07.005

Zijlmans, E. A. O., Tijmstra, J., van der Ark, L. A., & Sijtsma, K. (2019, January). Item-score reliability as a selection tool in test construction. *Frontiers in Psychology*, *9*, 2298. https://doi.org/10.3389/FPSYG.2018.02298/BIBTEX

10 Technological Competency Framework for Project Managers

10.1 Introduction

This chapter elaborates on the development and validation of the TCF for PMs to manage projects incorporating smart technologies. Previous chapters emphasised the importance of comprehensive knowledge in project management areas, adept project management skills and a positive attitude towards technologies for effective management of such projects. Simultaneously, a foundational understanding of smart technologies and certain personal characteristics as identified by Spencer and Spencer (1993), further corroborated by Dainty et al. (2004), were posited as crucial for PMs in the construction industry to be technologically proficient.

While these factors offer a solid explanation for PMs technological competency, the balance between comprehensiveness and parsimony is paramount in developing a robust yet accessible framework. A parsimonious model, as suggested by (Benitez et al., 2020), contributes to theory development by explaining significant phenomena with minimal complexity, thereby enhancing replicability and comprehensibility.

In line with this principle, this study employs PLS–SEM to elucidate the interrelationships among constructs affecting attitudes towards technologies, thereby refining the TCF for enhanced parsimony. This method enables a more nuanced understanding of the factors influencing one's attitude towards technologies, following the preliminary grouping of these factors through exploratory factor analysis. The subsequent sections will detail the development and validation of the TCF, culminating in a comprehensive yet accessible framework for technological competency in PMs.

10.2 Interrelationships of Factors Affecting Attitude Towards Technologies

10.2.1 *Exploratory Factor Analysis for Attitude Towards Technologies*

EFA was conducted for the factors that affect one's attitude towards technologies. Following the recommendations of Costello and Osborne (2005), the EFA was conducted using principal axis rotation as the data is not normally distributed; the promax rotation as the factors that affect attitude towards technologies may be correlated. The number

DOI: 10.1201/9781003462231-10

of factors to be retained was determined using the eigenvalues greater than 1.0, and by observing the results of the screen test (Costello & Osborne, 2005).

The EFA results are shown in Table 10.1. The KMO measure was found to be 0.836 and the results of the Bartlett's Test of Sphericity were found to have a *p*-value of 0.000. The results indicated that the data was suitable for EFA. Based on the EFA results, eight factor groupings were extracted. The results indicated that there are two factor groupings for the construct of TSE and for the construct of TAFF.

The first factor group for TSE involves the factors that reflect one's ability to complete work tasks using smart technologies with guidance available for the user, while the second factor group includes the factors that concerns one's ability to complete work tasks using smart technologies without guidance. Although the factors influencing TSE have been found to belong to one group in previous studies, this finding may signify a change in the perception of digital intelligence and evolution of the ability to use technologies in people over the years (Vaportzis et al., 2017).

Next, the first factor group for TAFF includes factors relating to one's enjoyment and liking to work with technologies and consideration that smart technologies are necessary in work settings. The second factor grouping includes factors that are related to an individual having fun or being interested in work that involves the use of smart technologies. While the factors representing TAFF were posited to be under the same construct, there may be distinctions perceived by the users as one is related to workplace settings while the other factor grouping may be more related to one's general interests in technologies in leisure contexts. This is consistent with Zhou and Feng (2017), who found that perceived enjoyment played a more important role in predicting intention to use video-calling technologies in leisure contexts while the predictive effect of PU is stronger in work settings.

10.2.2 Partial Least Squares–Structural Equation Modelling for Attitude Towards Technologies

Attitude towards technologies has been theorised to be made up of six exogenous variables: TSE, TAFF, TANX, PEU, PU and PIT. The six exogeneous variables were modelled as reflective while the construct of attitude towards technologies is modelled as a higher-order construct (Becker et al., 2012; Hair et al., 2017). Accordingly, the model is a reflective-formative model. The initial model for attitude towards technologies includes all identified factors for the six variables, as shown in Figure 10.1.

10.2.2.1 Assessment of the Initial Measurement Model

The measurement model was assessed using the factor analysis using an iteration of 1,000. Following the recommendations of Hair et al. (2017), the measurement model was assessed for its validity and reliability with the following criteria:

- Reflective indicator loadings should be more than 0.708
- CR of more than 0.70
- AVE of more than 0.50
- HTMT ratio of less than 0.90

Table 10.1 Summary of EFA Results for Attitude Towards Technologies

Factors	Factor Loadings							
	Group 1	Group 2	Group 3	Group 4	Group 5	Group 6	Group 7	Group 8
TSE1								0.855
TSE2			0.823					
TSE3			0.820					
TSE4			0.880					
TSE5			0.951					
TSE6			0.870					
TSE7			0.851					
TSE8			0.846					
TES9								0.904
TSE10			0.813					
TAFF1				0.811				
TAFF2				0.803				
TAFF3					0.873			
TAFF4					0.835			
TAFF5				0.850				
TAFF6					0.799			
TAFF7				0.801				
TANX1	0.807							
TANX2	0.625							
TANX3	0.860							
TANX4	0.954							
TANX5	0.843							
TANX6	0.918							
TANX7	0.949							
TANX8	0.887							
TANX9								
TANX10	0.896							
PEU1		0.967						
PEU2		0.896						
PEU3		0.782						
PEU4		0.872						
PEU5		0.888						
PEU6		0.972						
PEU7		0.940						
PU1				0.891				
PU2				0.720				
PU3				0.875				
PU4				0.926				
PU5				0.950				
PU6				0.797				
PIT1							0.903	
PIT2							0.682	
PIT3							0.734	
PIT4							0.782	
Initial eigenvalue	16.105	6.465	4.356	3.356	2.503	1.746	1.531	1.243
Extracted variance	36.248	14.207	9.523	7.189	5.210	3.455	3.046	2.416

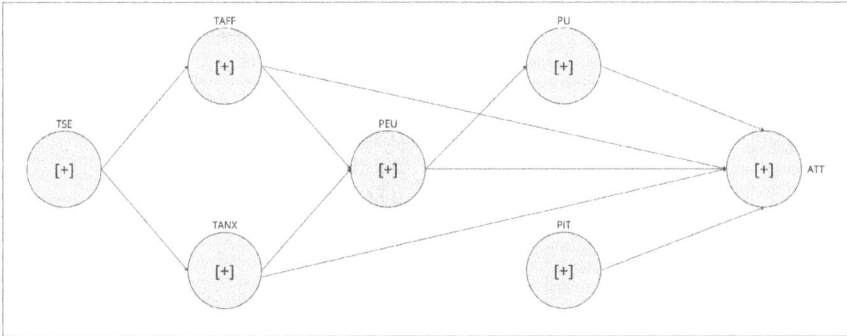

Figure 10.1 Initial PLS-SEM Model for Attitude towards Technologies.

Tables 10.2 and 10.3 show the summary of the results for the validity and reliability of the measurement model. According to the results, TSE1, TSE9, TANX2 and TANX9 had indicator loadings of less than 0.708. The CR, AVE and HTMT ratios were satisfactory.

While TSE1 and TSE9 had indicator loadings of less than 0.708, the two factors were found to have the highest mean and median. They were also found to be in a separate factor grouping from the other TSE factors, as mentioned in Section 10.2.1. For TANX2 and TANX9, the two factors were found to be in a separate factor group together with TANX1, which were found to be the factors that were perceived by respondents to contribute to one's attitude towards technologies the least. Hence, the measurement model was revised to include the factor groupings of the factors for analysis in the second round.

10.2.2.2 *Revision of the Measurement Model*

The measurement model was revised to include the factor groupings of the constructs TSE and TAFF according to the results of the factor analysis in Section 10.2.1. Factors TANX2 and TANX9 were removed from the model as the factor loadings were not significant despite being in the same factor group based on the EFA results. The revised model is shown in Figure 10.2.

Similarly, the measurement model was assessed for its reliability and validity following the recommendations of Hair et al. (2017). The results for the assessment of validity and reliability are shown in Tables 10.4 and 10.5. Based on the results, the factors were found to be reliable and valid to measure one's attitude towards technologies. However, as the main objective of reducing the number of factors has not been achieved, further revision of the measurement model has been made.

Table 10.2 Results of Assessment of the Initial Measurement Model

Variable	Indicator	Indicator Loading	CR	AVE
Technology Self-Efficacy	TSE1	0.154*	0.922	0.573
	TSE2	0.896		
	TSE3	0.825		
	TSE4	0.868		
	TSE5	0.868		
	TSE6	0.832		
	TSE7	0.869		
	TSE8	0.767		
	TSE9	0.239*		
	TSE10	0.787		
Technology Affect	TAFF1	0.905	0.936	0.678
	TAFF2	0.875		
	TAFF3	0.721		
	TAFF4	0.784		
	TAFF5	0.876		
	TAFF6	0.728		
	TAFF7	0.853		
Technology Anxiety	TANX1	0.810	0.956	0.689
	TANX2	0.566*		
	TANX3	0.835		
	TANX4	0.904		
	TANX5	0.872		
	TANX6	0.921		
	TANX7	0.869		
	TANX8	0.885		
	TANX9	0.668*		
	TANX10	0.899		
Perceived Ease of Use	PEU1	0.934	0.980	0.873
	PEU2	0.901		
	PEU3	0.907		
	PEU4	0.916		
	PEU5	0.947		
	PEU6	0.971		
	PEU7	0.965		
Perceived Usefulness	PU1	0.971	0.986	0.919
	PU2	0.957		
	PU3	0.955		
	PU4	0.960		
	PU5	0.958		
	PU6	0.951		
Personal Innovativeness in Technologies	PIT1	0.851	0.707	0.670
	PIT2	0.820		
	PIT3	0.743		
	PIT4	0.857		

* Indicator loading of less than 0.708.
CR, Composite Reliability; AVE, Average Variance Extracted.

Table 10.3 Results of HTMT Ratio of the Initial Measurement Model

	TSE	*TAFF*	*TANX*	*PEU*	*PU*
TAFF	0.327				
TANX	0.307	0.421			
PEU	0.249	0.319	0.336		
PU	0.247	0.409	0.526	0.728	
PIT	0.220	0.366	0.379	0.434	0.378

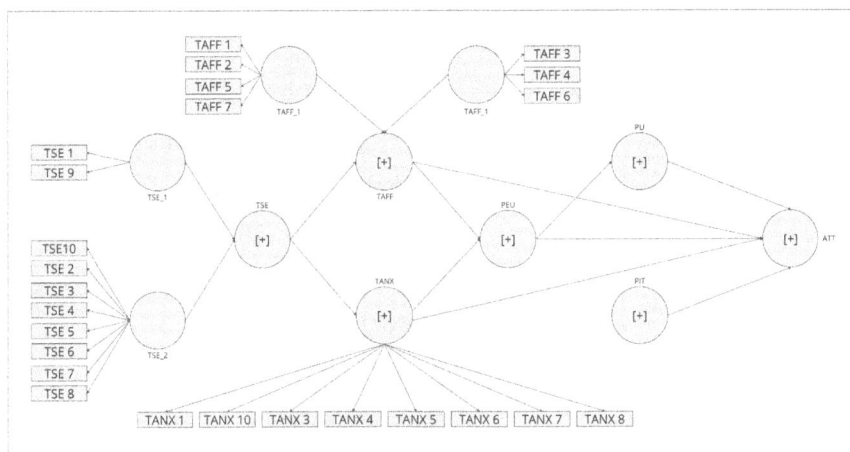

Figure 10.2 Revision 1 of PLS-SEM Model for Attitude towards Technologies.

10.2.2.3 Finalised Measurement Model

The measurement model was further refined by removing the factor grouping for TSE and TAFF with the lower mean factors based on the results discussed in Section 9.4.4. Interviewees A3, A4, A5, A6, V1 and V10 also agreed that the first TSE grouping consisting of TSE1 and TSE9 and the first TAFF grouping consisting of TAFF1, TAFF2, TAFF5 and TAFF7 were more representative of one's attitude towards technologies in the context of managing projects in smart technologies. As it was recommended that the number of indicators in lower-order constructs should be comparable to minimise bias (Becker et al., 2012; Matthews et al., 2018), subsequent iterations were made by removing one indicator at a time and comparing the AVE and CR with and without the indicator for each construct. As the purpose of PLS-SEM is to maximise the explained variance of the target construct, the objective for the iterations was to maximise the AVE of the construct (Hair et al., 2017; Riou et al., 2016). In addition, as single-item constructs were not recommended, the measurement model included at least two indicators and a maximum of four indicators per construct (Hair et al., 2017).

Table 10.4 Results of Assessment of the Revision 1 of the Measurement Model

Variable	Grouping	Indicator	Indicator Loading	CR	AVE
Technology Self-Efficacy	1	TSE1	0.994	0.944	0.895
		TSE9	0.896		
	2	TSE2	0.855	0.961	0.756
		TSE3	0.834		
		TSE4	0.902		
		TSE5	0.941		
		TSE6	0.899		
		TSE7	0.829		
		TSE8	0.854		
		TSE10	0.836		
Technology Affect	1	TAFF1	0.935	0.960	0.856
		TAFF2	0.926		
		TAFF5	0.938		
		TAFF7	0.901		
	2	TAFF3	0.945	0.963	0.897
		TAFF4	0.965		
		TAFF6	0.931		
Technology Anxiety	1	TANX1	0.771	0.966	0.780
		TANX3	0.821		
		TANX4	0.909		
		TANX5	0.903		
		TANX6	0.929		
		TANX7	0.886		
		TANX8	0.911		
		TANX10	0.923		
Perceived Ease of Use	1	PEU1	0.934	0.980	0.873
		PEU2	0.901		
		PEU3	0.907		
		PEU4	0.916		
		PEU5	0.947		
		PEU6	0.971		
		PEU7	0.965		
Perceived Usefulness	1	PU1	0.971	0.986	0.919
		PU2	0.957		
		PU3	0.955		
		PU4	0.960		
		PU5	0.958		
		PU6	0.951		
Personal Innovativeness in Technologies	1	PIT1	0.851	0.707	0.670
		PIT2	0.820		
		PIT3	-0.743		
		PIT4	0.857		

CR, Composite Reliability; AVE, Average Variance Extracted.

Table 10.5 Results of HTMT Ratio of Revision 1 of the Measurement Model

	TSE1	TSE2	TAFF1	TAFF2	TANX	PEU	PU
TSE2	0.173						
TAFF1	0.245	0.207					
TAFF2	0.078	0.316	0.582				
TANX	0.282	0.188	0.402	0.225			
PEU	0.395	0.157	0.386	0.138	0.366		
PU	0.260	0.190	0.418	0.277	0.550	0.728	
PIT	0.455	0.109	0.396	0.219	0.375	0.434	0.378

The finalised measurement model is shown in Figure 10.3 and the corresponding summary of the results for the reliability and validity assessment are shown in Tables 10.6 and 10.7.

10.2.2.4 Evaluation of the Structural Model

Following the finalisation of the measurement model, the structural model was further analysed using the path analysis and a bootstrapping sample of 5,000 (Hair et al., 2011). Based on the recommendations of Hair et al. (2017), the structural model was evaluated according to the following guidelines:

- Variance inflation factor of less than 3
- Q^2 value larger than 0 (0 depicting small predictive accuracy, 0.25 depicting medium predictive accuracy and 0.5 depicting large predictive accuracy of the PLS-SEM model)
- Q^2 predict values of more than 0
- Compare RMSE value with the LM value of each indicator and check if PLS-SEM analysis yields higher prediction errors in terms of RMSE for all (no predictive power), majority (low predictive power), minority or the same number (medium predictive power) or none of the indicators (high predictive power)

In addition to the aforementioned, the finalised model was evaluated using the CTA to check the measurement model, the I-PMA to identify the important components to focus on to improve one's attitude towards technologies and MGA according to one's experience in smart technologies to investigate if respondents with and without experience in smart technologies perceive the direction and strength of the relationships between the constructions differently (Gudergan et al., 2008; Hair et al., 2011; Ringle & Sarstedt, 2016).

The results of the evaluation of the structural model are shown in Tables 10.8 to 10.11. The Q^2 value of ATT was found to be 0.418. This represents a medium predictive accuracy of the model on one's attitude towards technologies. The Q^2 predict values of all the latent indicators were also larger than 0. As TSE1, TSE9, PIT2 and PIT4 are manifest variables, these factors are not included in the comparison

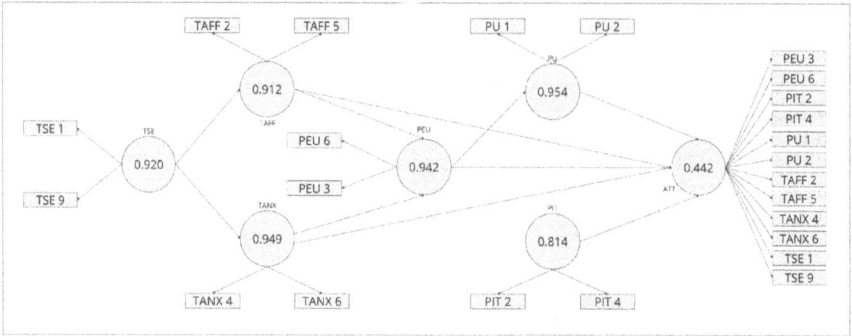

Figure 10.3 Finalised Measurement Model for Attitude Towards Technologies.

Table 10.6 Results of Assessment of the Finalised Measurement Model

Variable	Indicator	Indicator Loading	CR	AVE
Technology Self-Efficacy	TSE1	0.956	0.958	0.920
	TSE9	0.963		
Technology Affect	TAFF2	0.953	0.954	0.912
	TAFF5	0.956		
Technology Anxiety	TANX4	0.971	0.974	0.949
	TANX6	0.977		
Perceived Ease of Use	PEU3	0.972	0.970	0.942
	PEU6	0.969		
Perceived Usefulness	PU1	0.976	0.977	0.955
	PU2	0.978		
Personal Innovativeness in	PIT2	0.903	0.897	0.814
Technologies	PIT4	0.901		

Table 10.7 Results of HTMT Ratio of the Finalised Measurement Model

	TSE	TAFF	TANX	PEU	PU	PIT
TSE						
TAFF	0.248					
TANX	0.252	0.414				
PEU	0.395	0.434	0.344			
PU	0.253	0.418	0.549	0.766		
PIT	0.409	0.413	0.399	0.433	0.382	

between the RMSE of the PLS-SEM model and the LM values (García-Machado et al., 2021). In addition, the RMSE of the PLS-SEM model was compared with the LM values of each indicator, and an almost equal number of the indicators yielded lower prediction errors, indicating a medium predictive power of the PLS-SEM model. Hence, the results indicated that the finalised PLS-SEM for one's

Table 10.8 Summary of Evaluation of Structural Model

Paths	Path Coefficient	p-Value	MGA1	MGA2	MGA3	MGA4
TSE → TAFF	0.225	0.033*	0.573	0.302	0.708	0.712
TSE → TANX	−0.236	0.022*	0.648	0.850	0.776	0.887
TAFF → PEU	0.323	0.002*	0.891	0.142	0.914	0.182
TAFF → ATT	0.235	0.000*	0.393	0.345	0.813	0.337
TANX → PEU	−0.201	0.025*	0.095	0.221	0.041*	0.117
TANX → ATT	−0.269	0.000*	0.988	0.375	0.261	0.394
PEU → PU	0.724	0.000*	0.835	0.713	0.126	0.273
PEU → ATT	0.340	0.000*	0.253	0.196	0.312	0.965
PU → ATT	0.283	0.000*	0.762	0.321	0.869	0.437
PIT → ATT	0.221	0.000*	0.654	0.581	0.496	0.875

MGA1, multi-group analysis among respondents with and without experience in smart technologies; MGA2, multi-group analysis among consultant PMs vs contractor PMs; MGA3, multi-group analysis among consultant PMs versus developer PMs; MGA4, multi-group analysis among contractor PMs versus developer PMs.

* Significant at p-value <0.05.

Table 10.9 I-PMA Results

Variable	I-PMA (Effect on ATT)
Technology Self-Efficacy	0.189
Technology Affect	0.489
Technology Anxiety	−0.535
Perceived Ease of Use	0.600
Perceived Usefulness	0.359
Personal Innovativeness in Technologies	0.267

attitude towards technologies has medium predictive accuracy and power and does not suffer from multi-collinearity. In addition, the CTA-PLS results indicated that the model is a reflective measurement model, as the confidence interval for the tetrads included 0 (Gudergan et al., 2008). The results from the MGA analysis also indicated that respondents with and without experience in smart technologies did not perceive the direction and strength of the relationships between the constructs of attitude towards technologies differently. MGA was further conducted for the different organisation domains. It was found that there were no significant differences in the perceived significance of the relationships between the constructs of attitude towards technologies, except for the interrelationship between TANX and PEU between consultant PMs and developer PMs. Specifically, the path coefficient from TANX to PEU perceived by consultant PMs was 0.017, while the path coefficient perceived by developer PMs was −0.624, with a p-value of 0.044. In addition, the path coefficient perceived by contractor PMs was found to be −0.177, with a p-value of 0.366. Hence, the finding of the differences perceived by consultant PMs and developer PMs in the relationship between TANX and PEU was unexpected.

Table 10.10 Summary of Test for Predictive Power of Model

Indicator	Q^2 Predict (PLS)	RMSE (PLS)	RMSE (LM)	RMSE (PLS) < RMSE (LM)
PEU6	0.129	11.914	12.271	Y
TAFF2	0.084	11.398	11.744	Y
TSE9	0.218	12.465	NA	NA
PEU3	0.171	11.810	12.010	Y
TSE1	0.195	120.65	NA	NA
TAFF5	0.086	11.687	12.001	Y
PIT4	0.237	10.989	NA	NA
PIT2	0.251	10.894	NA	NA
PU1	0.094	10.829	11.263	Y
TANX4	0.092	9.828	10.004	Y
TANX6	0.097	9.189	9.421	Y
PU2	0.107	10.197	10.560	Y
PEU6	0.063	12.357	12.271	N
PEU3	0.063	12.557	12.010	N
PU2	0.029	10.631	10.560	N
PU1	0.023	11.248	11.263	Y
TAFF5	0.015	12.131	12.001	N
TAFF2	0.028	11.746	11.744	N
TANX6	0.038	9.485	9.421	N
TANX4	0.022	10.204	10.004	N

RMSE (PLS), Root Mean Square Error of each indicator in the PLS-SEM; RMSE (LM), Root Mean Square Error (RMSE) of each indicator in Linear Regression Model; Y, Yes; N, No; NA, Not Applicable.

Table 10.11 VIF Results

	TSE	TAFF	TANX	PEU	PU	PIT	ATT
TSE	1.000	1.000					
TAFF				1.170			1.341
TANX				1.170			1.536
PEU					1.000		2.309
PU							2.615
PIT							1.273
ATT							

One reason could be the possibility of conducting extensive research by consultant PMs or experimentation with the technologies to understand the technologies better when they feel anxious using them. However, future studies may be conducted to further understand the perceptions from different PMs. Finally, the I-PMA results demonstrated that PEU, TANX and TAFF should be focus areas for improvements to improve one's attitude towards technologies. The finalised PLS-SEM model for PMs' attitude towards technologies is shown in Figure 10.4.

The resulting PLS-SEM model is consistent with previous studies. It was found that TSE influences an individual's affective feelings towards technologies, which subsequently influences an individual's perception of the ease of use of technologies.

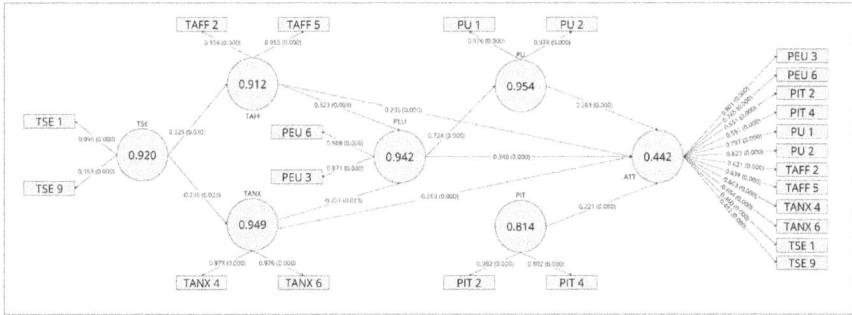

Figure 10.4 Finalised PLS-SEM Model for Attitude towards Technologies.

This is consistent with Venkatesh (2000), who found that computer self-efficacy, facilitating conditions, computer playfulness and computer anxiety form anchors that users employ when forming the PEU of new technologies. The PEU of technologies was found to influence PU, consistent with previous studies. The findings also confirm that both affective and cognitive components of attitude influence one's attitude towards technologies. Besides affective and cognitive components of attitude, an individual's underlying characteristics of willingness to experiment with new technologies was found to influence one's attitude towards technologies, consistent with Yi et al. (2006). The results also found that an individual's experience with smart technologies did not result in statistically significant differences in the PLS-SEM model. Experience in smart technologies was hypothesised to influence one's PEU through adjustments to one's TSE, TAFF and TANX and moderate one's PEU to PU (Venkatesh, 2000; Venkatesh & Bala, 2008). However, this study did not employ a longitudinal method that tracks one's TSE, TAFF, TANX, PEU and PU towards a specific technology. Hence, the cross-sectional collection of perception towards the factors affecting one's general attitude towards technologies may not reflect the influence of experience in technologies on these constructs. In addition, the use of the smart technologies in this study in the Singapore construction industry is still in its infancy. Interviewees P5, P6, V1, V4, V7 and V10 highlighted that many of these technologies are still in their experimentation stage and are not technologically mature for deployment in actual projects. Thus, experience in smart technologies may not influence one's perception of the factors affecting one's general attitude towards technologies.

10.3 Developed Technological Competency Framework for Project Managers

By integrating the finalised PLS-SEM model for attitude towards technologies with the other components of technological competency, the Technological Competency Framework for PMs is derived, as shown in Figure 10.5. All project management knowledge areas were found to be important in managing projects with smart

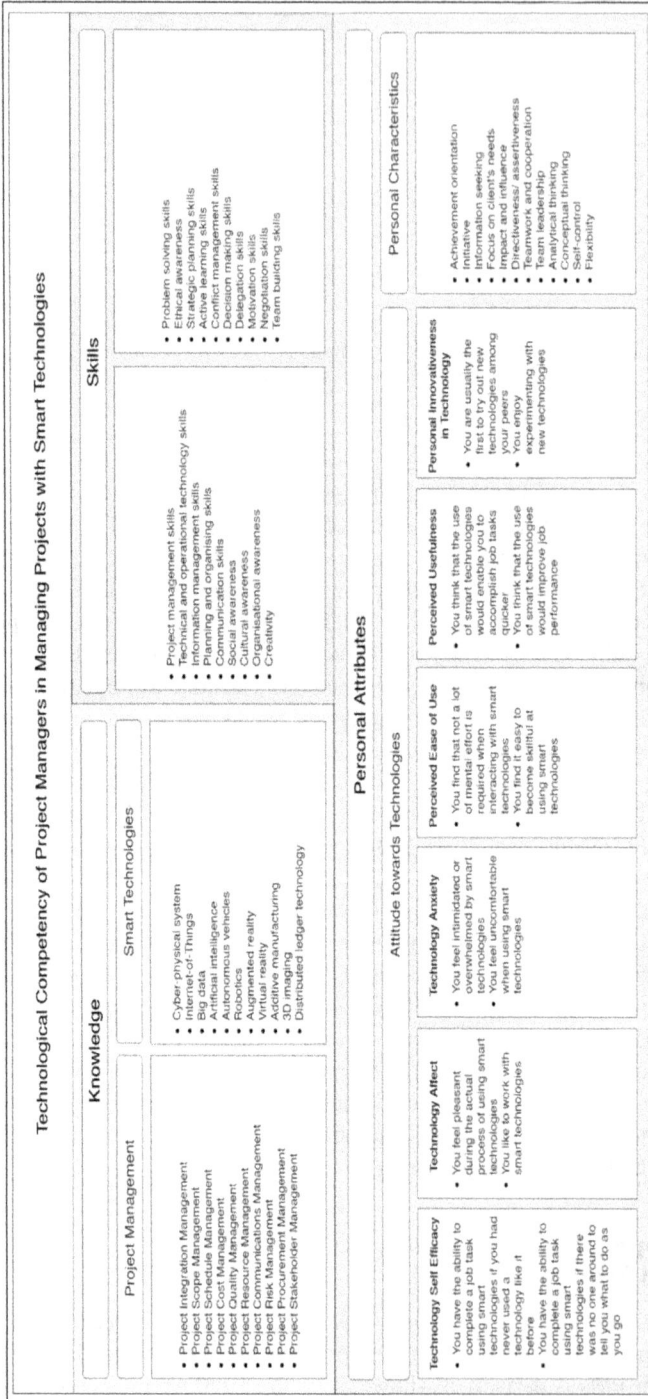

Figure 10.5 Technological Competency Framework for PMs.

Technological Competency of Project Managers in Managing Projects with Smart Technologies

Knowledge

Project Management
- Project Integration Management
- Project Scope Management
- Project Schedule Management
- Project Cost Management
- Project Quality Management
- Project Resource Management
- Project Communications Management
- Project Risk Management
- Project Procurement Management
- Project Stakeholder Management

Smart Technologies
- Cyber-physical system
- Internet-of-Things
- Big data
- Artificial intelligence
- Autonomous vehicles
- Robotics
- Augmented reality
- Virtual reality
- 3D imaging
- Additive manufacturing
- Distributed ledger technology

Skills
- Project management skills
- Technical and operational technology skills
- Information management skills
- Planning and organising skills
- Communication skills
- Social awareness
- Cultural awareness
- Organisational awareness
- Creativity

- Problem solving skills
- Ethical awareness
- Strategic planning skills
- Active learning skills
- Conflict management skills
- Decision making skills
- Delegation skills
- Motivation skills
- Negotiation skills
- Team building skills

Personal Characteristics
- Achievement orientation
- Initiative
- Information seeking
- Focus on client's needs
- Impact and influence
- Directiveness/ assertiveness
- Teamwork and cooperation
- Team leadership
- Analytical thinking
- Conceptual thinking
- Self-control
- Flexibility

Personal Attributes

Attitude towards Technologies

Technology Self Efficacy
- You have the ability to complete a job task using smart technologies if you had technologies if you had never used a technology like it before
- You have the ability to complete a job task using smart technologies if there was no one around to tell you what to do as you go

Technology Affect
- You feel pleasant during the actual process of using smart technologies
- You like to work with smart technologies

Technology Anxiety
- You feel intimidated or overwhelmed by smart technologies
- You feel uncomfortable when using smart technologies

Perceived Ease of Use
- You find that not a lot of mental effort is required when interacting with smart technologies
- You find it easy to become skilful at using smart technologies

Perceived Usefulness
- You think that the use of smart technologies would enable you to accomplish job tasks quicker
- You think that the use of smart technologies would improve job performance

Personal Innovativeness in Technology
- You are usually the first to try out new technologies among your peers
- You enjoy experimenting with new technologies

technologies and should be supported by the knowledge in smart technologies that will be used for the projects in order for the PM to have the necessary knowledge to integrate the use of smart technologies for effective project management. On top of the knowledge in project management and smart technologies, PMs should also have the necessary skills to manage the use of technologies in projects. The survey found that all identified skills remain relevant and important in managing construction projects with smart technologies. The findings also suggest that project management principles and objectives remain the same in projects with smart technologies (Paton & Hodgson, 2016). One's attitude towards technologies was also found to play a role in the level of technological competency, with all identified factors being perceived as important. A reduced set of factors for the various sub-constructs of attitude towards technologies was derived through PLS-SEM. Finally, the underlying personal characteristics of PMs identified through previous studies will continue to play a critical role in PM's work performance (Dainty et al., 2004; Spencer & Spencer, 1993).

10.4   Validation of the Technological Competency Framework

The TCF was validated through interviews with 10 experts with relevant experience in the construction industry and smart technologies. The interviewees perceived the three main components of knowledge, skills and personal attributes to be representative of one's technological competency to manage projects with smart technologies.

Specifically, the interviewees were in consensus that knowledge in all project management areas is important for managing projects, across the different project phases. While key project success factors still emphasised on project schedule, cost, quality and safety performances, Interviewees A2 and A3 shared that all project management knowledge areas must be managed in synergy in order to successfully manage projects. Next, the interviewees agreed that a basic knowledge of smart technologies will be required by PMs. While Interviewees P3 and A3 noted that PMs from different organisation domains will require different levels of proficiencies in the technologies due to the job scope, the PMs should have a basic knowledge of the smart technologies that will be implemented in the projects in order to manage the project with the technology. Interviewees P2, A8, V3 and V10 also shared that some of the technologies, such as DLT, may not be mature enough for deployment in projects currently, so it may not be essential for PMs to have the knowledge in the technology.

Moving on, the identified skills required to manage projects with smart technologies were found to contribute to a PM's technological competency to manage projects with smart technologies. The interviewees were of the view that the identified skills are all required for effective project management for all types of projects. In particular, Interviewees V2, V7 and V8 shared that the nature of the construction industry brings about uncertainties and requires management of the team regardless of whether the project uses smart technologies. Hence, the identified skills are essential considerations when assessing a PM's competency.

The list of personal characteristics that have been found to be critical in determining one's work performance in Spencer and Spencer (1993) and Dainty et al. (2004) were presented to the interviewees. The interviewees agreed that the presented characteristics are essential for one to manage people and uncertainties and perform well in the workplace. Interviewees V4, V9 and V10 further stressed that while the characteristics are essential for PMs to perform well in managing projects, too much of certain characteristics such as assertiveness/directiveness may backfire and sour relationships between the PMs and other team members.

Finally, the interviewees agree with the identified interrelationships of the sub-components affecting one's attitude towards technologies established through PLS-SEM. Interviewees A2, A3, A4, A7, A8, V2 and V6 noted that while two factors measuring each sub-component appear to be reasonably parsimonious, the factors under each sub-component may be interpreted as similar factors although they measure different aspects of the sub-component. Interviewees V2 and V6 further explained that many of the PMs in the current workforce are foreign talents or mature workers and may not be so proficient in English, which may result in an unclear understanding of the factors. However, the interviewees agreed that the factors cover the aspects of each sub-component and attitude towards technologies adequately.

10.5   Summary

This chapter presented the development of the TCF. Initial steps involved conducting an EFA to statistically determine groupings of the sub-factors under the attitude towards technologies. This was followed by the development of a PLS-SEM to illuminate the interrelationships among those sub-factors. A parsimonious model of 'attitude towards technologies' was generated after several iterations. The finalised PLS-SEM was subsequently incorporated into the TCF. Expert interviews were conducted to validate and finalise the TCF, ensuring its robustness and applicability. Table 10.12 shows the summary of the findings.

Table 10.12 Summary of Findings

	Hypothesis	Finding
H1	Knowledge of PMs is a significant determinant of PMs' technological competency in managing projects with smart technologies	Supported
H1.1	Knowledge in project management is a significant determinant of PMs' technological competency in managing projects with smart technologies	Supported
H1.2	Knowledge in smart technologies is a significant determinant of PMs' technological competency in managing smart technologies	Supported
H2	Skills of PMs is a significant determinant of PMs' technological competency in managing projects with smart technologies	Supported

	Hypothesis	Finding
H3	PMs' personal attributes are significant determinants of PMs' technological competency in managing projects with smart technologies	Supported
H3.1	The underlying personal characteristics are a significant determinant of PMs' technological competency in managing projects with smart technologies	Supported
H3.2	An individual's TSE has a positive relationship with one's TAFF	Supported
H3.3	An individual's TSE has a negative relationship with one's TANX	Supported
H3.4	An individual's TAFF has a positive relationship with one's PEU of technologies	Supported
H3.5	An individual's TANX has a negative relationship with one's PEU of technologies	Supported
H3.6	An individual's TAFF has a positive relationship with one's attitude towards technologies	Supported
H3.7	An individual's TANX has a negative relationship with one's attitude towards technologies	Supported
H3.8	An individual's PEU of technologies positively influences PU of technologies	Supported
H3.9	An individual's PEU of technologies has a positive relationship with one's attitude towards technologies	Supported
H3.10	An individual's PU of technologies has a positive relationship with one's attitude towards technologies	Supported
H3.11	An individual's PIT has a positive relationship with one's attitude towards technologies	Supported
H3.12	PMs' attitude towards technologies is a significant determinant of PMs' technological competency in managing projects with smart technologies	Supported

References

Becker, J. M., Klein, K., & Wetzels, M. (2012). Hierarchical latent variable models in PLS-SEM: Guidelines for using reflective-formative type models. *Long Range Planning*, *45*(5–6), 359–394. https://doi.org/10.1016/J.LRP.2012.10.001

Benitez, J., Henseler, J., Castillo, A., & Schuberth, F. (2020). How to perform and report an impactful analysis using partial least squares: Guidelines for confirmatory and explanatory IS research. *Information & Management*, *57*(2), 103168. https://doi.org/10.1016/J.IM.2019.05.003

Costello, A. B., & Osborne, J. (2005). Best practices in exploratory factor analysis: Four recommendations for getting the most from your analysis. *Research, and Evaluation Practical Assessment, Research, and Evaluation*, *10*, 7. https://doi.org/10.7275/jyj1-4868

Dainty, A. R. J., Cheng, M. I., & Moore, D. R. (2004). A competency-based performance model for construction project managers. *Construction Management and Economics*, *22*(8), 877–886. https://doi.org/10.1080/0144619042000202726

García-Machado, J. J., Sroka, W., & Nowak, M. (2021). R&D and innovation collaboration between universities and business – A pls-sem model for the Spanish province of Huelva. *Administrative Sciences, 11*(3), 83. https://doi.org/10.3390/ADMSCI11030083

Gudergan, S. P., Ringle, C. M., Wende, S., & Will, A. (2008). Confirmatory tetrad analysis in PLS path modeling. *Journal of Business Research, 61*(12), 1238–1249. https://doi.org/10.1016/j.jbusres.2008.01.012

Hair, J. F., Hult, G. T. M., Ringle, C. M., & Sarstedt, M. (2017). *A primer on partial least squares structural equation modeling (PLS-SEM)*. Sage.

Hair, J. F., Ringle, C. M., & Sarstedt, M. (2011). PLS-SEM: Indeed a silver bullet. *Journal of Marketing Theory and Practice, 19*(2), 139–152. https://doi.org/10.2753/MTP1069-6679190202

Matthews, L., Hair, J., & Matthews, R. (2018). PLS-SEM: The holy grail for advanced analysis. *The Marketing Management Journal, 28*(1), 1–13.

Paton, S., & Hodgson, D. (2016). Project managers on the edge: Liminality and identity in the management of technical work. *New Technology, Work and Employment, 31*(1), 26–40. https://doi.org/10.1111/NTWE.12056

Ringle, C. M., & Sarstedt, M. (2016). Gain more insight from your PLS-SEM results the importance-performance map analysis. *Industrial Management and Data Systems, 116*(9), 1865–1886. https://doi.org/10.1108/IMDS-10-2015-0449/FULL/PDF

Riou, J., Guyon, H., & Falissard, B. (2016). An introduction to the partial least squares approach to structural equation modelling: A method for exploratory psychiatric research. *International Journal of Methods in Psychiatric Research, 25*(3), 220. https://doi.org/10.1002/MPR.1497

Spencer, L. M., & Spencer, S. M. (1993). *Competence at work: Models for superior performance*. Wiley.

Vaportzis, E., Clausen, M. G., & Gow, A. J. (2017, October). Older adults perceptions of technology and barriers to interacting with tablet computers: A focus group study. *Frontiers in Psychology, 8*. https://doi.org/10.3389/FPSYG.2017.01687

Venkatesh, V. (2000). Determinants of perceived ease of use: Integrating control, intrinsic motivation, and emotion into the Technology Acceptance Model. *Information Systems Research, 11*(4), 342–365. https://doi.org/10.1287/isre.11.4.342.11872

Venkatesh, V., & Bala, H. (2008). Technology Acceptance Model 3 and a research agenda on interventions. *Decision Sciences, 39*(2), 273–315. https://doi.org/10.1111/j.1540-5915.2008.00192.x

Yi, M. Y., Fiedler, K. D., & Park, J. S. (2006). Understanding the role of individual innovativeness in the acceptance of IT-based innovations: Comparative analyses of models and measures. *Decision Sciences, 37*(3), 393–426. https://doi.org/10.1111/j.1540-5414.2006.00132.x

Zhou, R., & Feng, C. (2017, March). Difference between leisure and work contexts: The roles of perceived enjoyment and perceived usefulness in predicting mobile video calling use acceptance. *Frontiers in Psychology, 8*, 350. https://doi.org/10.3389/FPSYG.2017.00350/BIBTEX

11 Case Studies

11.1 Introduction

The construction industry is navigating a transformative period underscored by the integration of smart technologies and innovative practices. Amidst the paradigm shift towards digitalisation and smart technologies, understanding the experiences, challenges and opportunities presented by these changes is critical for development of TCF for PMs and provide recommendations for PMs to improve their technological competency levels.

In light of this, this chapter presents findings from three case studies involving key stakeholders in construction projects, namely developer, consultant and contractor firms. These firms have demonstrated successful integration of various smart technologies in their operations, including VR, Robotics, BIM, 3D Printing, Vision Analytics and Sensing technologies. Four experts from these firms were invited to contribute to the research through expert panel interviews. The interviewees also participated in the pilot interviews, which served to verify the completeness, validity and readability of the survey questionnaire.

Through these case studies, insights were generated regarding the application of smart technologies, the challenges encountered during implementation and the adaptations required in the knowledge, skills and personal attributes of PMs. Furthermore, the interviewees provided recommendations for improving technological competency among PMs, thus setting the stage for the development of a KBTCAIS.

11.2 Case Study I: Developer Firm in Singapore

The first case study was conducted through secondary data collection and interview of a developer firm in Singapore established for more than 60 years and is largely involved in developing industrial buildings. The developer firm is also an advocate for innovation and digitalisation and have been a pioneer in implementing new technologies in their projects.

11.2.1 Applications of Smart Technologies

The firm has started experimenting with almost all of the listed smart technologies in this study. More prominent applications of smart technologies in the developments

DOI: 10.1201/9781003462231-11

of the firm include the adoption of digitalised project delivery throughout the project lifecycle, VR collaboration system to review, analyse and simulate work processes to reduce clashes, the use of drones for progress monitoring and an automated fabricated process. These have been implemented in actual projects and have demonstrated the benefits of digitalising and automating work processes through time and cost savings, as well as improvements in safety and quality performance.

11.2.2   Challenges Faced During Implementation

According to Interviewee P3, one key challenge faced during the implementation of smart technologies was to get buy-ins. As a developer firm beyond the firm, buy-in must also be sought from consultants and contractors. Interviewee P3 shared a past experience on the implementation of an online planning and collaboration software which required the involvement of the project team to come together and develop project plans in terms of resources, schedule and the like. In order to do so, buy-in was first sought with the PMs within the developer firm. This was done through sharing of the benefits and training to familiarise the PMs with the software. Consultants, contractor PMs and technical staff were then trained in the usage of the collaboration tool. Interviewee P3 highlighted that more mature staff especially from the contractor firms displayed higher level of resistance towards the use of the collaboration tool. It was further explained that contractors are typically the ones who will be planning and executing the works, and mature workers have been executing the works in the same manner; hence it was reasonable for mature workers to display higher level of resistance when implementing a new technology.

Another key challenge highlighted by Interviewee P3 was the technological challenge. All technologies have their pros and cons with a great variety of software and hardware available. The selection of one specific software and/or hardware to be implemented was a challenge as bugs and technical issues are present in all software/hardware. Furthermore, technologies and software are always evolving. For example, in BIM, there are plenty of BIM software available but bugs are bound to occur when using any BIM software. This presents a challenge in selecting the best BIM software for implementation. Another aspect of technological challenge is related to how the output from using the technology can be utilised effectively. Specifically, although drones have been used to capture project information, how the collected information can be effectively used and analysed to provide meaningful outputs and impact to project performance presents a key challenge. Hence, Interviewee P3 was of the view that applying technologies in construction projects is still in its development stage.

11.2.3   Changes in Knowledge, Skills and Personal Attributes Required of Project Managers

Interviewee P3 was of the view that the principles of project management and the knowledge of project management remain the same even in projects with smart technologies. Specifically, it was emphasised that managing projects with smart technologies relies on the same project management principles, albeit through the use of different tools and methodologies. In addition, unless robots completely replace human workers, which is unlikely, Interviewee P3 emphasised the importance of the soft skills to manage projects with smart technologies. In terms of the knowledge of smart technologies, Interviewee P3 noted that different levels of proficiency in the smart technologies are required for different stakeholders depending on how the technology is applied in the specific areas of work. For example, PMs from the developer firm would not be required to have the same level of proficiency in drones compared to contractor PMs as contractors PMs will have to know how to operate and fly the drone while developer PMs will only be required to know that drones can be used to collect project progress information. Finally, Interviewee P3 shared that individuals with a more open attitude towards technologies were more likely to embrace the use of new technologies and took a shorter time to pick up new technologies and have performed quite well using the technologies, highlighting the role of attitude towards technologies in facilitating the use of technologies in projects with smart technologies.

11.2.4   *Recommendations to Facilitate Technology Adoption and Improve Technological Competency Level of Project Managers*

To facilitate technology adoption, Interviewee P3 emphasised the importance of a top–down approach to drive technology adoption within the firm and across the value chain. A top–down approach is essential as effective technology adoption requires resources to be allocated for pilot testing and experimentation of technology applications that will be useful for projects, training of all staff that will be involved in projects, specifying in project tenders and getting buy-ins from stakeholders across the value chain. It was also highlighted that the government plays a critical role in driving technology adoption. In Singapore, with the active push of the Building and Construction Authority to adopt digital technologies, contractors have shown more willingness to try out new technologies. Within the firm, Interviewee P3 shared that small-scale pilot testing played a critical role in facilitating technology adoption as it enabled the firm to identify technologies which were most useful in projects, and the features that could be effectively implemented using the technologies. With this understanding, it can assist in gaining buy-in of the project stakeholders as the benefits of the technology applications can be shared and understood, and proper training and familiarisation of the technology can take place.

11.3   Case Study II: Consultant Firm in Singapore

The second case study was conducted through secondary data collection and interview of a consultant firm in Singapore. The firm is a global consulting firm established for more than 70 years in end-to-end construction project delivery. The firm has been involved in projects across a diverse range of sectors such as aviation, healthcare, transport and energy sectors. The firm emphasises delivering sustainable solutions through innovation, problem solving and research and development.

11.3.1   *Applications of Smart Technologies*

Interviewee P4 shared that the firm has been focused on using digital twins, computational BIM and VR and have incorporated these technologies to all of their projects as the firm is driven towards the use of a digital project delivery approach in their projects. The firm is also extending the use of BIM in projects towards asset management and facilities management to ensure that BIM can be used throughout the entire building lifecycle. An example of how computational BIM was applied was shared by Interviewee P4. Computational BIM has been used for value engineering of a façade and by setting the design form and number of standard parts, the design of the façade was automatically regenerated and the cost of the façade could be reduced by half in a short period of time.

11.3.2   Challenges Faced During Implementation

Interviewee P4 highlighted three key challenges faced during the implementation of new technologies in the firm. First, the cost of new technologies was a concern. Implementing new technologies is costly due to the high capital costs and difficulty in ascertaining the rate of returns through productivity gains. Interviewee P4 further shared that while the top construction organisations in Singapore may already be starting their adoption of new technologies, smaller firms will require assistance in terms of costs and sharing of the benefits of new technologies to encourage the implementation of new technologies. In addition, the pandemic has further driven organisations to be operating on lean margins.

The second key challenge faced is changing human behaviour. Interviewee P4 highlighted that changing human behaviour is typically harder in more mature workers as they are more used to the conventional ways of working. In addition, the industry is largely fragmented and interoperability of software adopted also needs to be addressed when implementing new technologies. For example, a Bentley user may find it hard to switch to a Revit user and thus, the firm has different people specialising in different software to address the interoperability issue. Another aspect is related to the fear of workers getting displaced by technologies.

Finally, the need to have a good training plan is a key challenge to be overcome in order to drive the implementation of new technologies in the firm. A good training plan should address the skill gaps and concerns of workers in adopting the new technologies, However, the challenge lies in the variety of technologies available and used in the market.

11.3.3   Changes in Knowledge, Skills and Personal Attributes Required of Project Managers

Interviewee P4 was of the view that there will be significant changes in the competencies required of PMs. While Interviewee P4 emphasised that domain knowledge remains critical in managing projects, PMs will be required to understand how the specific technologies work, and how they can be applied in projects. Having a certain level of digital skills will also be essential for PMs to perform in the digital era. An example provided by Interviewee P4 was in e-tendering. Quantity surveyors will be required to understand how BIM works in order to conduct quantity take-off in the model, and they would have to understand how contracts will work in a digital environment. In addition to that, they should have a certain proficiency in programming in order to modify and interrogate the BIM model to conduct e-measurements for tendering purposes. Having an ecosystem of likeminded people who are willing to learn and are open to the use of technologies is also essential in encouraging PMs to use technologies in projects.

11.3.4 *Recommendations to Facilitate Technology Adoption and Improve Technological Competency Level of Project Managers*

To facilitate technology adoption, Interviewee P4 highlighted the importance of top management support. Top management support is required so that dedicated resources can be allocated to drive technology adoption in the firm. At the same time, in order for technology to be effectively implemented in projects, it is essential for the firm to conduct pilot testing and research to make sure that the specified technology works and the legalities of applying the technology in projects are taken care of. Hence, Interviewee P4 emphasised that a dedicated department should be set up to manage and specialise in the works that require the use of digital technologies. In addition, social coaching is one key practice that should be adopted to facilitate technology adoption. Interviewee P4 highlighted that as adults learn differently and may be concerned with being displaced by technologies, a coaching approach can be undertaken to identify the gaps and understand the needs of each worker, and to align them with the objectives of technology adoption. In line with this, it is also essential that technologies adopted have practical use cases and can solve targeted issues and proper training is provided for the staff to manage the technology. Finally, having an ecosystem of likeminded people will create an environment for staff to be encouraged to adopt new technologies, albeit from a sense of peer pressure, or having a support network.

11.4 Case Study III: Contractor Firm in Singapore

The third case study was conducted through secondary data collection and two interviews of a contractor firm in Singapore. The contractor firm is in operation for more than 60 years and believes in integrating robust planning procedures and innovations in their work. The firm has been involved in various projects in both the private and the public sectors, using both design-bid-build and design-and-build approaches.

11.4.1 *Applications of Smart Technologies*

Interviewee P5 shared that the firm has developed customised plugins for BIM to suit the organisation's needs. In particular, the plugins were developed to simplify and integrate work processes across the different work domains. Interviewee P6 further shared that VR is part of the practices for utilising BIM. VR allows the client to immerse in the simulated environment. One recent project that the firm was involved in conducted 10 sessions of VR immersion to allow the end users to provide feedback to improve the design of the project. Another use of VR was to simulate the design and to reduce the potential clashes in the design before work is executed. Apart from that, Interviewee P6 also highlighted that other technologies, such as HoloLens, were under trial and have shown potential to improve project performance. However, the AR technology has not been developed to a level in which it can be deployed full scale in actual projects sustainably.

Another segment of smart technology applications is the use of field tools. Mobile devices can be used to collect project data at source for processing and analysis. Interviewee P5 shared that RFID tags, sensors and beacons are tagged to materials and equipment which provide the information to a dashboard when they pass through designated gateways. This provides managers with key information including the location, person in charge, time and the activity conducted. With this information, further analysis can be conducted to understand the work progress and identify areas for improvement. Another example was to use mobile devices to log defects and notify the relevant stakeholders in real time. The collected data can also be used to support decision making in future project tenders.

Apart from that, the firm has also been involved in automated prefabrication using robotic arms and 3D printing. However, Interviewee P6 highlighted that 3D printing is currently at the stage of experimentation before being deployed full scale. Beyond that, the firm is also experimenting the use of robots, body cameras and video analytics to be deployed in projects.

11.4.2   *Challenges Faced During Implementation*

Some of the key challenges faced during the implementation of smart technologies include difficulty in getting buy-in, time pressures faced in projects, costs of investment, change in work processes and technology challenges. Interviewee P5 stated that the buy-in during the initial digitalisation journey was very slow. One reason for the slow buy-in was the time pressures faced in projects. In particular, since conventional ways of executing the work could already produce the project deliverables, there was no need for a new way of working, which may require more time for project delivery. In addition, as buildings can only be built from bottom up in a typical manner, it may be faster to adopt conventional ways to build instead of spending time to experiment with the new technologies. Apart from that, new technologies require high initial investment costs and training for staff. Together with the uncertainty in returns of investment, it was hard to gain buy-in for new technologies. Interviewee P5 further emphasised that as the majority of the workforce on site is low wage, they may not be able to afford mobile data plans and may be unwilling to use their own mobile data plans for work tasks. Since real-time monitoring and control of the value chain is enabled by network connectivity, the firm also needs to provide tablets for use in the field. Interviewee P6 also highlighted that workers are concerned of being displaced by technologies. Hence, it is critical to maintain communication with all stakeholders to understand their concerns and to address these concerns.

With the collection of data from the field, the challenge of how to effectively use the data collected arises. While plenty of data can be collected, not all collected data can provide useful analysis to improve the performance of projects. Hence, studies need to be conducted to determine the critical data that will be useful for the firm to improve their performance. At the same time, Interviewee P5 highlighted the importance of maintaining control of the data collected as data is most critical for analysis to enhance the performance of the projects.

Next, as integrating new technologies requires changes in work processes, another key challenge was to make sure that the use of new technologies does not create more work for workers, and instead assists workers in their jobs. Interviewee P5 also highlighted that simplifying the data collection process into two or three steps and making data collection interesting are important ways to gain buy-in and reduce resistance of workers to use new technologies.

Finally, apart from the human challenges, both interviewees also highlighted the challenges with the existing technologies. This includes the hardware and software limitations of the technologies. For example, in HoloLens, some of the limitations faced include the capacity of data that can be loaded and how long the battery of the HoloLens can last. Another example provided was the use of body cameras, where ergonomics becomes an issue for prolonged use.

11.4.3 Changes in Knowledge, Skills and Personal Attributes Required of Project Managers

Interviewee P6 was of the view that the fundamentals of project management remain the same regardless of whether projects are incorporated with smart technologies or not. However, a new set of capabilities to think out of the box and not be confined to usual practices will be required. At the same time, the ability to understand what the new technologies can offer and how to integrate the technologies into the workflow to effectively use the technologies is critical in projects with smart technologies. It was emphasised by Interviewee P6 that it is important for workers to be learning continuously to be more empowered as more technologies are being introduced in the workplace. Both Interviewees P5 and P6 also recognised that the attitude towards technologies is more important than the skills and knowledge in smart technologies and that it is essential for workers to embrace new technologies. Interviewee P5 shared that with an open mind-set towards technologies, workers will take the initiative to learn new skills and knowledge of the new technologies.

11.4.4 Recommendations to Facilitate Technology Adoption and Improve Technological Competency Level of Project Managers

To facilitate technology adoption, Interviewee P5 shared that the firm made it compulsory for all staff to turn digital. The new hires of the firm must be data ready, data driven or data savvy. With this, the firm can create an ecosystem of likeminded people and cultivate an organisation culture that encourages the adoption of new technologies. This can only be done with top management support. Interviewee P6 was also of the view that with top management support, it is easier to implement new technologies as the staff will also be in support of the new technologies. In the case where the staff are unwilling to adopt the new technologies, it may be indicative of the high costs that reduce their project budget, of minimal practical benefits that may be derived from the technologies or of too many steps or changes in the work processes in order to integrate the technology. Hence, it is important to integrate the technology into existing work processes and ensure that the technology is easy to implement.

While top management support within the organisation is important, government push is also essential in getting the entire value chain to adopt new technologies. Financial support from the government also plays an important role in encouraging organisations, especially smaller firms, in trying out new technologies. Both the government and the organisation's support towards the use of technologies are essential to the top–down approach in driving technology adoption. Interviewee P6 further shared that while the top–down approach is essential in driving technology adoption, a bottom–up approach can also play an important role in driving technology adoption in the firm. Workers who are involved on site have a better understanding of the problems faced during the execution of works, and with an open mind-set towards technologies, workers on the ground may suggest new ways of working or new technologies that may be adopted to improve work performance. Interviewee P6 was of the view that a bottom–up approach to technology adoption is suitable for cheaper technologies while a top–down approach to technology adoption may be more suitable for more expensive technologies.

Beyond costs, both interviewees highlighted the importance of understanding the use cases of new technologies. New technologies should only be implemented if they can solve specific problems faced in projects. This can also facilitate the gaining of buy-in from the stakeholders as clear benefits could be derived from the technologies. Communication is also essential when getting buy-in from the stakeholders. Apart from the firm, as many stakeholders are involved in construction projects, Interviewee P6 also mentioned the importance of helping the entire value chain improve their competency in digital technologies. Interviewee P6 shared the example of training the subcontractors in the use of BIM during its roll out to ensure that BIM can be effectively implemented in their projects.

Finally, the availability of a dashboard and transparency of the performance can help to improve performance of the firm and encourage technology adoption. As data is being collected using RFID tags, sensors, beacons and cameras, analysis of this data can help to evaluate performance of the different stakeholders. The visibility of the performance of stakeholders can motivate people to improve their performance. With this motivation to improve performance, it may also encourage people to implement technologies in their projects.

11.5   Discussions

The three case studies were compared to identify (i) the most common smart technology applications in projects; (ii) common challenges faced by organisations during the implementation of new technologies; (iii) changes to the knowledge, skills and personal attributes required of PMs to manage projects with smart technologies; and (iv) the effective strategies to facilitate technology adoption.

11.5.1   *Common Smart Technology Applications in Projects*

All three organisations have adopted real-time monitoring and control on site and along the value chain, through the use of RFID tags and sensors and a digital model. This was complemented by the use of drones for progress monitoring. The increased visibility of the activities in the project allows PMs to have a more holistic monitoring of the construction activities, enabling changes to the project schedule and resource allocation to be made to optimise the resources. At the same time, the collected data could be used to improve future projects and for supplier, vendor or contractor evaluation.

The organisations also highlighted the use of BIM as the foundation for other add-ins such as integration capabilities, computational capacity and collaboration in their projects. The three organisations have also adopted AR and/or VR collaboration tools in their projects. This is reflective of the importance of collaboration in the fragmented construction industry. All interviewees have highlighted that improved collaboration has helped to improve the productivity, quality, safety and performance of the projects. Other key benefits from utilising AR and/or VR collaboration and simulation tools were found to be the early identification of clashes in the design simulation, and for client immersion in the design to improve client satisfaction.

The common applications in projects reflected the nature of fragmentation in the construction industry and have demonstrated the potential to overcome the challenges resulting from the fragmentation of the construction industry. Specifically, improving access and availability to project information provides a more holistic view of the project progress and enhances collaboration among the project stakeholders. This subsequently results in improved project performance.

11.5.2 *Common Challenges Faced During Implementation*

The interviewees highlighted common challenges related to the human resources, integration in work processes, utilisation of the outputs, costs of investment and technological challenge. First, the fear of being displaced by technologies needs to be addressed. This could be done through communication and coaching of the employees. In addition to that, training of the employees to have the relevant digital skills is critical in empowering employees to embrace new technologies. However, developing a suitable training plan was highlighted as another challenge due to the wide variety of technologies available.

Gaining buy-in for the technology adoption was also another key challenge faced by the organisations. Apart from the fear of being displaced by technologies, other reasons that cause resistance towards the use of technologies include the complexities of integrating the technology use into existing workflows, unclear or minimal practical benefits of adopting new technologies and difficulty in changing human behaviour. Similarly, a good training plan, identification of practical use cases of technologies and rigorous research and experimentation prior to deployment are essential to facilitate the gaining of buy-in.

Apart from human-related and work-process-related challenges, challenges related to the existing level of maturity of the smart technologies were also highlighted by all interviewees. Specifically, hardware and software limitations of the technologies were emphasised by the interviewees. In particular, ergonomics, data capacity, battery life and interoperability of the technologies were some of the key concerns of the interviewees. The wide variety of available software and hardware also presents the challenge of selecting the optimal technology to adopt. On top of the existing maturity level of the technologies, the cost of investment was another key challenge that was highlighted by all interviewees. Costs of investment include the initial costs of hardware and software, training of staff and maintenance costs. The interviewees also emphasised the limited capacities of smaller firms downstream in adopting new technologies, which require more assistance in order for them to adopt new technologies.

Finally, the interviewees also highlighted the difficulties in utilising the outputs from the technologies to improve performance of projects and the firm. Specifically, while a large variety, volume and velocity of data could be collected with real-time monitoring and control, how the collected data should be utilised to produce meaningful results and improve performance in terms of productivity, safety, quality and costs needs to be studied extensively.

11.5.3   *Changes in Knowledge, Skills and Personal Attributes Required of Project Managers*

The cases demonstrated that the fundamentals of project management knowledge and skills remain the same in both conventional projects and projects with smart technologies. The interviewees noted that the integration of smart technologies in projects will be used as tools in assisting PMs to manage the project and will only change the method or tools adopted instead of the principles of project management.

While the principles of project management do not change, all interviewees highlighted an increasing need for soft skills, digital skills and an open attitude that embraces new technologies. The cases also demonstrated the role an individual's attitude towards technologies plays in adopting new technologies. With a mind-set that embraces new technologies, individuals were found to be more willing to pick up new knowledge and skills in relation to smart technologies.

11.5.4   *Effective Strategies to Facilitate Technology Adoption and Improve Technological Competency Level of Project Managers*

To overcome the common challenges faced by organisations during the implementation of smart technologies, several strategies were adopted in all three cases to facilitate technology adoption. First, top–down approach driven by the government and the organisation's top management is needed. Government support is essential in terms of providing financial support especially for small and medium enterprises and may further drive technology adoption through mandating the use of the technologies. Mandating the use of technologies has been effective in driving the adoption of technologies, like in the case of BIM in Singapore. While the government support can drive technology adoption on a higher level, the support of an organisation's top management plays an important role in driving the adoption of technologies within the organisation. Specifically, the organisation can benefit from setting up a dedicated department to conduct pilot testing and experimentation of new technologies, study the use cases of smart technologies, study how the outputs of the technologies may be effectively utilised and how to simplify and integrate the use of technologies into the work processes, gain buy-of stakeholders into the new technologies and train the employees in the use of the new technologies.

In order to facilitate technology adoption, it is critical for the organisation to investigate the problems faced by workers and study how technologies can be adopted to overcome a targeted problem. Having practical use cases for the technologies will facilitate the buy-in of stakeholders and increase the willingness of employees to implement the technologies. Beyond the identification of practical use cases of the technologies, pilot testing and experimentation of the technologies in projects should be conducted to understand the limitations of the technologies, how to better integrate the technologies into the workflow and to understand how the outputs of the technologies can be best utilised. Having conducted extensive research and experimentation, the benefits of utilising the technology can be shared among the staff to gain buy-in and can facilitate the training of the staff in using the technology. Along the same line, having a good training plan is essential in driving the adoption of new technologies in the firm. While new hires of the firm may already have the relevant digital skills and mind-set, the existing employees of the firms need to be trained to manage projects with new technologies. Although the training plan needs to be personalised according to the individual's scope of works and specified technology, the basic building blocks of the training plan should include knowledge of the specified technology, digital skills such as programming, and can benefit from coaching and understanding of the worker's concerns and aligning the objectives of technology adoption with the worker's concerns.

Creating an ecosystem of likeminded individuals and an organisation culture that embraces technologies is the cornerstone of facilitating technology adoption. This would require top management support. Having an organisation culture that embraces technologies will provide a support network for individuals to experiment with new technologies and may also create peer pressure to encourage one another to continuously learn new technologies. This can help to create an environment that constantly innovates and experiments with new technologies, facilitating technology adoption.

Finally, as construction projects involve various stakeholders along the value chain, it is important to improve the technological competency of organisations and workers along the entire value chain. Specifically, organisations upstream with the capacity can assist in the training of workers downstream, or financial support may be provided by the government to increase the level of technological competency of all organisations in the value chain. This is essential in order to overcome the issues of interoperability of technologies and fragmentation through effective collaboration, which can only be possible if all project stakeholders are on board and have the relevant skills to manage the smart technologies.

11.6 Summary

This chapter presented the case studies that aimed to investigate (i) the most common smart technology applications in projects, (ii) common challenges faced by organisations during the implementation of new technologies, (iii) changes to the knowledge, skills and personal attributes required of PMs to manage projects with smart technologies and (iv) the effective strategies to facilitate technology adoption. Table 11.1 summarises the findings of the three case studies. These findings can be used to support the development of the TCF and the identification of recommendations to improve the level of technological competency of PMs.

Table 11.1 Findings of the Case Studies

Objectives	Main Findings
Smart Technology Applications in Projects	Real-time monitoring and control through the use of RFID tags, sensors, beacons and drones
	Big data to analyse the collected data and visualise on a dashboard
	AR and/or VR for simulation to identify potential clashes
	AR and/or VR as a collaboration and immersion tool
Common Challenges Faced During the Implementation of New Technologies	Workers' fear of getting displaced by technologies
	Complexities of integrating the technology into existing workflows
	Unclear or minimal practical benefits of adopting new technologies
	Getting buy-in to change conventional ways of work
	Utilisation of outputs from technologies that are meaningful in improving performance
	Costs of technologies
	Developing a suitable training program for employees
	Hardware and software limitations of technologies
	Interoperability of technologies
	Selection of optimal technology solution
Changes to Knowledge, Skills and Personal Attributes Required of PMs	Fundamentals of project management do not change
	Increasing importance of soft skills and digital skills
	Importance of having an open mind-set that embraces the use of technologies
	Require different level of proficiency in smart technologies according to scope of works
Effective Strategies to Facilitate Technology Adoption and Improve Technological Competency Level of PMs	Government support in terms of financial incentives
	Top management support
	Setting up a dedicated department that manages smart technologies
	Identify practical use cases and targeted problems to be solved by the technology
	Extensive pilot testing and experimentation of new technologies before deployment
	Simplify and integrate the use of technologies into existing workflows
	Gain buy-in of stakeholders through sharing of benefits
	Personalised training plan
	Coaching approach to understand concerns of workers and to align objectives of technology adoption with workers' concerns
	Create an ecosystem of likeminded individuals who are open to technologies
	Help organisations along the value chain to adopt new technologies

12 Developing a Knowledge-Based Analytics and Innovations System for Technological Competency in Project Managers

12.1 Introduction

The KBTCAIS is a Microsoft Excel–based analytics platform that has been designed to assess a PM's technological competency. This system offers a visualisation of assessment results, provides customised recommendations to improve the technological competency of the PM and generates a downloadable report of the technological competency assessment.

The KBTCAIS is made up of three key components: a knowledge base, a GUI and a DSS. The knowledge base, established from extensive literature reviews and case studies, and further endorsed by expert input, houses recommendations to improve PMs' technological competency levels. Factor weightings are determined through the AHP using the AtIP-GMM.

This system was validated by 10 industry experts experienced in both the construction industry and the application of smart technologies. The GUI offers a user-friendly platform for interaction with the KBTCAIS, while the DSS processes assessment inputs and calculates the technological competency level based on pre-established weightings assigned to various factors.

Finally, depending on the assessment outcome, a tailored technological competency report inclusive of custom recommendations is generated. This allows PMs to improve their competency in managing projects involving the use of smart technologies.

12.2 Objectives of the Knowledge-Based Technological Competency Analytics and Innovations System

This research aims to develop the KBTCAIS to improve the level of technological competency of PMs in managing construction projects to ensure that projects with smart technologies are effectively managed. To achieve the aim, the KBTCAIS is developed with the following specific objectives:

(i) Assess PMs technological competency to manage construction projects with smart technologies
(ii) Visualise the technological competency level of PMs to manage construction projects with smart technologies

DOI: 10.1201/9781003462231-12

(iii) Provide tailored recommendations according to the PM's technological competency level to improve the level of technological competency
(iv) Provide recommendations to organisations to provide a conducive environment to improve the technological competency of PMs
(v) Allow organisations to allocate or recruit suitable PMs for projects of different technological complexity levels
(vi) Generate a printable technological competency assessment report

Users will be asked for their inputs to each of the technological competency factor by selecting the level that corresponds to the description provided. Using the input data and pre-determined weightages, the KBTCAIS will compute the overall level of technological competency and the level of competency for each of the technological competency component. The computed level of overall technological competency and technological competency components will be visualised in a bar chart with its corresponding score and quartile displayed. Following that, the KBTCAIS will select the set of recommendations corresponding to the level of competency for each of the technological competency components and provide recommendations for the organisation. To allow for easy review, the users can print the technological competency report and recommendations from the KBTCAIS.

By using the KBTCAIS, PMs can better understand their level of technological competency and receive recommendations to improve their level of technological competency. Furthermore, the KBTCAIS allows PMs to understand the factors that contribute to one's technological competency and provide PMs with the key factors they should work on to increase their level of technological competency. Aside from improving the technological competency of PMs, organisations can also use the results from the KBTCAIS for recruitment purposes, and to allocate suitable PMs for projects of different technological complexities.

Other users of KBTCAIS may include university or polytechnic students, and individuals looking to have a career switch to become PMs. These users may use the KBTCAIS to obtain a general assessment of their technological competency to supplement their portfolio and support their job applications and to identify areas of improvements to further develop to be technologically competent and consistently deliver successful projects with smart technologies.

12.3   Architecture and Tools of the Knowledge-Based Technological Competency Analytics and Innovations System

The KBTCAIS is made of three main components: (i) the knowledge base; (ii) graphical user interface and (iii) decision support system. The architecture of the KBTCAIS is shown in Figure 12.1.

12.3.1   Knowledge Base

The knowledge base in the KBTCAIS consists of (i) the technological competency factors; (ii) the weightages of the technological competency factors determined

Input	Decision Support System	Output
· Assessment of level of proficiency in identified factors affecting one's technological competency	· Calculation of the technological competency level · Visualisation of the technological competency level · Select appropriate recommendations for PMs · Generate competency report **Knowledge Base** · Technological competency factors · Weightage of technological factors · Recommendations to improve technological competency	· Assessed level of technological competency, including quartile and score for the overall technological competency and sub-components · Visualised assessment results · Tailored recommendations to improve level of technological competency · Printable reports

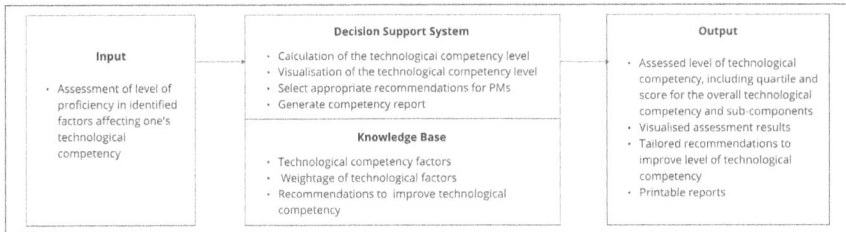

Figure 12.1 Architecture of the KBTCAIS.

by experts and (iii) recommendations to improve the technological competency of PMs. The weightages of the technological competency factors were determined using the analytical hierarchy process with experts and the recommendations to improve the technological competency level were developed based on literature review and expert interviews.

12.3.2   Graphical User Interface

The GUI provides the platform for users to interact with the KBTCAIS. The GUI of the KBTCAIS consists of the (i) introduction interface; (ii) user guide interface; (iii) assessment interface; (iv) report interface and (v) recommendations interface. The user will first be presented with the introduction interface, which provides a brief description of the KBTCAIS and allows users to input their information. The user will then be directed to the user guide, where a more detailed description of technological competency and its components and the instructions on how to use the KBTCAIS are provided. The assessment interface displays each of the technological competency factor, categorised according to the various technological competency components. The assessment utilises a 7-point Likert scale (1 = strongly disagree/not proficient at all, 4 = neutral, 7 = strongly agree/very proficient), and provides a description of the corresponding level of competency to the users. After submitting the assessment, the user will be shown the report interface, which comprises of the computed overall technological competency score and quartile, a bar chart displaying the computed score and quartile of each technological competency component, the key technological competency factors to work on under each technological competency component and the breakdown of the responses submitted. The user will then be guided to the recommendations interface, where the user will receive recommendations to improve one's overall technological competency level, each of the technological competency component and a set of recommendations for the organisations to improve the technological competency of PMs. The KBTCAIS interface also allows the user to print the report and recommendations for easy reference.

12.3.3 Decision Support System

The decision support system transforms the assessment inputs and calculates the technological competency level according to the pre-determined factor weightages. The user's technological competency level will be categorised into four quartiles. At the same time, the corresponding scores and quartiles are visualised in the bar chart. The decision support system also retrieves the recommendations according to the user's technological competency level according to the rules coded as shown in Table 12.1.

12.3.4 Tools to Develop the Knowledge-Based Technological Competency Analytics and Innovations System

The KBTCAIS was developed using Microsoft Excel VBA. Microsoft Excel VBA was selected as the platform to develop the KBTCAIS as Microsoft Excel is an essential office software that is widely used in the industry (B.-G. Hwang et al., 2018; Ngo et al., 2020). This means that the KBTCAIS can be easily accessed by the public. In addition, Microsoft VBA has been highly optimised to support rapid application development in Microsoft Office programs (Hwang et al., 2018; Ngo et al., 2020). Furthermore, since Microsoft VBA is built within Microsoft Excel, the KBTCAIS will be automatically updated when Microsoft Excel is updated, reducing the need for a separate maintenance plan for the KBTCAIS (Hwang et al., 2018; Ngo et al., 2020).

12.4 Determination of Factor Weightages

With the different importance of each technological competency factor in contributing to one's technological competency to manage construction projects with smart technologies, one key component of the KBTCAIS is the factor weightages which will be used to compute the user's technological competency level. This study adopted the AHP approach to determine the weightages of factors in contributing to the overall technological competency level of the PM. Nine PMs with various years of experience in the construction industry and in smart technologies have been invited as experts to participate in the AHP.

12.4.1 Analytical Hierarchical Process

The AHP structure adopted in this study is shown in Figure 12.2. The AHP structure consists of four hierarchy levels. The first level contains the goal of the decision

Table 12.1 Rules of Selecting Recommendations in the DSS

Rules	Score	Corresponding Quartile	Recommendation Level
Rule 1	≤0.25	Quartile 4	Low to Medium
Rule 2	0.26 to 0.50	Quartile 3	Low to Medium
Rule 3	0.51 to 0.75	Quartile 2	Medium to High
Rule 4	≥ 0.76	Quartile 1	High to Very High

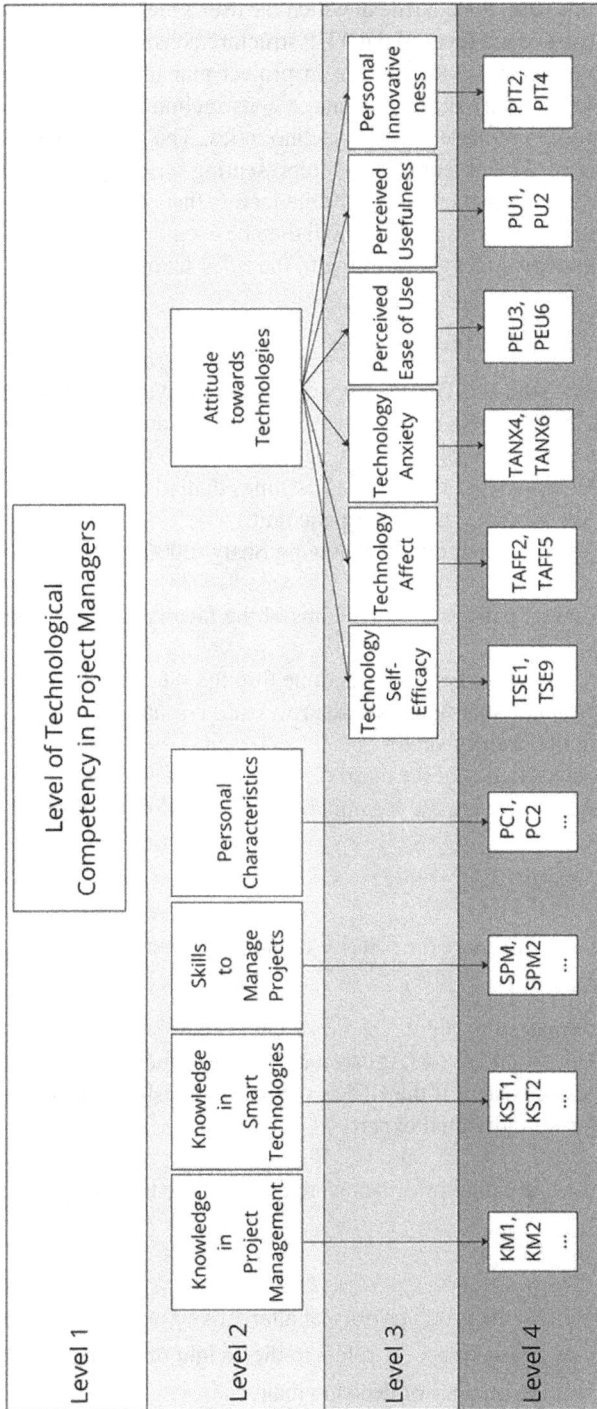

Figure 12.2 AHP Model for Technological Competency in PMs.

problem, and in this case, the quartile in which the user's technological competency level falls into. The second level of the AHP structure consists of the technological competency components: (i) knowledge in project management, (ii) knowledge in smart technologies; (iii) skills to manage construction projects, (iv) personal characteristics and (v) attitude towards technologies. The third level of the AHP structure consists of the sub-components representing attitude towards technologies. Finally, the fourth level comprises of the factors that contribute to one's technological competency. The respondents will then be categorised into the respective technological competency level according to the rules set out in Table 12.1.

12.4.2   Deriving Consensus Among Experts

The AIP computed with the GMM was adopted to derive consensus among the experts (Forman & Peniwati, 1998). In AIP, the individual priorities were first derived for individual experts and then aggregated to obtain the priorities as a group (Aguarón et al., 2019). This method assumes that all experts interviewed are individuals who do not work as a synergistic unit.

Individual priorities were derived following Saaty (2006) as follows:

1. Individuals conduct pairwise comparisons of the factors using the fundamental scale of 1 to 9.
2. Computation of the priority vectors is done through the formation of the matrix of ratio comparisons, summing over the rows and normalising by its geometric mean to obtain the priority vector.
3. Assessing the consistency of the matrix is made by first establishing the geometric consistency index using the formula (Aguarón & Moreno-Jiménez, 2003):

$$GCI_n = \frac{1}{(n-1)(n-2)} \sum_{i,j=i}^{n} \log^2 e_{ij}$$

where $e_{ij} = a_{ij}\omega_j / \omega_i$ and ω are the priority vectors obtained using the row geometric method $\omega = (\omega_i) = (\prod_{k=1}^{n} a_{ik}^{1/n})$.

4. This is then compared to the threshold of GCI = 0.31, GCI = 0.35 and GCI = 0.37 for $n = 3$, $n = 4$ and $n > 4$ (Aguarón & Moreno-Jiménez, 2003). The matrix is considered as consistent if the GCI is within the threshold values.
5. This is done for all individual experts.

Following that, the individual priorities were then aggregated using the formula

$$P_g(A_j) = \prod_{i=1}^{n} P_i(A_j)^{wi}$$

where $P_g(A_j)$ refers to the group priority of alternative j, $P_i(A_j)$ refers to the individual i's priority of alternative j, w_i refers to the weight of the individual i, where $\sum_{i=1}^{n} w_i = 1$ and n is the number of decision makers.

Finally, the consistency of the group matrix was calculated following the same procedure.

12.4.3 Finalisation of Weightages

Following the AtIP-GMM, Table 12.2 shows the finalised weightages of the technological competency factors which are adopted in the KBTCAIS. The finalised weightages for the components of technological competency indicated that the fundamentals of project management remain important in determining one's technological competency to manage projects with smart technologies, with high weightages given to knowledge in project management, skills to manage projects and personal characteristics. This finding is consistent with previous findings where it was found that project management principles and processes remain the same (Paton & Hodgson, 2016). Essentially, this shows that smart technologies are used to assist PMs in project management. In addition, Interviewees A2 and A3 emphasised that the role of PMs is to deliver the project to meet its set objectives, and PMs have to utilise the smart technologies that are specified to be used in the project contract. Hence, the PMs' attitude towards technologies and knowledge in smart technologies contributes less to the PMs' technological competency, albeit being important.

Equal weightages were allocated to each project management knowledge area. Although the survey findings indicated that project communication management, project cost management and project risk management were most important when managing projects with smart technologies, it was emphasised by Interviewees A3 and A4 that each project management knowledge area is equally important as each plays a role in different phases of a project. In particular, the interviewees highlighted that while project schedule, cost and quality performance are key metrics used to evaluate project success, the other project management knowledge areas must be managed well in order to successfully complete the project. Similarly, under the context where all smart technologies are to be used in projects, it is important for the PM to have at least a basic understanding of each smart technology including their potential use cases and limitations and how each smart technology can be used to complete the pre-determined work task. Hence, equal weightages have been allocated to knowledge in each smart technology. While the KBTCAIS allocated equal weightages to knowledge in each smart technology, it is recognised that different organisations and projects will require a different level of proficiency in each technology. Hence, organisations may adjust the weightages of knowledge in each smart technology according to organisational needs within the KBTCAIS.

Next, project management skills, problem-solving skills and planning and organising skills were found to be the three most important skills in determining one's technological competency to manage projects with smart technologies. This is consistent with the survey findings where these skills were found to be most important in managing construction projects, with or without smart technologies as these are key task-related skills required of PMs to manage projects (Ahadzie et al.,

Table 12.2 Finalised Weightages of Technological Competency Factors

Hierarchy Level 2: Components of Technological Competency	Finalised Weightage (%)	Hierarchy Level 3: Sub-Components of Attitude Towards Technologies	Finalised Weightage (%)	Hierarchy Level 4: Factors Affecting Technological Competency	Finalised Weightage (%)
Knowledge in Project Management	30.02	NA	NA	Project integration management	10.00
				Project scope management	10.00
				Project schedule management	10.00
				Project cost management	10.00
				Project quality management	10.00
				Project communication management	10.00
				Project procurement management	10.00
				Project risk management	10.00
				Project resource management	10.00
				Project stakeholder management	10.00
Knowledge in Smart Technologies	8.94	NA	NA	Cyber-physical system	9.09
				Internet-of-Things	9.09
				Big data	9.09
				Artificial intelligence	9.09
				Robotics	9.09
				Autonomous vehicle	9.09
				Augmented reality	9.09
				Virtual reality	9.09
				Additive manufacturing	9.09
				3D imaging	9.09
				Distributed ledger technology	9.09
Skills to Manage Projects	23.37	NA	NA	Social awareness	2.47
				Organisational awareness	4.29
				Cultural awareness	1.85
				Ethical awareness	2.06

	Strategic planning skills			7.35
	Delegation skills			4.84
	Conflict management skills			6.17
	Motivation skills			3.96
	Negotiation skills			3.49
	Team-building skills			6.83
	Project management skills			10.00
	Technical and operational technology skills			4.42
	Information management skills			5.84
	Planning and organising skills			9.22
	Creativity			1.32
	Problem-solving skills			9.48
	Decision-making skills			8.68
	Active learning skills			1.36
	Communication skills			6.39
Personal Characteristics	Achievement orientation	27.67	NA	10.40
	Initiative		NA	11.18
	Information seeking			9.71
	Focus on client's needs			7.23
	Impact and influence			5.29
	Directiveness/assertiveness			7.31
	Teamwork			6.29
	Team leadership			7.64
	Analytical thinking			11.76
	Conceptual thinking			7.58
	Self-control			5.58
	Flexibility			10.03

(Continued)

Table 12.2 (Continued)

Hierarchy Level 2: Components of Technological Competency	Finalised Weightage (%)	Hierarchy Level 3: Sub-Components of Attitude Towards Technologies	Finalised Weightage (%)	Hierarchy Level 4: Factors Affecting Technological Competency	Finalised Weightage (%)
Attitude Towards Technologies	10.00	Technology self-efficacy	14.00	Ability to complete a job using smart technologies if you had never used a technology like it before	59.05
				Ability to complete a job using smart technologies if there was no one around to tell you what to do as you go	40.95
		Technology affect	11.26	You feel pleasant during the actual process of using smart technologies	45.54
				You like to work with smart technologies	54.46
		Technology anxiety	7.00	You feel intimidated by smart technologies	61.77
				You feel uncomfortable when using smart technologies	38.23
		Perceived ease of use of technology	25.64	Not a lot of mental effort is required when interacting with smart technologies	55.86
				It is easy to become skilful at using smart technologies	44.14
		Perceived usefulness of technology	27.54	You think that the use of smart technologies will enable you to accomplish job tasks quicker	60.42
				You think that the use of smart technologies will improve job performance	39.58
		Personal innovativeness in technology	14.56	You are usually first to try new technologies	38.23
				You enjoy experimenting with new technologies	61.77

2008; Alvarenga et al., 2019). This is followed by decision-making skills and strategic planning skills, which were found to be more important when managing projects with smart technologies due to the uncertainties and novelty of integrating the smart technologies into existing work processes (Sniazhko, 2019). Following the importance of the key task-related skills, soft skills, including team-building skills, communication skills and conflict management skills were found to also play relatively important roles in managing projects with smart technologies. This is in alignment with Ahadzie et al. (2008), who found that task-related behaviours contribute to about half of managerial performance while contextual behaviours contribute to about 30 per cent of managerial performance. While information management skills were found to be the second most important skill in managing projects with smart technologies from the survey, Interviewees A2, A6 and A9 shared that PMs should be able to manage the project and the project team and resolve problems using the outputs from the technologies. Thus, higher importance was placed on task-related skills compared to information management skill, although the ability to manage the information input and output of the technologies is also essential in order to support decision making. Similarly, while technical and operational technology skills are important in managing projects with smart technologies, Interviewees A1, A3 and A8 highlighted that the execution of the technologies in projects will largely be conducted by workers downstream. Hence, PMs need not be very proficient in operating the technologies. The findings highlight the importance of PMs' ability to interpret and utilise the output from the smart technologies and communicate with project stakeholders in order to improve the performance of projects (Bilal et al., 2016). Cultural awareness, active learning skills and creativity were found to have the lowest weightages in determining one's technological competency. This is consistent with the survey findings where cultural awareness and active learning were perceived to be least important in contributing to one's technological competency. Interviewee A6 highlighted that cultural awareness may not play a more critical role in managing projects with smart technologies as the existing level of technology maturity and organisational work processes do not support the formation of virtual project teams across different geographical locations. In addition, Interviewees A5 and A9 shared that active learning may be less important as extensive testing and experimentation of the technology will be conducted prior to the adoption in construction projects. Finally, although creativity was found to be more important in managing projects with smart technologies due to the relatively new way of work, Interviewees A5 and A9 noted that the integration of technologies in project works are typically tested and experimented by a dedicated team before being deployed into actual projects managed by PMs. Hence, it is reasonable that creativity plays a less active role in determining one's technological competency to manage projects with smart technologies.

Analytical thinking, initiative, achievement orientation, flexibility and information-seeking characteristics made up almost half of one's personal characteristics that contribute to one's technological competency to manage projects with smart technologies. While this finding is in contrast with Dainty et al. (2004), who found that composure and team leadership were most important in determining a construction PM's competency, Interviewee A2 highlighted the need for PMs to be able to troubleshoot problems when they arise during the course of work. This requires analytical thinking to systematically break down the problems faced into smaller parts to be resolved. The findings are consistent with Low et al. (2021), who argued that entrepreneurial thinking, pursuing convictions, vision, adaptability and curiosity were key competencies required of projects and facility management graduates to be future-ready in the era of 4IR. Specifically, in Low et al. (2021), the concept of entrepreneurial thinking is consistent with the concept of initiative, which refers to an entrepreneurial mind-set to spot and seize opportunities by working out solutions to existing problems and to improve performance of projects independently (Low et al., 2021). In the case of managing projects with smart technologies, initiative may be important for PMs to identify opportunities to apply the technologies to improve the performance of the project. In addition, Interviewee A4 shared that as problems are likely to arise due to the relative new way of works, PMs with initiative are more likely to anticipate the potential problems that may arise from the use of smart technologies and try to resolve them beforehand. Achievement orientation was also found to be essential in determining one's technological competency to manage projects with smart technologies. This coincides with pursuing convictions and vision, which refers to pursuing of one's beliefs and seeking to contribute and improve situations of oneself and others. It was highlighted that having vision is important in order to stand out from others, especially when AI has the abilities to self-learn and execute technical tasks (Low et al., 2021). This can differentiate a superior PM from an average PM when done in moderation and hence is important in determining one's technological competency to manage projects with smart technologies. Flexibility was also found to be important in managing smart technologies as the relatively new ways of work with the smart technologies may bring about unexpected issues and may require PMs to adjust thoughts, behaviours and plans according to the changing circumstances (Sirotiak et al., 2019; Teerajetgul & Chareonngam, 2008). Next, information-seeking characteristic is similar to curiosity, which was found to influence an individual's ability to develop solutions to overcome problems encountered, explore further in unfamiliar situations and to find out how one can do better in one's work performance (Low et al., 2021). Similarly, since the use of smart technologies in projects is still in its infancy, PMs may have to gather information on their own in terms of how to use the technologies, interpret the outcomes or resolve problems arising from the use of technologies. This was highlighted by Interviewee A4, who shared that PMs managing projects with smart technologies are frequently faced with unexpected problems arising from unfamiliarity with the technology, and hence need to gather information on the issues each project team member faced with the technology in order to resolve the problems encountered. The other personal characteristics are more focused on team leadership, the ability to think of the big picture, lead and influence others to achieve the objectives of the project. These are essential in the management of both conventional projects and projects with smart technologies, and hence may be perceived as less important by the interviewees in managing projects with smart technologies. Interviewee A2 highlighted that as projects are typically limited by time constraints, it is more important for PMs to be able to overcome the problems encountered before they can focus on leading the project team to effectively adopt the smart technologies as the PMs can then show by example on how the smart technologies can be used.

Lastly, the interviewees perceived the cognitive beliefs of technologies to play a greater role in determining one's technological competency. This was also highlighted by several interviewees who shared that workers are willing to use new technologies only when the technology can overcome a targeted problem at work and can be easily integrated into the existing workflows. Hence, more importance was placed on the PEU and usefulness of technologies. This is followed by the PIT and TSE of individuals. Interviewees A2, A7 and A8 shared that when one has high TSE, this will naturally make the user to be less anxious when using technologies and be more willing to experiment and use the technologies. Hence, a higher weightage was given to TSE and PIT compared to how one feels towards technologies. Furthermore, Interviewees A2, A5, A7, A8 and A9 noted that the time and cost pressures of projects put greater focus on how the technology can be used to improve the project performance and how easily it can be executed, thus resulting in a relatively lower importance of how one feels towards technologies in contributing to one's technological competency to manage projects with smart technologies.

12.5 Recommendations to Improve Technological Competency

To improve the level of technological competency of PMs to manage projects with smart technologies, recommendations to improve the technological competency were first developed through literature review and case studies. Learning theories formed the basis on which the recommendations were developed. There are nine key learning theories: (i) cognitive learning theory; (ii) behaviourism learning theory, (iii) constructivism learning theory, (iv) humanism learning theory, (v) connectivism learning theory, (vi) transformative learning theory, (vii) social learning theory, (viii) experiential learning theory and (ix) andragogy.

Cognitive learning theory understands learning as 'acquiring knowledge and skills, and having them in memory so that you can make sense of future problems and opportunities' (G. Brown, 2004); behaviourism learning theory views learning as the building of relationships between behaviours and responses which are shaped with reinforcement (G. Brown, 2004); constructivism learning theory understands learning as fitting new information into what the learner already knows, in alignment with one's experiences, beliefs and attitudes (Bada & Olusegun, 2015; Kimmons & Caskurlu, 2020); humanism learning theory views learners as persons with their own aspirations and believes that learners should have the freedom and responsibility to learn to achieve their own goals (G. Brown, 2004); connectivism learning theory views learning as knowledge that is acquired when the learner is connected to and participates in a learning community (Goldie, 2016; Kop & Hill, 2008); transformative learning theory understands learning as the development of a critical reflection in learners which transforms the learner's perspectives of reality (G. Brown, 2004; Mezirow, 2000); social learning theory considers learning through the interaction of environmental and cognitive factors, where learners learn by observing and modelling how others act and respond to the action (Bandura, 1986); experiential learning theory posits that learning occurs when learners learn by reflecting on their experience, develop a theoretical understanding of the experience and apply it in new situations (G. Brown, 2004); andragogy is focused on adult learning, where learning is self-directed and based on actual problems faced by adults and are perceived to be relevant to their needs (G. Brown, 2004).

Apart from the learning theories, Brown et al. (2016) studied learner control and e-learning and highlighted the importance of understanding the learner's level of motivation to learn and level of prior knowledge before designing a suitable training programme for the learner. The authors also recommended that learners with prior knowledge should be given more autonomy to decide the magnitude of control in e-learning. In addition, the authors differentiated the methods that are suitable for training for closed and open skills. Closed skills refer to recall of knowledge and skills which are required depending on the environmental information for subsequent action and is performed in a largely predictable environmental setting while open skills refer to the determination of a current situation through identification of general patterns of information and to establish if the state is different from what is desired, which is affected by the environment (K. G. Brown et al., 2016). Training of closed skills was recommended to be conducted in training environments that are similar to the transfer environment to minimise variability in employee behaviours while open skills are better trained when learners are first familiarised with the basic concepts and skills necessary, and adaptively structured to introduce the active learning of guided exploration (K. G. Brown et al., 2016). Post-training interventions such as encouraging practice and providing opportunities to perform should also be employed by organisations to ensure effective transfer behaviours (K. G. Brown et al., 2016).

In another study, Eby and Robertson (2020) studied the psychology of workplace mentoring relationships through attachment theory, interdependence theory, self-expansion theory, Rhode's model of formal youth mentoring and the working alliance. While workplace mentoring has been found to be essential in employee development and in improving the performance of employees, it is important to understand the relationships of workplace mentoring in order to maximise the benefits for both the mentor and the protégé (Eby & Robertson, 2020). First, the authors recommended that the goals of the mentorship need to first be clarified and how the mentor and protégé plan to achieve these goals should be established. Care must also be taken in arranging for the matches among the mentor and protégé, as too much similarity may limit the self-expanding potential of the mentorship (Eby & Robertson, 2020). Along this note, the unique skills and perspectives of individuals such as areas of expertise, organisational units or cultural and ethnic backgrounds can be taken into consideration when arranging for the match in mentoring programs (Eby & Robertson, 2020).

Next, Edmondson (1999) developed a model of team learning to understand how psychological safety and team efficacy can impact learning and performance in organisational work teams. The study found that learning behaviour is a significant predictor of team performance (Edmondson, 1999). In the study, learning behaviour consists of activities such as seeking feedback, sharing information and asking for help (Edmondson, 1999). Learning on the group level was conceptualised as a continual process of reflection and action, characterised by asking questions, seeking feedback, experimenting, reflecting and discussing the outcomes of actions (Edmondson, 1999). Performance was also found to benefit from seeking feedback by individual managers (Ashford & Tsui, 1991).

Loon et al. (2020) explores how human resource can facilitate the development of key needed capabilities for organisations to thrive in the new normal. The authors recommended that recruitment and selection of new hires should focus on creative abilities, analogical reasoning, having high disposition towards learning, ability to manage and operate in the new business model and experience in managing large-scale complex projects. While these recommendations target new hires, training for existing staff should focus on multi-skilling, technical abilities and cross-functional capabilities to develop flexibility and adaptability in existing staff (Loon et al., 2020). Apart from that, job redesigns and rotations can also be implemented to allow for autonomy and encourage collaboration across the organisation (Loon et al., 2020; Rouleau, 2005). The organisation should also align performance targets with rewards and developmental opportunities to encourage learning (Loon et al., 2020).

With a focus on motive acquisition, McClelland (1965) studied how achievement motive can be developed and found that who perceive a new motive or goal as being relevant and important to their self-image, in alignment with prevailing cultural values, or as a signifier of membership in a new reference group, are more likely to succeed in educational efforts to achieve that motive or goal. Linking the newly acquired knowledge or skill to existing related actions can improve the endurance and likelihood of changes in both thoughts and actions of the individual (McClelland, 1965). In addition, the author found that keeping records of progress towards the goal increased the likelihood that the newly formed motive can influence one's future thoughts and actions.

Based on the aforementioned, the preliminary list of recommendations to improve the level of technological competency of PMs can be summarised as follows:

- Goal setting with the PMs to identify areas for improvements
- Formal training courses for PMs to gain new knowledge or skills
- Case studies or project-based learning to apply new knowledge into simulated similar situations
- Simulation or games to gain new knowledge or skills
- Familiarisation with newly acquired knowledge or skills through practice, simulation and on-the-job training
- Job shadowing, on-the-job training and receiving mentoring to observe how specific knowledge and skills are applied in actual projects
- Cross-training of PMs to facilitate collaboration

- Seeking feedback and regular progress meeting to evaluate and reflect on progress
- Organisations involving PMs in job design
- Organisations creating an environment that will facilitate the use of technologies
- Creating a support network and cultivating a digital culture in the organisation

Goal setting with the PMs is in alignment with humanism learning theory and andragogy and with the recommendations of Eby and Robertson (2020), which requires the PM to set a goal to be achieved through the training program. This will be complemented by formal and informal training courses, case studies, project-based learning, simulation or games to allow PMs to gain new knowledge and skills. Cross-training of PMs follows the recommendations of Loon et al. (2020), to develop flexible and adaptable PMs that can collaborate with other project team members. Apart from these, some learners learn through observation (Bandura, 1986; G. Brown, 2004). Hence, job shadowing is another approach that can be undertaken by PMs to increase their level of technological competency. These newly acquired skills can then be reinforced through on-the-job training, so that PMs can be familiarised and understand the outcome of certain actions (Bandura, 1986; G. Brown, 2004). As highlighted by Eby and Robertson (2020), workplace mentoring is essential to improving the performance of employees. This is also supported by Interviewee P4, who emphasised that a softer coaching approach was critical in facilitating the digital transformation within his organisation. Next, having a feedback loop is an underlying characteristic of the learning theories and recommendations. Hence, it is an essential step for PMs to take in order to evaluate and reflect on their progress so that future learning may be adjusted. Organisations can also take a more active role by involving PMs in job design and creating an environment that encourages the use of technologies (Loon et al., 2020).

These recommendations are then categorised and tailored for the different level of technological competency. The recommendations are intended to help PMs improve their level of competency from a very low or low level of technological competency to a medium level and a medium level to a high level of technological competency and from a high to a very high level of technological competency. A very low or low level corresponds to quartile four and quartile three respectively, a medium level corresponds to quartile two in technological competency level and a high level corresponds to quartile one in technological competency level. The rationale behind having a combined set of recommendations for a very low and low level of technological competency was that improving the level of technological competency from very low to low may not be meaningful.

The preliminary set of recommendations was presented to 10 experts with relevant experience in managing projects and in smart technologies to solicit insightful comments and additional action plans. The recommendations were revised according to the feedback from the experts. Table 12.3 shows the finalised set of recommendations to improve PMs level of technological competency. The interviewees were of the view that the recommendations need not be broken down for each specific factor as the factors contribute to the same construct of competency and need

Table 12.3 Recommendations to Improve PMs' Level of Technological Competency

Technological Competency Component	Recommendation Level		
	Low to Medium (0 to 50%)	*Medium to High (51 to 75%)*	*High to Very High (76 to 100%)*
Overall	• Goal setting to understand the importance of knowledge, skills and personal characteristics and identify areas for improvements • Attend formal and hands-on training courses on project management, smart technologies, skills and personal characteristics in projects • Practice and familiarisation of project management processes, smart technologies and activities that demonstrate personal characteristics effective for project management through hands-on and on-the-job training, simulation, shadowing, mentorship and case-based learning • Shadowing to observe how more experienced colleagues manage projects and smart technologies in projects • On-the-job training with guidance and mentorship	• Goal setting to understand the importance of knowledge, skills and identify areas for improvements • Attend formal training courses on project management, smart technologies, skills and personal characteristics in projects • Familiarisation with project management processes, smart technologies, skills and activities that demonstrate personal characteristics effective for management through small-scale implementation, on-the-job training, mentorship and case-based learning • On-the-job training with guidance and mentorship	• Goal setting to understand the importance of knowledge, skills and personal characteristics and identify areas for improvements • Attend formal training and cross-training courses on project management, smart technologies, skills and personal characteristics in projects • Implementation in actual projects with no guidance • Provide mentorship to others
Knowledge in Project Management	• Goal setting to understand why project management knowledge is important and to identify key knowledge areas to work on • Attend formal training courses to pick up project management knowledge area processes and principles • Simulation or project management games to practice and familiarise yourself with project management processes	• Goal setting to identify key knowledge areas to work on and to specialise in • Attend formal training courses for intermediate project management processes and principles and best practices to apply the project management processes	• Goal setting to identify project management areas you like to specialise in • Pursue further studies on project management • Involvement in projects with autonomy and no guidance to manage actual projects

(Continued)

Table 12.3 (Continued)

Technological Competency Component	Recommendation Level		
	Low to Medium (0 to 50%)	Medium to High (51 to 75%)	High to Very High (76 to 100%)
	• Case studies or projects to learn the basic project management principles • Job shadowing to observe how project management is conducted by others • On-the-job training with guidance and mentorship to practice and familiarise yourself with project management processes • After the training, practice and familiarisation with project management processes, seek feedback on areas that have been done well and areas for improvement	• Pursue further studies on project management • Involvement in projects and on-the-job training with less guidance/more autonomy to apply learned project management processes into actual projects • Be mentored to learn from project management experts • Seek feedback on areas that have been done well and areas for improvement	• Serve as a mentor to others • Seek feedback on areas that have been done well and areas for improvement
Knowledge in smart technologies	• Goal setting to understand why knowledge in smart technologies is important and to identify key smart technologies to pick up • Attend formal training courses to learn about specified smart technologies including the basic applications in construction projects • Attend hands-on training courses to experience the use of the specified smart technologies • Simulation or games to learn about the basic use of the specified smart technologies • Case studies or projects to learn about the smart technologies in general • Shadowing to observe how the specified smart technologies are applied and managed in actual projects	• Goal setting to identify key smart technologies to pick up • Attend formal training workshops on the specified smart technologies • Pursue further studies in the specified smart technologies • Case studies or projects to learn more in depth about the specified smart technologies' use in project management and best practices • On-the-job training with guidance to familiarise yourself with the application of specified smart technologies in projects and how to manage them	• Goal setting to identify key smart technologies to specialise in • Attend training workshops or seminars on advanced use cases and operations of smart technologies in complex projects • Pursue further studies in the specified smart technologies • Case studies or projects that involve the use of smart technologies in complex project management and best practices • Implementation of specified smart technologies in actual projects • Provide mentorship to others

Competency			
(continued from previous page)			• Small-scale pilot testing of specified smart technologies to familiarise yourself with the use of the smart technologies • Be mentored by experts with experience in the specified smart technologies in projects
Skills to Manage Projects	• Goal setting to identify skill gap and skills to improve on • Attend formal training courses on identified skills including case studies and projects to apply identified skills and for familiarisation • Play simulation games that require the use of the identified skills to practice and familiarise yourself with the applications of the skills • Shadowing to observe how project management is conducted by more experienced colleagues • On-the-job training and mentorship to apply the identified skills in actual projects with guidance • Seek feedback on skills that have been applied well and skills to improve on	• Goal setting to identify skill gap and skills to improve on • Attend training workshops for intermediate application of skills required in project management and best practices • Cross-training in other disciplines to facilitate interdisciplinary collaboration in projects • On-the-job training with less guidance to apply the skills in actual projects • Be mentored to learn from experts in project management • Seek feedback on skills that have been applied well and skills to improve on	• Goal setting to identify skills for improvement • Attend training workshops on applications of the identified skills in complex projects • Pursue further studies on project management • Cross-training in other disciplines to facilitate interdisciplinary collaboration in projects • Involvement in projects with no guidance and autonomy to manage projects • Provide mentorship to others • Seek feedback on skills that have been applied well and skills to improve on
Personal Characteristics	• Goal setting to identify career path and personal characteristics to work on • Attend formal training courses on the identified personal characteristics and how they can be applied in project management • Be mentored to learn from experts in project management	• Goal setting to identify career path and personal characteristics to work on • Attend training workshops on the identified personal characteristics and how they can be applied in project management	• Goal setting to identify career path and personal characteristics to work on • Attend training workshops on the identified personal characteristics and how they can be applied in the management of complex projects

(Continued)

Table 12.3 (Continued)

Technological Competency Component	Recommendation Level		
	Low to Medium (0 to 50%)	*Medium to High (51 to 75%)*	*High to Very High (76 to 100%)*
	• Job shadowing to observe how project management is conducted by others and the demonstration of personal characteristics in project management • On-the-job training with guidance and mentorship to practice and familiarise yourself with demonstration of personal characteristics that are effective for project management • Seek feedback on personal characteristics that have been applied well in project management and personal characteristics to improve on	• Cross-training in other disciplines to facilitate interdisciplinary collaboration in projects • On-the-job training with less guidance and mentorship to practice and familiarise yourself with demonstration of personal characteristics that are effective for project management • Seek feedback on personal characteristics that have been applied well in project management and personal characteristics to improve on	• Cross-training in other disciplines to facilitate collaboration in projects • Involvement in projects with no guidance and autonomy to manage projects • Provide mentorship to others • Seek feedback on personal characteristics that have been applied well in project management and personal characteristics to improve on
Attitude towards technologies	• Goal setting to understand why it is important to adopt smart technologies and identify aspects to work on • Attend formal training courses on specified smart technologies • Attend hands-on training courses on the specified smart technologies • Case studies or projects to learn about smart technologies in general • Simulation to practice and familiarise yourself with smart technologies • Shadowing to observe how the specified smart technologies are applied and managed in actual projects	• Goal setting to identify career path and identify aspects to work on • Attend training workshops on more in-depth knowledge of the specified smart technologies • Cross-training in other disciplines to facilitate interdisciplinary collaboration in projects • On-the-job training with guidance and mentorship to familiarise yourself with the application of specified smart technologies in projects and how to manage them	• Goal setting to identify career path and identify aspects to work on • Cross-training in other disciplines to facilitate interdisciplinary collaboration in projects • Implementation of specified smart technologies in actual projects • Involvement in projects with no guidance and autonomy to manage projects

- On-the-job training with guidance and mentorship to familiarise yourself with the application of specified smart technologies in projects and how to manage them
- Seek feedback on aspects of attitude towards technologies that are effective to manage projects with technologies and areas to improve on

- Small-scale pilot testing of specified smart technologies to familiarise yourself with the use of the smart technologies
- Seek feedback on aspects of attitude towards technologies that are effective to manage projects with technologies and areas for improvement

- Provide mentorship to others
- Seek feedback on aspects of attitude towards technologies that are effective to manage projects with technologies and areas for improvement

Recommendations

For Organisations

Conduct goal setting with PMs. A goal-setting session should help to identify the PM's aspired career path and areas of interests, and establish goals for the PM's career development. This session should also provide a platform to understand the concerns of the PM and convey the purpose of the proposed training and development plan, especially in terms of digitalisation and uptake of technologies. The intervals between each progress check and feedback session can also be established during this session.

Involve PM in job design. Considering the PM's strengths and weaknesses, job design may be conducted to utilise the PM's strengths. Some examples could be to assign more technologically competent PMs to manage projects with the use of technologies or to automate specific work processes that PMs are relatively weaker in.

Create an environment that facilitates use of technologies. Strong top-down leadership is required for an organisation to digitally transform. The organisation can also hire people with a digital mind-set and skills to integrate the use of technologies into everyday work processes and mentor the other workers who are less inclined towards technologies. This could provide a support network within the organisation. At the same time, it is important that the organisation encourages experimentation of technologies. This could be done through the provision of avenues for small-scale testing, allow for mistakes through a no-blame culture and to set up an incentive system that rewards employees to pick up new technology skills. The organisation can also provide support for PMs with high technological competency in terms of autonomy and resources to implement new technologies in their work.

A support network can play an essential role in creating a digital culture in the organisation. For those who are less inclined towards the use of technologies, a support network can serve as a platform to seek advice and help and to address their concerns in the use of technologies.

Regular progress check and feedback sessions should be conducted with the PMs to track their progress in learning and development and to address concerns of PMs faced during the course of training and development. Adjustments to the training and development plan can also be made in these regular progress checks.

Source: (Ashford & Tsui, 1991; Bada & Olusegun, 2015; Bandura, 1986; G. Brown, 2004; K. G. Brown et al., 2016; Eby & Robertson, 2020; Edmondson, 1999; Goldie, 2016; Kimmons & Caskurlu, 2020; Kop & Hill, 2008; Loon et al., 2020; McClelland, 1965)

to be used in synergy in order to successfully manage projects with smart technologies. Hence, the recommendations were consolidated for each component of technological competency. In addition, all interviewees highlighted the importance of the organisation's support and commitment to improving the technological competency of PMs in order for the recommendations to be effective and actionable.

12.6   Demonstration of the Knowledge-Based Technological Competency Analytics and Innovations System

The KBTCAIS is a Microsoft Excel–based tool, consisting of the introduction interface, user guide interface, assessment interface report interface and recommendations interface. The users will be required to have Microsoft Excel installed in their computer with the Windows operating system. To start, the user has to open the KBTCAIS macro-enabled file.

The introduction interface will be presented, providing the user with a brief introduction of the KBTCAIS and an input table to gather the user's name, job role, email and date of assessment. The user needs to click 'Next' to be directed to the user guide interface. The introduction interface is shown in Figure 12.3.

Next, the user guide interface is shown to the user. Some examples of the contents that can be found in the user guide interface are shown in Figures 12.4 and 12.5. The user guide interface provides more details on technological competency and its components and instructions on how to use the KBTCAIS.

The user will be guided to the input rating interface after clicking on 'Next'. The input rating interface allows users to rate their level of proficiency in the identified factors of affecting technological competency in PMs, categorised by the technological competency component. The input rating interface is shown in Figures 12.6 to 12.10. Each factor is rated on a 7-point Likert scale, with the description of the levels provided in the description box on the right, which will be changed when the cells relating to the factor are clicked on by the user. The user will click on the cell that corresponds to the description that describes the user's level of proficiency in the identified factor best. Upon completion, the user should click 'Submit' to submit the assessment. The user will then be directed to the output reporting interface.

In the case where not all factors have been assessed by the user, the user will receive a pop-up notification to notify the user that the assessment is incomplete. The user will be redirected to the input rating interface upon clicking 'OK'.

In the output reporting interface, the user receives the report of his results on the assessed technological competency level and the overall score. The first page includes a cover page with the user's name, job role, email and date of assessment. Next, a bar chart is provided to visualise the level of competency in each technological competency component and the corresponding quartile of technological competency. Each technological component is colour coded. Below the bar chart are shown the top factors under each technological competency component the user should focus on to improve the level of technological competency, sorted from the lowest scored to highest score factor. The breakdown of the assessment responses is also provided in the output reporting interface. An example of the report is shown in Figure 12.11.

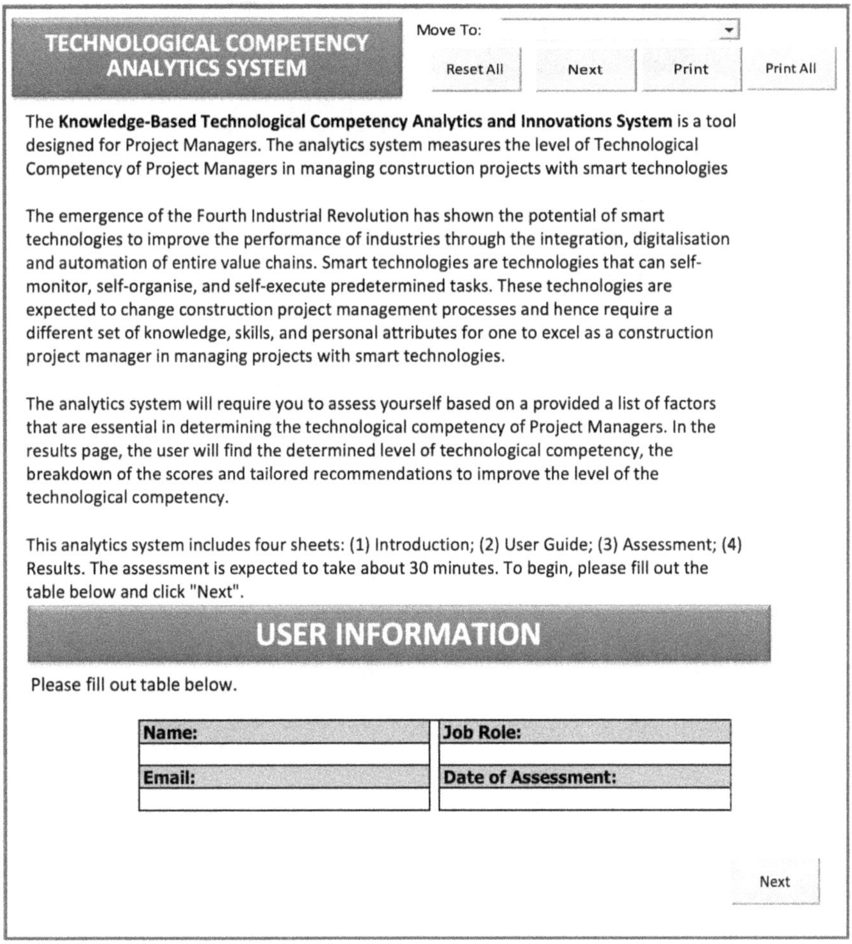

Figure 12.3 Introduction Interface of the KBTCAIS.

The user should click 'Next' to go to the recommendations interface. The recommendations interface first shows the recommendation report page and provides the user with the recommendations to improve the overall technological competency, knowledge in project management, knowledge in smart technologies, skills to manage projects, personal characteristics, attitude towards technologies and recommendations for the organisations. The recommendations are extracted based on the quartile and score of the assessed overall technological competency level and for each component. A sample of the recommendation interface is shown in Figure 12.12. The user may choose to print the report by clicking on the 'Print' or 'Print All' buttons.

USER GUIDE

Move To: | Home | Previous | Next | Print | Print All

Knowledge-Based Technological Competency Analytics and Innovations System

Introduction and Purpose

The Knowledge-Based Technological Competency Analytics and Innovations System is an Excel based platform designed to assess the level of technological competency of Project Managers in managing construction projects with smart technologies. It can be used to help Project Managers and their organisations in the construction industry for the following purposes:

1. To develop a training and develop plan for project managers to improve their level of technological competency
2. Determine the most suitable project managers for construction projects of different technological complexities

Background Information

In order to use the analytics system effectively, it is important to understand the concept of technological competency and the key smart technologies associated with the Fourth Industrial Revolution.

Technological Competency

Technological competency refers to the underlying characteristics of an individual that results in superior work performance. Technological competency of project managers in managing construction projects with smart technologies consists of five main components:

(i) Knowledge in project management: Information that an individual has in project management principles and processes

(ii) Knowledge in smart technologies: Information that an individual has in smart technologies

(iii) Skills in managing construction projects: Ability of the individual to perform a certain task to manage construction projects

(iv) Personal characteristics: The individual's underlying personal characteristics (for e.g. personality traits) that result in superior work performance

(v) Attitude towards technologies: How an individual feels towards technologies and their beliefs of technologies

Smart Technologies

Smart technologies are technologies that can self-monitor, self-organise, and self-execute predetermined tasks. They can be used in synergy to digitalise, automate and integrate entire value chains, eventually improving the performance of industries. The key smart technologies that can influence construction project management significantly include cyber-physical system, Internet-of-Things, big data, artificial intelligence, robotics, autonomous vehicles, augmented reality, virtual reality, additive manufacturing, 3D imaging, and distributed ledger technology.

Content of the Knowledge-Based Technological Competency Analytics and Innovations System

Upon entering the system, the user will find the following five sheets:

1	2	3	4	5
Introduction	User Guide	Input Rating	Output Reporting	Recommendations

1. Introduction: this sheet contains an introduction to th system as well as a user information table. The user information table should be completed accordingly.
2. User Guide: this sheet provides the user instructions on how to use the system and output information.
3. Input Rating: this primary sheet is where the user will score each of the technological competency determinants
4. Output Reporting: this sheet provides the user with the assessment of the technological competency level
5. Recommendations: this sheet provides the user with the tailored recommendations to improve the user's technological competency level

Scoring

There is a built-in algorithm in the tool that calculates the technological competency of the project manager. Each factor is assigned a weightage after consultation with industry experts. Based on the self-assessed level of competency of the users and the pre-determined factor weightages, the user's technological competency level will be calculated. The assessed technological competency will be categorised into "Quartile 1", "Quartile 2", "Quartile 3", and "Quartile 4", with Quartile 1 being the most technologically competent and Quartile 4 being the least technologically competent. The pre-determined weightages can be found in the "Data Source" sheet and can be modified if required.

Figure 12.4 User Guide – Background Information.

Using the Tool

Step 1. After reviewing the Introduction and User Guide, please proceed to the Input Rating sheet in the system. This sheet is organised by the different technological competency components. The subfactors under each component are indicated by a combination of letters and numbers (i.e. KPM1 Project Integration Management). A brief description of the subfactor is also provided.

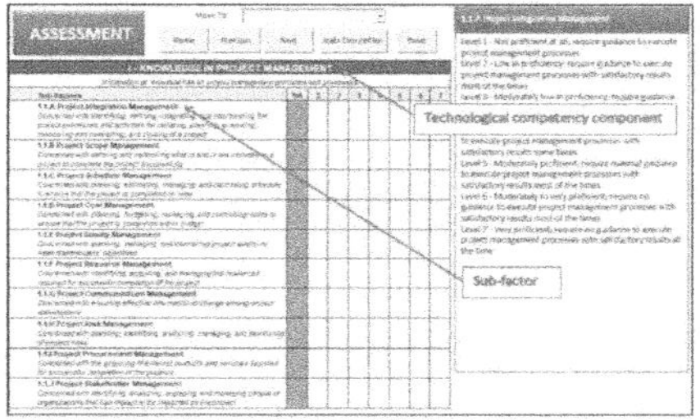

Step 2. Beside the description column, there are seven columns designated for scoring each sub-factor. The corresponding level of competency for each sub-factor can be measured against the statements provided in the dialogue box. The dialogue box will be displayed when the sub-task cell is clicked on.

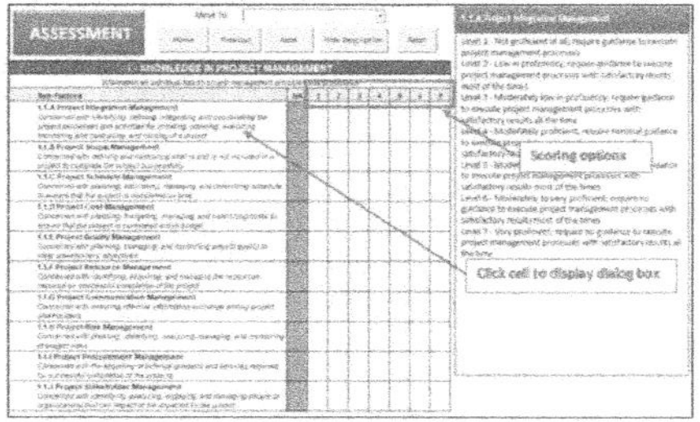

Step 3. After scoring the subfactors for all five technological competency components, the user may click on the "Submit" button. The user will be redirected to the Output Reporting sheet as seen below in Figure 4.

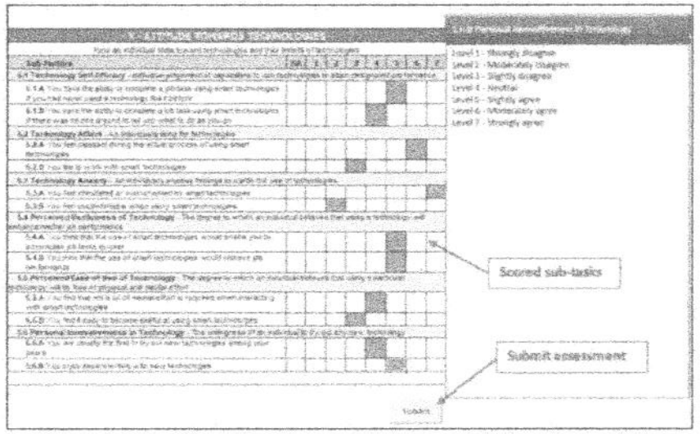

Step 4. The Output Reporting sheet displays the user's assessed overall technological competency level and the competency level of each technological competency component. The top sub-factors under each technological competency component that should be improved on are also highlighted in this sheet. Following that, a breakdown of the responses provided by the user is summarized.

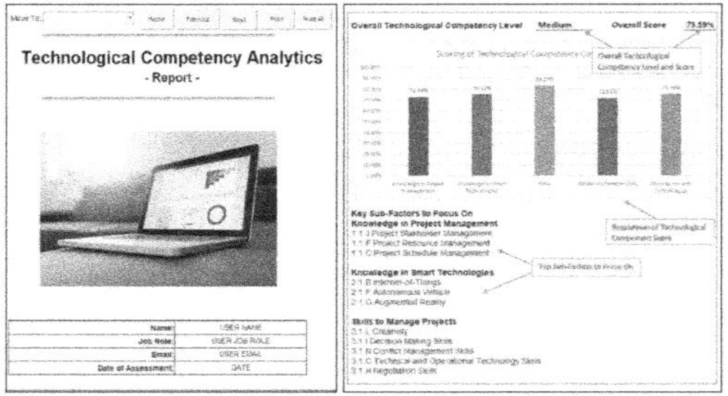

Step 5. In the Recommendations sheet, the user will receive tailored recommendations to improve the user's technological competency level. Recommendations will also be provided for the organizations towards the end of the sheet.

Figure 12.5 User Guide – Steps.

ASSESSMENT　Home　Previous　**Next**　Hide Description　Reset

1.1.F Project Resource Management

Level 1 - Not proficient at all; require guidance to execute project management processes
Level 2 - Low in proficiency; require guidance to execute project management processes with satisfactory results most of the times
Level 3 - Moderately low in proficiency; require guidance to execute project management processes with satisfactory results all the time
Level 4 - Moderately proficient; require minimal guidance to execute project management processes with satisfactory results some times
Level 5 - Moderately proficient; require minimal guidance to execute project management processes with satisfactory results most of the times
Level 6 - Moderately to very proficient; require no guidance to execute project management processes with satisfactory results most of the times
Level 7 - Very proficient; require no guidance to execute project management processes with satisfactory results all the time

I - KNOWLEDGE IN PROJECT MANAGEMENT
Information an individual has on project management principles and processes

Sub-Factors	NA	1	2	3	4	5	6	7
1.1.A Project Integration Management *Concerned with identifying, defining, integrating, and coordinating the project processes and activities for initiating, planning, executing, monitoring and controlling, and closing of a project*								
1.1.B Project Scope Management *Concerned with defining and controlling what is and is not included in a project to complete the project successfully*								
1.1.C Project Schedule Management *Concerned with planning, estimating, managing, and controlling schedule to ensure that the project is completed on time*								
1.1.D Project Cost Management *Concerned with planning, budgeting, managing, and controlling costs to ensure that the project is completed within budget*								
1.1.E Project Quality Management *Concerned with planning, managing, and controlling project quality to meet stakeholders' objectives*								
1.1.F Project Resource Management *Concerned with identifying, acquiring, and managing the resources required for successful completion of the project*								
1.1.G Project Communication Management *Concerned with ensuring effective information exchange among project stakeholders*								
1.1.H Project Risk Management *Concerned with planning, identifying, analyzing, managing, and monitoring of project risks*								
1.1.I Project Procurement Management *Concerned with the acquiring of external products and services required for successful completion of the projects*								
1.1.J Project Stakeholder Management *Concerned with identifying, analyzing, engaging, and managing people or organizations that can impact or be impacted by the project*								

Figure 12.6 Assessment for Knowledge in Project Management.

2.1.H Virtual Reality

Level 1 - Not proficient at all; have not heard of the smart technology before and require guidance to execute work processes with the smart technology
Level 2 - Low in proficiency; have basic knowledge of what the smart technology is and require guidance to execute work processes with the smart technology
Level 3 - Moderately low in proficiency; require guidance to execute work tasks with the smart technology with satisfactory results most of the times
Level 4 - Slightly proficient; require minimal guidance to execute work tasks with the smart technology with satisfactory results some times
Level 5 - Moderately proficient; require minimal guidance to execute work tasks with the smart technology with satisfactory results most of the times
Level 6 - Moderately to very proficient; require no guidance to execute work tasks with the smart technology with satisfactory results most of the times
Level 7 - Very proficient; require no guidance to execute work processes with the smart technology with satisfactory results all the time

II - KNOWLEDGE IN SMART TECHNOLOGIES
Information that an individual has in smart technologies

Sub-Factors	NA	1	2	3	4	5	6	7
2.1.A Cyber-Physical System *A system that integrates the physical and virtual environments to form situation-integrated analytical system that can respond intelligently to dynamic changes of real world conditions*								
2.1.B Internet-of-Things *A system that integrated various devices equipped with sensing, identification processing, communication and networking capabilities that can be monitored and controlled in real-time through the Internet*								
2.1.C Big Data *Software tools and techniques that capture, store, manage, analyze and visualize huge volume, velocity and variety of data*								
2.1.D Artificial Intelligence *Machines that simulate human intelligence processes that drives the optimization of processes and self-adaptation of machines*								
2.1.E Robotics *A form of machine designed and programmed to execute one or more tasks with little or no human intervention (for e.g. painting robots)*								
2.1.F Autonomous Vehicle *Another form of machine designed and programmed to execute one or more tasks with little or no human intervention (for e.g. drones)*								
2.1.G Augmented Reality *Display of a real-world scene and virtual information into the user's view*								
2.1.H Virtual Reality *Display of an environment generated from a model to allow for user experience*								
2.1.I Additive Manufacturing/ 3D Printing *Process of building up successive laters of materials based on a computer-aided drawing model to make objects*								
2.1.J 3D Imaging *Collection of 3D geometric as-built information using technologies such as laser scanners or photogrammetry*								
2.1.K Distributed Ledger Technology/ Blockchain *A data structure consisting of an ordered list of blocks that contain transactions, logged and added in chronological order that is permanent*								

Figure 12.7 Assessment for Knowledge in Smart Technologies.

III – SKILLS IN MANAGING CONSTRUCTION PROJECTS								
Ability of an individual to perform a certain task to manage the project								
Sub-Factors	NA	1	2	3	4	5	6	7
3.1.A Project Management Skills *Skills required to manage the project to achieve project objectives (as highlighted in PMBOK)*								
3.1.B Planning and Organizing Skills *Skills required to plan, organize, and coordinate work tasks to achieve project objectives*								
3.1.C Technical and Operational Technology Skills *Skills required to use technology to accomplish job tasks*								
3.1.D Information Management Skills *Skills required to use technology to search, select, evaluate, and organize information to accomplish job tasks*								
3.1.E Strategic Planning Skills *Skills that are required to use the technology as a means to achieve project objectives*								
3.1.F Organizational Awareness *Skills to show understanding of organization systems and culture and ability to work within it*								
3.1.G Ethical Awareness *Skills to behave in a socially responsible way, demonstrating awareness and knowledge of legal and ethical aspects when using technologies*								
3.1.H Negotiation Skills *Ability to negotiate win-win agreements*								
3.1.I Decision Making Skills *Ability to identify key issues and pick the optimal solution with available information and alternatives in a timely manner*								
3.1.J Problem Solving Skills *Skills required to conceive, analyze, and reason project issues to be able to resolve them and make appropriate decisions to achieve the project objectives. This includes critical thinking, analytical thinking, and conceptual thinking skills.*								
3.1.K Active Learning Skills *Skills required to learn new knowledge effectively to be applied into the project*								
3.1.L Creativity *Skills required to generate new ideas or treat familiar ideas in a different way and transform such ideas into a novel way of doing things*								
3.1.M Cultural Awareness *Skills to show cultural understanding and respect other cultures*								
3.1.N Conflict Management Skills *Ability to negotiate and resolve disagreements in projects and resolve conflicts in a satisfactory manner*								
3.1.O Communication Skills *Skills required for information exchange, in an effective and accurate manner, and the ability to actively listen, persuade, and understand what others mean*								
3.1.P Motivation Skills *Ability to motivate the project team to accomplish project tasks to fulfil project objectives*								
3.1.Q Team Building Skills *Ability to build effective relationships with project team members and stakeholders, to enable the project team to work together cooperatively to achieve project objectives*								
3.1.R Social Awareness *Ability to read the emotions of others*								
3.1.S Delegation Skills *Ability to allocate project tasks to team members based on their strengths and weaknesses, and to provide opportunities for development*								

3.1.N Conflict Management Skills

Level 1 - Not proficient at all; require guidance in executing skill to manage project

Level 2 - Low in proficiency; require guidance in executing skill to manage project with satisfactory results most of the times

Level 3 - Moderately low in proficiency; require guidance in executing skill to manage project with satisfactory results all the time

Level 4 - Moderately proficient; require minimal guidance in executing skill to manage project with satisfactory results most of the times

Level 5 - Moderately proficient; require minimal guidance in executing skill to manage project with satisfactory results all the time

Level 6 - Moderately to very proficient; require no guidance in executing skill to manage project with satisfactory results most of the times

Level 7 - Very proficient; require no guidance in executing skill to manage project with satisfactory results all the time

Figure 12.8 Assessment for Skills to Manage Projects.

IV – PERSONAL CHARACTERISTICS								
An individual's underlying personal characteristics (for e.g. personality traits) that result in superior work performance								
Sub-Factors	NA	1	2	3	4	5	6	7
4.1.A Achievement Orientation *Concern for working well or for competing against a standard of excellence*								
4.1.B Initiative *Taking proactive actions to avert problems to improve project performance and prevent problems*								
4.1.C Information Seeking *Making an effort to get more information, and not accepting situations at face value*								
4.1.D Focus on Client's Needs *A desire to meet the needs of clients*								
4.1.E Impact and Influence *Ability to influence, persuade, and convince people to support their agenda and complete project tasks*								
4.1.F Directiveness/ Assertiveness *Ability to use personal power or power of the individual's position for the long term good of the project*								
4.1.G Teamwork *Ability to work cooperatively with the project team and project stakeholders*								
4.1.H Team Leadership *Ability to lead groups to achieve project objectives by providing direction, vision, alignment of project stakeholders toward project success through communication, the motivation of project team members to achieve project objectives, and mentoring of project team to develop team members*								
4.1.I Analytical Thinking *Ability to break down a situation and organize the parts of a problem in a systematic way*								
4.1.J Conceptual Thinking *Ability to understand a situation by identifying patterns between situations or key and underlying issues in complex situations*								
4.1.K Self-control *Ability to keep emotions under control including stress management, and when faced with hostility*								
4.1.L Flexibility *Ability to adapt one's thinking, attitude, or behaviour to changing situations*								

4.1.J Conceptual Thinking

Level 1 - Not at all like me; I think in a very definitive manner

Level 2 - Not a lot like me; I use "rules of thumb" and past experiences to identify problems. I identify the similarities between current and past situations.

Level 3 - A bit like me; I observe discrepancies, trends and interrelationships in data and identify crucial differences between current situation and past situations.

Level 4 - Somewhat like me; I apply past knowledge to different situations and apply learned concepts (such as root-cause analysis) to the situation appropriately

Level 5 - Quite like me; I put together ideas, issues and observations into a single concept and can identify a key issue in a complex situation

Level 6 - Very much like me; I identify relationships among complex data from seemingly unrelated areas

Level 7 - Exactly like me; I create new models or theories a complex situation and can reconcile discrepant data

Figure 12.9 Assessment for Personal Characteristics.

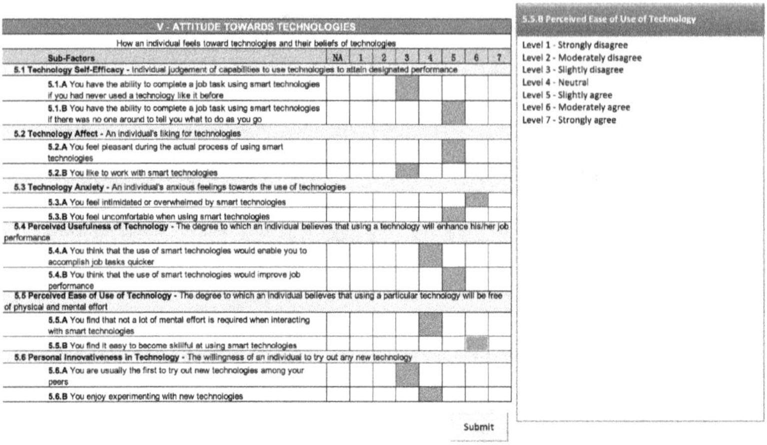

Figure 12.10 Assessment for Attitude towards Technologies.

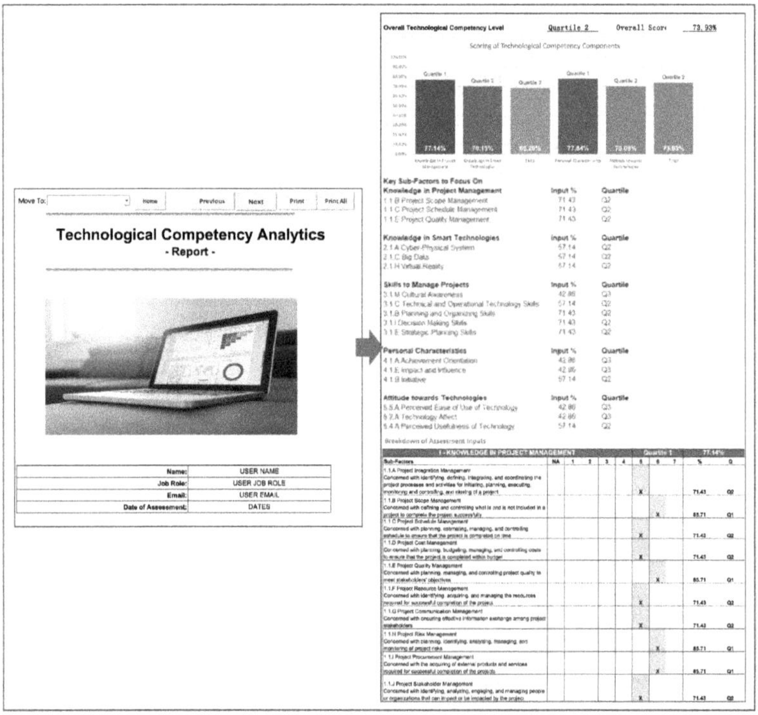

Figure 12.11 Technological Competency Report.

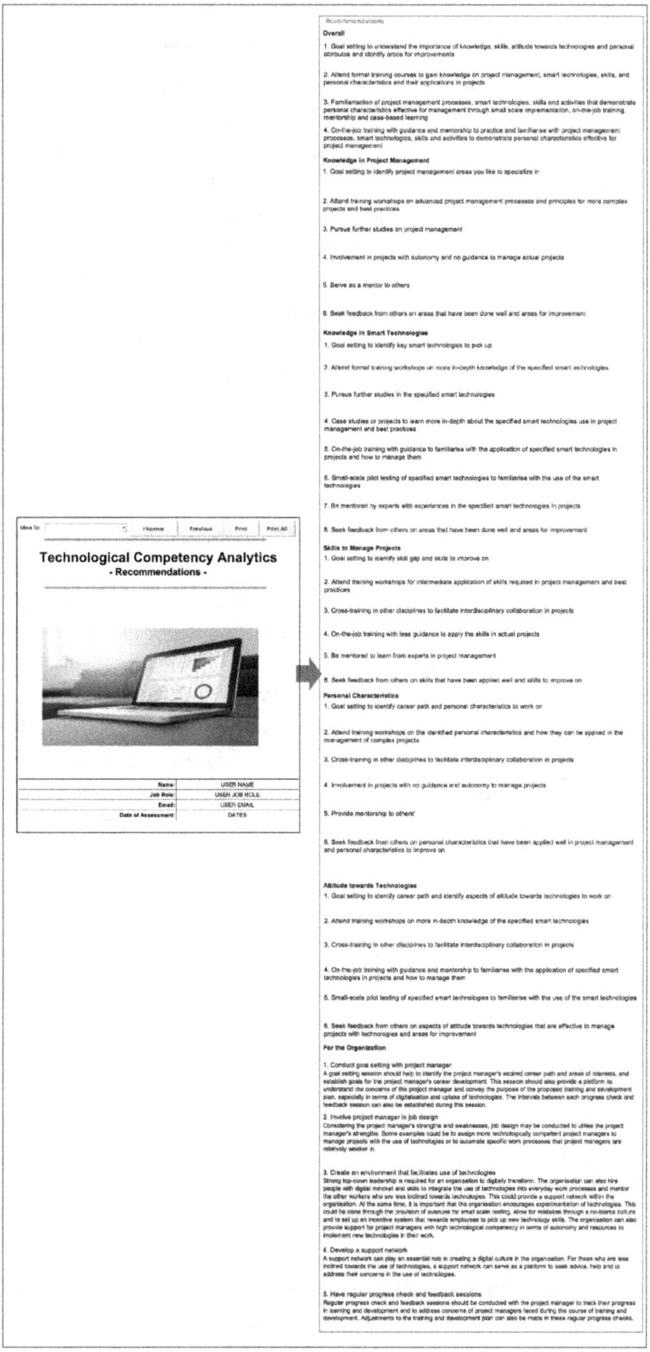

Figure 12.12 Tailored Recommendations for the User.

12.7   Validation of the Knowledge-Based Technological Competency Analytics and Innovations System

A computer-based tool should be validated to ensure that the tool has been developed to serve its purpose, is accurate and produces the same results consistently (Boehm, 1984; Huff & Sireci, 2001). In Huff and Sireci (2001), the authors found that validity of computer-based testing should acknowledge that (i) tests must be evaluated with respect to a particular purpose; (ii) the validation should be in respect of the inferences derived from the test scores; (iii) the evaluation of inferences made from the test score should comprise of qualitative and quantitative evidence; (iv) evaluation of the validity of inferences is a continuous process and (v) test users must evaluate the evidence for the validity of the inferences to ensure that it supports the purpose for which the test is used for.

The validation of the KBTCAIS consists of three steps. In the first step, the users were asked to self-assess their level of proficiency in each technological competency component and overall technological competency level. The user was then asked to use the KBTCAIS to perform the assessment in the second step. Upon completion of the assessment using the KBTCAIS, the user was asked to fill in his assessed technological competency level based on each technological competency component and overall technological competency level in step three. The user was also asked to evaluate the comprehensiveness of the factors presented in the KBTCAIS in assessing one's technological competency level, the usefulness and actionability of the recommendations provided by the KBTCAIS to improve one's technological competency level, the user friendliness of the KBTCAIS and whether the user is likely to use the KBTCAIS in his projects or organisation. This forms the qualitative evidence of the validity of the KBTCAIS. The quantitative evidence of the validity of the KBTCAIS is collected through the calculation of the PE, MPE and MAPE between the self-assessed and assessed technological competency levels. This approach follows Zhao et al. (2016) and Lim et al. (2012). The formulas of PE, MPE and MAPE are as follows:

$$PE = \left(\frac{\text{Actual value} - \text{forecasted value}}{\text{Actual value}} \right) \times 100\%$$

$$MPE = \frac{\sum PE_i}{n}$$

$$MAPE = \left| \frac{\sum PE_i}{n} \right|$$

where n is the number of experts interviewed.

The MPE was used to assess if the results of the KBTCAIS tends to be higher (positive sign) or lower (negative sign) when compared to the self-assessed technological competency level of experts, while MAPE indicates the magnitude of model errors. A lower MAPE indicates a lower magnitude of errors, and hence higher accuracy of the KBTCAIS. While self-assessment is typically associated with overrated assessment of one's ability, providing specific definitions for each factor was found to result in the gradual agreement of self-rated assessment and assessment results rated by others (Dunning et al., 2004). Specific definitions for each factor have been provided in KBTCAIS, minimising the risk of overestimation of one's ability. Furthermore, since the weightages of the factors are aggregated from expert interviews through AHP, the use of MPE and MAPE can further corroborate the results from the different experts in terms of the validity of the KBTCAIS.

Table 12.4 shows the summary of the PE, MPE and MAPE of the experts' technological competency assessment results. The PE values ranged from −16.08 per cent to 18.57 per cent while the MPE values ranged from −3.80 per cent to 5.34 per cent. The MPE results indicated that the model was likely to provide a higher assessment for the technological competency components compared to the self-assessment, except for attitude towards technologies. However, the magnitude of errors remains relatively low, demonstrating that the results from the KBTCAIS are fairly consistent with the user's assessed technological competency level. Next, the MAPE values ranged from 3.32 per cent to 9.02 per cent. This suggests that the accuracy of the KBTCAIS in assessing one's technological competency is higher than 90 per cent, which reflects a good validity of the KBTCAIS.

The interviewees were asked to evaluate the comprehensiveness of the factors presented in the KBTCAIS in assessing one's technological competency, the usefulness and actionability of the recommendations provided by the KBTCAIS to improve one's technological competency level and the user friendliness of the KBTCAIS and if they are likely to use KBTCAIS in their projects or organisation.

All 10 interviewees agree that the factors presented in the KBTCAIS were comprehensive enough for its purpose, to conduct a general assessment of the level of technological competency of PMs in managing projects with smart technologies. Interviewees V2, V5, V6 and V8 highlighted that different project types that use different smart technologies will require a different specific skillset, which needs to be broken down into the specific work processes. Besides that, Interviewees V2 and V6 shared that as the demographics of workers in the construction industry may not be fluent in English, the factors may sometimes be viewed as too comprehensive, as the interpreted meaning of certain factors such as the factors under TAFF or TANX may be perceived similarly. However, Interviewee V2 shared that the description provided under the factors can help in mitigating this risk, and that the factors provided are still relevant to the construct of technological competency.

Next, the 10 interviewees were also in consensus that the recommendations provided by the KBTCAIS are generally useful and actionable if organisation support is available. Interviewees V3 and V10 further highlighted that the usefulness of workplace mentoring depends greatly on the ability and the compatibility of the mentor and the protégé. Organisation support in terms of resources provided, goal setting and job design processes is also the cornerstone which determines if the recommendations are useful in improving one's technological competency level. In addition, Interviewees V2 and V8 agreed that the gradual increment of complexities and autonomy of applying the newly acquired knowledge and skills

Table 12.4 Summary of Percentage Errors in Technological Competency

Prediction Errors Between Self-Assessed and KBTCAIS Assessment Results

	Knowledge in Project Management (%)			Knowledge in Smart Technologies (%)			Skills to Manage Projects (%)			Personal Characteristics (%)			Attitude Towards Technologies (%)			Overall (%)		
	S	K	D	S	K	D	S	K	D	S	K	D	S	K	D	S	K	D
V1	90	94.29	4.29	50	40.26	-9.74	80	81.69	1.69	75	88.95	13.95	60	43.92	-16.08	80	80.00	0.00
V2	80	94.29	14.29	20	24.68	4.68	80	78.61	-1.39	75	83.80	8.80	40	34.89	-5.11	60	75.56	15.56
V3	75	78.57	3.57	40	38.96	-1.04	75	73.77	-1.23	75	72.41	-2.59	75	66.06	-8.94	70	70.95	0.95
V4	70	74.29	4.29	40	40.26	0.26	60	65.68	5.68	60	54.36	-5.64	70	69.27	-0.73	60	63.22	3.22
V5	80	90.00	10.00	51	49.35	-1.65	75	75.48	0.48	75	82.66	7.66	70	69.32	-0.68	75	78.87	3.87
V6	90	98.57	8.57	60	64.94	4.94	85	92.21	7.21	90	100.00	10.00	80	78.51	-1.49	80	92.47	12.47
V7	80	72.86	-7.14	20	29.87	9.87	70	61.48	8.52	70	64.76	-5.24	30	27.19	-2.81	60	59.55	-0.45
V8	100	100.00	0.00	25	25.97	0.97	80	80.11	0.11	80	75.04	-4.96	30	18.22	-11.78	70	73.65	3.65
V9	70	78.57	8.57	40	41.56	1.56	70	75.89	5.89	60	77.23	17.23	70	80.59	10.59	65	74.47	9.47
V10	60	64.29	4.29	10	28.57	18.57	60	58.97	-1.03	50	64.16	14.16	50	49.03	-0.97	55	58.29	3.29
Total MPE		5.07			2.84			0.89			5.34			-3.80			5.20	
Total MAPE		6.50			5.33			3.32			9.02			5.92			5.29	

D, difference between self-assessed and KBTCAIS-assessed results; K, KBTCAIS-assessed results; S, self-assessed results.

into projects, and the further improvement in technological competency level through mentoring others is an appropriate approach to improving the level of technological competency of PMs, instead of directly allocating the PMs to manage projects with smart technologies with no guidance and familiarisation phase provided.

The interviewees also agreed that the Microsoft Excel–based KBTCAIS is generally user friendly, except for Interviewee V6, who highlighted that the font size can be made bigger for ease of reading. However, Interviewee V6 noted that this can easily be solved by increasing the zoom size in Microsoft Excel. Other than that, Interviewees V1, V2, V5, V7 and V8 commended the ease of use enabled by the clicking of the corresponding cell, which reduces the time required for the assessment.

Overall, the 10 interviewees shared that they are likely to use KBTCAIS in their projects or organisations to have a general understanding of one's technological competency level. Interviewees V1 and V4 highlighted that possible use cases for KBTCAIS can include the allocation of suitable PMs for projects with different technological competency. On the other hand, Interviewees V2, V5, V6 and V8 shared that more specific knowledge, skillsets and their impact on work processes can be developed for each smart technology and project type. This can greatly help in the allocation of PMs to specific projects.

12.8 Summary

This chapter detailed the development of the KBTCAIS for technological competency of PMs in managing projects with smart technologies. Developed as a Microsoft Excel–based analytics system, the KBTCAIS can assess the technological competency of individual PMs, visualise the assessment results, provide tailored recommendations to improve technological competency of PMs and generate a printable technological competency assessment report. The KBTCAIS consists of a knowledge base, GUI and DSS. Within the knowledge base, the recommendations to improve the technological competency level of PMs were derived from literature review and case studies, which were further validated by experts. The factor weightages were also derived through the AHP using AtIP-GMM method. The developed KBTCAIS was then validated by 10 experts with relevant experience in the construction industry and smart technologies. Based on the validation results, the developed KBTCAIS was found to be valid and user friendly and the recommendations were useful and actionable.

References

Aguarón, J., Escobar, M. T., Moreno-Jiménez, J. M., & Turón, A. (2019). AHP-group decision making based on consistency. *Mathematics*, *7*(3), 242. https://doi.org/10.3390/math7030242

Aguarón, J., & Moreno-Jiménez, J. M. (2003). The geometric consistency index: Approximated thresholds. *European Journal of Operational Research*, *147*(1), 137–145. https://doi.org/10.1016/S0377-2217(02)00255-2

Ahadzie, D. K., Proverbs, D. G., & Olomolaiye, P. (2008). Towards developing competency-based measures for construction project managers: Should contextual behaviours be distinguished from task behaviours? *International Journal of Project Management*, *26*, 631–645. https://doi.org/10.1016/j.ijproman.2007.09.011

Alvarenga, J. C., Branco, R. R., Guedes, A. L. A., Soares, C. A. P., & E Silva, W. D. S. (2019). The project manager core competencies to project success. *International Journal of Managing Projects in Business*. https://doi.org/10.1108/IJMPB-12-2018-0274

Ashford, S. J., & Tsui, A. S. (1991). Self-regulation for managerial effectiveness: The role of active feedback seeking. *Academy of Management Journal*, *34*(2), 251–280. https://doi.org/10.5465/256442

Bada, S. O., & Olusegun, S. (2015). Constructivism learning theory: A paradigm for teaching and learning. *IOSR Journal of Research & Method in Education*, *5*(6).

Bandura, A. (1986). *Social foundations of thought and action: Social cognitive theory*. Prentice Hall.

Bilal, M., Oyedele, L. O., Qadir, J., Munir, K., Ajayi, S. O., Akinade, O. O., Owolabi, H. A., Alaka, H. A., & Pasha, M. (2016). Big Data in the construction industry: A review of present status, opportunities, and future trends. *Advanced Engineering Informatics*, *30*, 500–521. https://doi.org/10.1016/j.aei.2016.07.001

Boehm, B. W. (1984). Verifying and validating software requirements and design specifications. *IEEE Software*, *1*(1), 75–88. https://doi.org/10.1109/MS.1984.233702

Brown, G. (2004). How do students learn? In *Giving a lecture*. https://www.researchgate.net/publication/299402228_How_do_students_learn

Brown, K. G., Howardson, G., & Fisher, S. L. (2016). Learner control and e-learning: Taking stock and moving forward. *Annual Review of Organizational Psychology and Organizational Behavior*, *3*(1), 267–291. https://doi.org/10.1146/annurev-orgpsych-041015-062344

Dainty, A. R. J., Cheng, M. I., & Moore, D. R. (2004). A competency-based performance model for construction project managers. *Construction Management and Economics*, *22*(8), 877–886. https://doi.org/10.1080/0144619042000202726

Dunning, D., Heath, C., & Suls, J. M. (2004). Flawed self-assessment implications for health, education, and the workplace. *Psychological Science in the Public Interest, Supplement*, *5*(3), 69–106. https://doi.org/10.1111/j.1529-1006.2004.00018.x

Eby, L. T., & Robertson, M. M. (2020). The psychology of workplace mentoring relationships. *Annual Review of Organizational Psychology and Organizational Behavior*, *7*, 75–100. https://doi.org/10.1146/annurev-orgpsych-012119-044924

Edmondson, A. (1999). Psychological safety and learning behavior in work teams. *Administrative Science Quarterly*, *44*(2), 350–383. https://doi.org/10.2307/2666999

Forman, E., & Peniwati, K. (1998). Aggregating individual judgments and priorities with the analytic hierarchy process. *European Journal of Operational Research*, *108*(1), 165–169. https://doi.org/10.1016/S0377-2217(97)00244-0

Goldie, J. G. S. (2016). Connectivism: A knowledge learning theory for the digital age? *Medical Teacher*, *38*(10), 1064–1069. https://doi.org/10.3109/0142159X.2016.1173661

Huff, K. L., & Sireci, S. G. (2001). Validity issues in computer-based testing. *Educational Measurement: Issues and Practice*, *20*(3), 16–25. https://doi.org/10.1111/J.1745-3992.2001.TB00066.X

Hwang, B.-G., Shan, M., & Looi, K.-Y. (2018). Knowledge-based decision support system for prefabricated prefinished volumetric construction. *Automation in Construction*, *94*, 168–178. https://doi.org/10.1016/j.autcon.2018.06.016

Kimmons, R., & Caskurlu, S. (2020). The students' guide to learning design and research. In *EdTechBook.org*. EdTech Books. https://edtechbooks.org/studentguide

Kop, R., & Hill, A. (2008). Connectivism: Learning theory of the future or vestige of the past? *International Review of Research in Open and Distance Learning*, *9*(3). https://doi.org/10.19173/irrodl.v9i3.523

Lim, B. T. H., Ling, F. Y. Y., Ibbs, C. W., Raphael, B., & Ofori, G. (2012). Mathematical models for predicting organizational flexibility of construction firms in Singapore. *Journal of Construction Engineering and Management, 138*(3), 361–375. https://doi.org/10.1061/(ASCE)CO.1943-7862.0000439

Loon, M., Otaye-Ebede, L., & Stewart, J. (2020). Thriving in the new normal: The HR microfoundations of capabilities for business model innovation. An integrated literature review. *Journal of Management Studies, 57*(3), 698–726. https://doi.org/10.1111/JOMS.12564

Low, S. P., Gao, S., & Ng, E. W. L. (2021). Future-ready project and facility management graduates in Singapore for Industry 4.0: Transforming mindsets and competencies. *Engineering, Construction and Architectural Management, 28*(1), 270–290. https://doi.org/10.1108/ECAM-08-2018-0322

McClelland, D. C. (1965). Toward a theory of motive acquisition. *The American Psychologist, 20,* 321–333. https://doi.org/10.1037/h0022225

Mezirow, J. (2000). Learning to think like an adult: Core concepts of transformation theory. In *Learning as transformation. Critical perspectives on a theory in progress* (pp. 3–33). Jossey-Bass.

Ngo, J., Hwang, B.-G., & Zhang, C. (2020). Factor-based big data and predictive analytics capability assessment tool for the construction industry. *Automation in Construction, 110,* 103042. https://doi.org/10.1016/j.autcon.2019.103042

Paton, S., & Hodgson, D. (2016). Project managers on the edge: Liminality and identity in the management of technical work. *New Technology, Work and Employment, 31*(1), 26–40. https://doi.org/10.1111/NTWE.12056

Rouleau, L. (2005). Micro-practices of strategic sensemaking and sensegiving: How middle managers interpret and sell change every day*. *Journal of Management Studies, 42*(7), 1413–1441. https://doi.org/10.1111/J.1467-6486.2005.00549.X

Saaty, T. L. (2006). There is no mathematical validity for using fuzzy number crunching in the analytic hierarchy process. *Journal of Systems Science and Systems Engineering, 15*(4), 457–464. https://doi.org/10.1007/s11518-006-5021-7

Sirotiak, T., Asce, M., & Sharma, A. (2019). *Problem-based learning for adaptability and management skills.* https://doi.org/10.1061/(ASCE)

Sniazhko, S. (2019). Uncertainty in decision-making: A review of the international business literature. *Cogent Business and Management, 6*(1). https://doi.org/10.1080/23311975.2019.1650692

Teerajetgul, W., & Chareonngam, C. (2008). Tacit knowledge utilization in Thai construction projects. *Journal of Knowledge Management, 12*(1), 164–174. https://doi.org/10.1108/13673270810852467

Zhao, X., Hwang, B.-G., & Low, S. P. (2016). An enterprise risk management knowledge-based decision support system for construction firms. *Engineering, Construction and Architectural Management, 23*(3), 369–384. https://doi.org/10.1108/ECAM-03-2015-0042

13 Conclusions and Recommendations

13.1 Research Findings and Conclusions

This research investigation aimed to answer two main research questions: (1) how construction project management will be impacted by smart technology applications in the 4IR? (2) how technologically competent PMs can be developed to manage projects with smart technologies to consistently deliver successful projects. To answer the research questions, we achieved the following three objectives: (i) assessing the changes to the knowledge and skillsets required to manage projects with smart technologies, (ii) developing a TCF for PMs to be competent in managing projects with smart technologies and (iii) developing a KBTCAIS to assess and improve the technological competency of PMs.

13.1.1 Changes to Knowledge and Skills Required of Project Managers

The first research objective of this book aims to assess the changes to the knowledge and skills required by PMs to manage projects with smart technologies. Through the survey with 97 respondents, it was found that top three project management knowledge areas required to manage projects with smart technologies were project communication management, project cost management and project risk management. Similar rank-order of the importance of the project management knowledge areas was found in managing conventional projects and projects with smart technologies. While the knowledge in project management remains important in both conventional projects and projects with smart technologies, project integration, stakeholder, scope and procurement management need to be emphasised further in order to effectively manage projects. The survey also found that problem-solving skills, information management skills, communication skills, project management skills and planning and organisation skills were the top skills required of PMs to manage projects with smart technologies. The survey found that technical and operational technology skills, information management skills, creativity, ethical awareness, strategic planning skills, active learning skills, decision-making skills and motivation skills have grown in importance when managing projects with smart technologies. While those skills were found to have grown in importance when managing projects with smart technologies, the rank-order of the skills required

DOI: 10.1201/9781003462231-13

to manage both conventional projects and projects with smart technologies were found to be similar. Thus, the first research objective was fulfilled.

13.1.2 Proposed Technological Competency Framework for Project Managers

The second research objective of this book was achieved through the development of the TCF for PMs in managing projects with smart technologies. Through the literature review, the key components of technological competency were identified to include knowledge in project management, knowledge in smart technologies, skills to manage projects, personal characteristics and attitude towards technologies. More specifically, this research posited that attitude towards technologies comprises of technology self-efficacy, technology affect, technology anxiety, perceived ease of use, perceived usefulness and personal innovativeness in technologies. The survey questionnaire aimed to identify the factors affecting one's attitude towards technologies. Furthermore, EFA was conducted to identify the factor groupings of the factors affecting attitude towards technologies. PLS-SEM was then conducted to investigate the interrelationships between the different sub-components which affect one's attitude towards technologies. Based on the survey findings and PLS-SEM results, the TCF was developed and validated by 10 experts. The TCF was used as the basis to develop the KBTCAIS. Hence, the second research objective was achieved.

13.1.3 Knowledge-Based Analytics and Innovations System for Technological Competency in Project Managers

The third research objective was to develop a KBTCAIS to assess and improve the technological competency of PMs. The KBTCAIS was developed based on the validated TCF. Through the literature review and case studies, 11 key recommendations to improve the level of technological competency have been identified and further tailored for the different technological competency levels. The developed KBTCAIS can assess a PM's technological competency and visualise the technological competency level of the PM. PMs will also receive recommendations to improve their technological competency level, according to their assessed technological competency level. The KBTCAIS was developed as a Microsoft Excel–based tool, consisting of a knowledge base, GUI and DSS. The knowledge base consisted of the factors affecting PMs' technological competency, the weightages of these factors and the identified recommendations to improve the technological competency level. The DSS can compute the technological competency level according to the weightages given to these factors. The DSS also selects the relevant recommendations to be provided to the user. The validation of the KBTCAIS reflected an accuracy of more than 90 per cent in the assessment of one's technological competency level. In addition, the validated results also indicated that the recommendations provided by the KBTCAIS to improve the level of technological competency were useful and actionable, with the support of the organisation. Hence, the third research objective was achieved.

13.2   Contributions of the Research

13.2.1   *Contributions to Knowledge*

With the advent and the rapid proliferation of the 4IR and its technologies, there has been increasing emphasis on understanding the impact of the smart technology applications on one's competency in various industries such as education, healthcare and manufacturing. In the context of the construction industry, existing studies typically focus on the applications of the smart technologies in specific areas of projects, or on improving the performance of the smart technologies. This book fills the gap and contributes to knowledge by providing a better understanding of the impact of smart technology applications on PMs' competency in order for PMs to be technologically competent to consistently deliver successful projects with smart technologies.

The first contribution to knowledge is achieved by providing an understanding of the changes to project management knowledge and skills required to manage projects with smart technologies. While the basics of project management knowledge and skills have been posited to remain the same, certain project management knowledge and skills were found to require more focus when managing projects with smart technologies compared to the management of conventional projects. This finding enriches existing literature on the impact of smart technology applications on construction project management.

The second contribution to knowledge is the proposed TCF for PMs. The TCF integrates the competency theory and technology acceptance models to develop the model of technological competency for PMs, extending the existing competency theory. The TCF also takes into account the key personal characteristics that have been found to result in superior work performance. Hence, the TCF is a holistic model encompassing one's knowledge, skills, attitude towards technologies and underlying personal characteristics that are essential for PMs to be successful in managing projects with smart technologies. Despite the ever-evolving landscape of new technologies, certain core attributes remain unchanged to a PM's technological competency, including the domain-specific knowledge and skills, an open-minded ATT adoption and personal traits that foster a willingness to learn, experiment and integrate emerging technologies. These constants form the foundation for technological competency, enabling PMs to navigate and adapt to the continuous advent of new technologies. This should be complemented with upskilling for PMs to be proficient in the knowledge and skills to manage specific new technologies.

This research also contributes to knowledge in terms of the proposed recommendations for PMs to improve their level of technological competency. This research applied the learning theories in the development of recommendations to improve the level of technological competency in PMs, demonstrating the applicability of learning theories to training and development programs of PMs.

13.2.2   *Contributions to the Practices*

On top of the contributions to knowledge, this research also contributes to the practices. Firstly, through the identification of the changes in project management knowledge and skills required to manage projects with smart technologies, practitioners can better understand the key knowledge and skills to focus on during digital transformation process. This can serve as guidelines for training and developing PMs who have the relevant knowledge and skills to manage projects with smart technologies.

The second contribution to the practices is the development of KBTCAIS. The KBTCAIS operationalises the concept of technological competency, allowing practitioners to assess their level of technological competency to manage projects with smart technologies through an easy-to-use Microsoft Excel–based tool. Additionally, the tailored recommendations supplied by the KBTCAIS function as a comprehensive guide for the training and professional development of PMs, facilitating their effective management of projects that integrate smart technologies. The recommendations also serve as a benchmark for organisations in the development of training programs for PMs. At the same time, the factors that contribute to PMs' technological competency provided in the KBTCAIS can provide organisations and individual PMs with the focal areas in which training efforts could be placed on. Users may also conduct the assessment on a regular basis to track their progress. Besides that, organisations may also use the KBTCAIS in the recruitment process to select PMs with the required technological competency level which corresponds to the technology complexity of the project. Similarly, organisations may also use the KBTCAIS to allocate the PMs with varying technological competency level to the corresponding projects with different technology complexities.

On a broader level, improvements in PMs' technological competency can have a social impact by reducing the skills disparity among PMs in managing projects with smart technologies. The findings can facilitate the digital transformation of the construction industry, ultimately improving the performance of the construction industry through the increased adoption of smart technologies.

Although this book is focused on the construction industry, the developed TCF and KBTCAIS can also be modified for other industries and geographical regions by adopting a similar research approach, further extending the contributions of this research.

13.3 Limitations

While the objectives of this book have been achieved, this study has some limitations. First, the factors influencing one's technological competency, as well as recommendations for enhancing PMs' technological competency, were identified through an exhaustive literature review and expert panel interviews, which may not remain comprehensive as time progresses.

Furthermore, as this research developed the TCF and KBTCAIS for managing projects with smart technologies in general, knowledge and skills required for specific project types or smart technologies were not integrated in the TCF and KBTCAIS.

Next, as the survey questionnaire solicited the subjective opinions of respondents, the responses may be subject to individual perceptions and past experiences within the construction industry. Nevertheless, the findings from this research contribute to both knowledge and the practices through the developed TCF and KBTCAIS to improve the level of technological competency of PMs in managing projects with smart technologies.

13.4   Recommendations

13.4.1   *Recommendations for Future Research*

This research sets a foundation for future research on technological competency of PMs in managing projects with smart technologies. Future research is recommended to be conducted in the following areas.

First, this research studied the general technological competency of PMs in managing smart technology-aided projects. However, given that different project types require varying knowledge, skills and personal attributes of PMs, further research could investigate more specific competencies. Accordingly, the impact of specific smart technologies on project management processes may also be investigated in future research in this area.

Next, future research can evaluate the long-term predictive validity of the developed KBTCAIS. As this research was conducted when the adoption of smart technologies in the construction industry is not widespread, the long-term predictive validity of the developed KBTCAIS could not be assessed. Hence, future research may be conducted to assess the long-term predictive validity of the KBTCAIS. In addition, the effectiveness of the recommendations to improve the level of technological competency of PMs may also be studied in the future.

Furthermore, future research can be conducted to develop a technological competency benchmarking system for PMs. This will encompass the establishment of a database containing the technological competency scores collected from a large number of PMs; the benchmarking system could be embedded into the KBTCAIS, which will then enable users to compare their technological competency level to their peers'.

Future research may also be conducted to understand the perceptions from different PMs and to develop tailored KBTCAIS with differentiated weightages for different job roles when more concrete use cases with defined processes are developed and deployed.

As new technologies are continuously being developed, future studies may also be conducted to develop an additional feature embedded in the KBTCAIS to automatically filter newly developed smart technologies that are applicable for construction projects using pre-determined criteria, so that the KBTCAIS can reflect new technological developments without requiring manual review of the technologies by experts.

Finally, it is recognised that smart technology applications will impact organisation processes, culture and systems. At the same time, governmental policies will also be required to change in order to create an environment that encourages technology adoption. Hence, future research may be conducted on the organisational processes, systems and readiness that are essential to develop technologically competent PMs. This can further improve the level of technological competency of PMs to consistently deliver successful projects with smart technologies. Another key focus area could be assessing the impact of smart technology applications on organisational and governmental processes and policies and how organisational and governmental processes and policies affect adoption of smart technologies by PMs.

13.4.2   *Recommendations for the Industry*

This work identified project management knowledge areas and skills that need to be emphasised in managing projects with smart technologies. Hence, industry practitioners may place more focus on these knowledge areas and skills. Specifically, more emphasis should be placed on project integration, stakeholder, scope and procurement management and technical and operational technology skills, information management skills, creativity, ethical awareness, strategic planning skills, active learning skills, decision-making skills and motivation skills for PMs to be competent in managing projects with smart technologies.

Furthermore, PMs may adopt the recommendations to improve their technological competency level to be prepared in the management of projects with smart technologies. In particular, PMs should set goals prior to the training by pinpointing the key areas for training. PMs with low technological competency levels can first observe how the management of projects with smart technologies is conducted through job shadowing and familiarise themselves with the processes involved. When the PM is more familiar with the processes, the learning can be reinforced by doing and getting regular feedback. More technologically competent PMs should provide mentorship to others and may specialise in specific areas to become an expert in managing projects with smart technologies.

Finally, there is a need for organisations to create a suitable organisation environment and culture that encourages technology adoption to improve the technological competency of PMs. This can be done through a top–down approach driving the digital transformation of the organisation, by creating incentive systems that encourage workers to pick up new technology skills, conduct rigorous pilot testing to instil confidence in the technologies and demonstrate the benefits of the smart technologies, as well as creating a support network within the organisation to encourage technology adoption. With these, individuals may be more inclined towards the use of technologies, which can motivate them to improve their technological competency level.

Appendices

Appendix A

Data Analysis Results

Appendix A-1 Summary of Data Pre-processing Results for Knowledge in Project Management

Project Management Knowledge Areas	For Conventional Projects				For Projects with at least one Smart Technology			
	Cronbach's Alpha	Cronbach's Alpha if item is deleted	Corrected total-item correlation	Shapiro-Wilk Test	Cronbach's Alpha	Cronbach's Alpha if item is deleted	Corrected total-item correlation	Shapiro-Wilk Test
Project Integration Management	0.945	0.917	0.817	0.001*	0.970	0.965	0.909	0.001*
Project Scope Management		0.940	0.749	0.001*		0.967	0.854	0.001*
Project Schedule Management		0.939	0.789	0.001*		0.967	0.844	0.001*
Project Cost Management		0.939	0.777	0.001*		0.965	0.900	0.001*
Project Quality Management		0.941	0.723	0.001*		0.966	0.870	0.001*
Project Resource Management		0.936	0.840	0.001*		0.966	0.866	0.001*
Project Communication Management		0.938	0.786	0.001*		0.967	0.845	0.001*
Project Risk Management		0.938	0.802	0.001*		0.966	0.858	0.001*
Project Procurement Management		0.940	0.747	0.001*		0.968	0.806	0.001*
Project Stakeholder Management		0.942	0.724	0.001*		0.967	0.842	0.001*

* Significant at p-value <0.05.

Appendix A-2 Summary of Data Pre-processing Results for Skills to Manage Projects

Skills to manage projects	For Conventional Projects				For Projects with at least one Smart Technology			
	Cronbach's Alpha	Cronbach's Alpha if item is deleted	Corrected total-item correlation	Shapiro-Wilk Test	Cronbach's Alpha	Cronbach's Alpha if item is deleted	Corrected total-item correlation	Shapiro-Wilk Test
Technical and operational technology skills		0.943	0.258	0.001*		0.959	0.607	0.001*
Project management skills		0.940	0.544	0.001*		0.958	0.685	0.001*
Information management skills		0.942	0.412	0.001*		0.959	0.536	0.001*
Planning and organising skills		0.940	0.515	0.001*		0.958	0.664	0.001*
Communication skills		0.940	0.543	0.001*		0.958	0.731	0.001*
Social awareness		0.938	0.685	0.001*		0.958	0.679	0.001*
Cultural awareness		0.938	0.721	0.001*		0.958	0.727	0.001*
Organisational awareness		0.939	0.613	0.001*		0.958	0.731	0.001*
Creativity		0.942	0.421	0.001*		0.959	0.602	0.001*
Problem-solving skills	0.942	0.942	0.411	0.001*	0.960	0.958	0.692	0.001*
Ethical awareness		0.939	0.607	0.001*		0.958	0.675	0.001*
Strategic planning skills		0.938	0.675	0.001*		0.957	0.772	0.001*
Active learning skills		0.945	0.239*	0.001*		0.961	0.405	0.001*
Conflict management skills		0.937	0.783	0.001*		0.957	0.754	0.001*
Decision-making skills		0.939	0.622	0.001*		0.957	0.765	0.001*
Delegation skills		0.938	0.703	0.001*		0.958	0.635	0.001*
Motivation skills		0.938	0.750	0.001*		0.958	0.688	0.001*
Negotiation skills		0.938	0.722	0.001*		0.959	0.561	0.001*
Team-building skills		0.937	0.709	0.001*		0.957	0.771	0.001*

* Significant at p-value <0.05.

Appendix A-3 Summary of Data Pre-processing Results for Attitude towards Technologies

Attitude Towards Technologies	Applicability Cronbach's Alpha	Cronbach's Alpha if item is deleted	Corrected total-item correlation	Shapiro-Wilk Test	Impact Cronbach's Alpha	Cronbach's Alpha if item is deleted	Corrected total-item correlation	Shapiro-Wilk Test
TSE1	0.825	0.857	0.095*	0.001*	0.903	0.926	0.143*	0.001*
TSE2		0.788	0.750	0.001*		0.883	0.825	0.001*
TSE3		0.788	0.720	0.001*		0.882	0.832	0.001*
TSE4		0.800	0.613	0.001*		0.884	0.817	0.001*
TSE5		0.787	0.736	0.001*		0.886	0.778	0.001*
TSE6		0.793	0.695	0.001*		0.890	0.723	0.001*
TSE7		0.785	0.754	0.001*		0.883	0.830	0.001*
TSE8		0.806	0.545	0.001*		0.888	0.747	0.001*
TSE9		0.876	0.028*	0.001*		0.915	0.325	0.001*
TSE10		0.790	0.695	0.001*		0.891	0.699	0.001*
TAFF1	0.880	0.852	0.768	0.001*	0.924	0.907	0.823	0.001*
TAFF2		0.863	0.668	0.001*		0.912	0.771	0.001*
TAFF3		0.875	0.568	0.001*		0.916	0.733	0.001*
TAFF4		0.858	0.709	0.001*		0.909	0.795	0.001*
TAFF5		0.859	0.698	0.001*		0.914	0.746	0.001*
TAFF6		0.873	0.610	0.001*		0.914	0.751	0.001*
TAFF7		0.862	0.677	0.001*		0.917	0.722	0.001*
TANX1	0.944	0.928	0.786	0.001*	0.948	0.943	0.776	0.001*
TANX2		0.946	0.591	0.001*		0.952	0.550	0.001*
TANX3		0.939	0.759	0.001*		0.941	0.837	0.001*
TANX4		0.933	0.871	0.001*		0.938	0.877	0.001*
TANX5		0.935	0.832	0.001*		0.943	0.768	0.001*
TANX6		0.933	0.882	0.001*		0.938	0.881	0.001*
TANX7		0.934	0.862	0.001*		0.941	0.825	0.001*
TANX8		0.935	0.843	0.001*		0.940	0.840	0.001*
TANX9		0.953	0.453	0.001*		0.949	0.628	0.001*
TANX10		0.935	0.843	0.001*		0.940	0.836	0.001*

PU1		0.958	0.916	0.001*		0.969	0.936	0.001*
PU2		0.959	0.918	0.001*		0.974	0.884	0.001*
PU3	0.967	0.964	0.863	0.001*		0.972	0.907	0.001*
PU4		0.958	0.914	0.001*	0.975	0.969	0.930	0.001*
PU5		0.962	0.883	0.001*		0.968	0.949	0.001*
PU6		0.961	0.887	0.001*		0.971	0.909	0.001*
PEU1		0.966	0.833	0.001*		0.959	0.891	0.001*
PEU2		0.959	0.917	0.001*		0.969	0.761	0.001*
PEU3		0.967	0.811	0.001*		0.961	0.867	0.001*
PEU4	0.967	0.963	0.858	0.001*	0.965	0.960	0.876	0.001*
PEU5		0.960	0.905	0.001*		0.958	0.903	0.001*
PEU6		0.959	0.926	0.001*		0.955	0.941	0.001*
PEU7		0.957	0.949	0.001*		0.958	0.907	0.001*
PIT1		0.764	0.744	0.001*		0.810	0.741	0.001*
PIT2	0.838	0.825	0.597	0.001*	0.861	0.835	0.676	0.001*
PIT3		0.818	0.640	0.001*		0.837	0.688	0.001*
PIT4		0.773	0.735	0.001*		0.810	0.743	0.001*

* Significant at *p*-value <0.05.

Appendix A-4 Respondent's Self-Assessed Level of Technological Competency

	Mean (1 = not technologically competent at all, 7 = very technologically competent)	Standard deviation	Median	Shapiro-Wilk	One sample Wilcoxon signed rank test
Respondent's self-assessed level of technological competency	3.07	0.181	3	0.001*	0.001*

* Significant at *p*-value <0.05.

Appendix A-5 Summary of Results for Perceived Importance of Knowledge in Project Management by Respondent Age

Project Management Knowledge Areas	25 to 34 years old			35 to 44 years old			45 to 54 years old			55 to 64 years old			Kruskal Wallis
	Mean	SD	Rank	Mean	SD	Rank	Mean	SD	Rank	Mean	SD	Rank	
Project Integration Management	6.31	0.918	2	6.00	1.304	6	6.06	1.289	5	6.55	0.934	6	0.488
Project Scope Management	6.06	0.966	9	5.90	1.261	8	5.88	1.204	8	6.27	1.191	9	0.615
Project Schedule Management	6.27	1.151	6	6.19	1.401	3	6.19	1.377	3	6.64	0.924	2	0.683
Project Cost Management	6.31	1.084	4	6.14	1.352	4	6.25	1.342	1	6.64	0.924	2	0.683
Project Quality Management	6.29	1.061	5	6.19	1.289	1	6.06	1.340	6	6.64	0.924	2	0.449
Project Resource Management	6.10	1.005	7	5.90	1.338	9	5.81	1.223	9	6.45	1.214	7	0.200
Project Communication Management	6.47	0.960	1	6.19	1.327	2	6.19	1.328	2	6.73	0.905	1	0.326
Project Risk Management	6.31	1.025	3	6.14	1.352	4	6.19	1.377	3	6.64	0.924	2	0.661
Project Procurement Management	5.86	1.061	10	5.71	1.271	10	5.81	1.223	9	6.00	1.414	10	0.838
Project Stakeholder Management	6.10	1.104	8	6.00	1.304	6	5.94	1.289	7	6.36	1.206	8	0.619

SD, standard deviation.

Appendix A-6 Summary of Results for the Perceived Importance of Skills to Manage Projects by Respondent Age

Skills to Manage Projects	25 to 34 years old			35 to 44 years old			45 to 54 years old			55 to 64 years old			Kruskal Wallis
	Mean	SD	Rank	Mean	SD	Rank	Mean	SD	Rank	Mean	SD	Rank	
Technical and Operational Technology Skills	5.71	1.118	12	6.14	0.964	3	5.87	1.088	6	6.36	1.120	4	0.178
Project Management Skills	6.16	0.921	4	5.95	1.244	7	6.19	0.911	1	6.27	1.104	7	0.863
Information Management Skills	6.27	1.114	2	6.10	1.261	4	5.94	1.389	5	6.27	1.009	6	0.865
Planning and Organising Skills	6.04	1.079	7	6.05	0.973	5	6.06	1.124	3	6.18	1.079	9	0.960
Communication Skills	6.24	0.990	3	6.24	0.944	2	5.75	1.612	8	6.09	0.944	10	0.809
Social Awareness	5.43	1.307	17	5.05	1.244	18	5.00	1.211	19	5.91	1.044	13	0.174
Cultural Awareness	5.22	1.177	18	4.95	1.024	19	5.13	1.147	17	5.45	1.293	19	0.763
Organisational Awareness	5.63	1.131	14	5.24	1.136	17	5.44	1.315	13	5.64	1.120	16	0.710
Creativity	5.82	1.054	11	5.81	0.873	9	5.56	1.209	10	6.27	0.905	5	0.414
Problem-Solving Skills	6.29	0.866	1	6.33	0.796	1	6.13	1.025	2	6.45	0.820	3	0.859
Ethical Awareness	5.57	1.258	15	5.57	1.363	14	5.38	1.500	14	6.09	0.944	10	0.597
Strategic Planning Skills	6.12	0.949	5	5.86	1.236	8	5.50	1.366	11	6.45	0.688	2	0.168
Active Learning Skills	5.18	1.202	19	5.33	1.065	16	5.31	1.352	15	5.55	1.368	17	0.796
Conflict Management Skills	6.00	0.979	9	5.62	1.244	11	5.81	1.377	7	6.55	0.820	1	0.158
Decision-Making Skills	6.12	1.148	6	6.00	1.414	6	5.94	1.340	4	6.09	1.136	12	0.992
Delegation Skills	5.65	1.110	13	5.57	1.326	13	5.19	1.167	16	5.91	1.136	14	0.441
Motivation Skills	5.90	1.159	10	5.48	1.078	15	5.50	1.438	12	5.55	1.635	18	0.472
Negotiation Skills	5.45	1.138	16	5.71	1.309	10	5.06	1.289	18	5.82	0.982	15	0.291
Team-Building Skills	6.02	1.145	8	5.62	1.359	12	5.69	1.352	9	6.27	1.104	7	0.391

SD, standard deviation.

Appendix A-7 Summary of Results for Perceived Differences in Factors Affecting Attitude towards Technologies by Respondent's Age

Factors Affecting Attitude Towards Technologies

Applicability

	25 to 34 Years old				35 to 44 years old				45 to 54 years old				55 to 64 years old				KW
	Mean	SD	RW	RA	Mean	SD	RW	RA	Mean	SD	RW	RA	Mean	SD	RW	RA	
TSE1	5.45	1.555	2	13	6.05	0.865	2	2	6.00	1.211	1	1	5.73	1.272	2	8	0.465
TSE2	5.16	1.048	7	24	5.33	1.017	6	28	5.06	0.854	6	32	5.64	1.027	4	12	0.466
TSE3	4.84	1.124	10	36	5.29	0.956	9	32	5.13	0.885	4	29	5.45	1.036	7	23	0.169
TSE4	5.27	1.095	4	20	5.43	0.978	4	24	5.06	0.72	5	31	5.64	0.924	3	11	0.427
TSE5	5.16	1.124	7	25	5.33	1.065	7	29	4.88	0.806	10	40	5.55	1.036	6	18	0.367
TSE6	5.27	1.016	4	19	5.38	1.024	5	27	4.94	0.854	9	38	5.36	0.809	9	27	0.565
TSE7	5.08	1.115	9	29	5.33	1.065	7	29	5.25	0856	3	25	5.64	1.027	4	12	0.337
TSE8	5.20	1.136	6	21	5.14	1.014	10	39	5.06	0.854	6	32	5.36	0.505	8	26	0.878
TSE9	5.37	1.728	3	16	6.38	0.973	1	1	5.94	1.181	2	2	5.09	1.921	10	32	0.079
TSE10	5.78	1.165	1	1	5.76	1.044	3	14	5.00	0.816	8	35	6.00	1.000	1	3	0.052
TAFF1	5.43	1.000	4	14	5.62	1.244	2	19	5.88	1.088	1	3	5.64	1.206	4	15	0.454
TAFF2	5.45	1.022	3	11	5.48	1.209	3	22	5.81	1.047	2	4	5.73	1.272	3	8	0.635
TAFF3	5.06	1.126	6	30	5.24	1.221	5	34	5.00	0.816	6	35	4.18	1.168	7	42	0.148
TAFF4	5.10	1.104	5	28	5.19	1.289	6	37	5.06	0.854	5	32	4.55	1.440	6	40	0.693
TAFF5	5.57	1.242	2	7	5.48	1.401	4	23	5.81	1.223	3	5	6.09	0.944	1	1	0.588
TAFF6	4.98	1.421	7	33	5.19	1.601	7	38	4.94	1.289	7	39	4.73	1.489	5	36	0.803
TAFF7	5.65	1.393	1	2	5.71	1.271	1	15	5.62	1.258	4	11	6.00	0.894	2	3	0.946
TANX1	4.65	1.110	7	41	5.05	1.465	8	42	4.50	1.211	8	42	4.18	0.751	8	41	0.336
TANX2	4.88	1.317	1	34	5.05	1.396	7	41	4.37	1.360	9	43	3.91	1.375	10	44	0.141
TANX3	4.76	1.392	4	38	4.90	1.546	9	43	4.88	1.408	7	41	4.00	1.265	9	43	0.446
TANX4	4.86	1.414	2	35	5.24	1.758	5	36	5.25	1.732	4	28	4.91	2.023	3	33	0.546
TANX5	4.37	1.439	9	43	5.24	1.609	4	35	5.00	1.713	6	37	4.82	1.722	5	35	0.127
TANX6	4.76	1.392	4	38	5.33	1.494	2	31	5.13	1.544	5	30	4.91	1.300	4	34	0.315
TANX7	4.78	1.403	3	37	5.43	1.690	1	26	5.31	1.493	1	24	4.73	1.555	6	37	0.178

(Continued)

Appendix A-7 (Continued)

Factors Affecting Attitude Towards Technologies

Applicability

	25 to 34 Years old				35 to 44 years old				45 to 54 years old				55 to 64 years old				KW
	Mean	SD	RW	RA	Mean	SD	RW	RA	Mean	SD	RW	RA	Mean	SD	RW	RA	
TANX8	4.73	1.351	6	40	5.29	1.586	3	33	5.25	1.483	3	26	5.36	1.433	1	28	0.192
TANX9	4.33	1.360	10	44	4.71	1.586	10	44	4.06	1.482	10	44	4.64	1.859	7	39	0.542
TANX10	4.63	1.236	8	42	5.10	1.609	6	40	5.25	1.571	2	27	5.09	1.514	2	31	0.283
PU1	5.63	1.035	1	3	5.95	1.024	4	6	5.38	1.088	6	21	6.00	1.095	4	3	0.292
PU2	5.61	0.909	2	4	5.90	0.995	5	8	5.56	0.964	2	12	5.64	1.027	6	12	0.614
PU3	5.51	1.210	6	9	6.00	1.049	2	4	5.75	0.931	1	6	6.00	1.000	3	5	0.336
PU4	5.59	1.135	4	6	5.95	0.973	3	5	5.44	1.263	4	18	5.73	1.272	5	8	0.599
PU5	5.55	1.191	5	8	5.90	1.179	6	10	5.44	1.263	4	18	6.09	1.136	1	2	0.369
PU6	5.59	1.098	3	5	6.00	1.000	1	3	5.56	0.964	2	12	6.00	0.894	2	3	0.377
PEU1	5.18	1.439	5	23	5.67	1.494	6	18	5.50	1.211	6	15	5.45	1.572	5	24	0.484
PEU2	5.47	1.309	1	10	5.95	1.024	1	6	5.63	1.088	2	8	5.55	1.214	3	19	0.562
PEU3	5.20	1.291	4	22	5.76	0.995	4	13	5.62	1.147	4	10	5.27	1.272	7	29	0.306
PEU4	5.16	1.344	6	26	5.71	1.271	5	15	5.63	1.204	3	9	5.55	1.368	3	20	0.339
PEU5	5.16	1.359	7	27	5.62	1.499	7	20	5.56	1.153	5	14	5.45	1.572	5	24	0.401
PEU6	5.39	1.239	3	15	5.86	1.108	3	11	5.50	1.211	6	15	5.64	1.206	1	15	0.501
PEU7	5.45	1.209	2	12	5.90	1.044	2	9	5.69	1.078	1	7	5.64	1.206	1	15	0.543
PIT1	5.29	1.258	2	18	5.81	1.123	1	12	5.44	1.459	2	20	5.45	0.820	1	21	0.453
PIT2	4.98	1.250	4	32	5.43	1.326	4	25	5.31	1.352	4	23	4.64	1.120	4	38	0.283
PIT3	5.02	1.479	3	31	5.67	1.390	2	17	5.50	1.461	1	17	5.27	1.954	3	30	0.568
PIT4	5.35	1.200	1	17	5.57	1.165	3	21	5.31	1.138	3	22	5.45	0.820	1	21	0.836

Impact

	25 to 34 Years old				35 to 44 years old				45 to 54 years old				55 to 64 years old				KW
	Mean	*SD*	*RW*	*RA*	*Mean*	*SD*	*RW*	*RA*	*Mean*	*SD*	*RW*	*RA*	*Mean*	*SD*	*RW*	*RA*	
TSE1	5.78	1.403	1	1	5.95	1.024	2	6	5.81	1.424	1	1	5.64	1.629	7	14	0.993
TSE2	4.82	1.253	9	35	5.62	1.161	5	21	4.75	1.000	5	36	5.73	1.104	5	8	0.020*
TSE3	4.84	1.313	8	34	5.43	0.978	7	24	4.75	1.183	7	38	5.73	1.272	6	10	0.078
TSE4	5.14	1.173	4	22	5.71	1.056	3	17	4.81	1.167	4	34	5.82	1.079	2	2	0.045*
TSE5	5.00	1.354	6	27	5.38	1.071	8	27	4.69	0.946	8	39	5.73	1.009	3	4	0.107
TSE6	5.04	1.274	5	26	5.48	0.928	6	23	4.75	1.125	6	37	5.27	1.104	10	27	0.346
TSE7	4.88	1.218	7	30	5.38	1.161	9	28	4.88	1.147	3	32	5.73	1.009	3	4	0.113
TSE8	4.73	1.511	10	38	5.24	1.261	10	34	4.38	1.455	10	44	5.55	1.036	8	15	0.150
TSE9	5.76	1.331	2	2	6.19	1.078	1	1	5.69	1.250	2	5	5.55	1.572	9	20	0.476
TSE10	5.29	1.258	3	15	5.71	1.056	3	17	4.63	1.025	9	40	5.91	1.136	1	1	0.017*
TAFF1	5.14	1.339	4	24	5.43	1.248	2	25	5.75	1.065	1	2	5.36	1.362	3	26	0.439
TAFF2	5.14	1.275	3	23	5.24	1.261	5	34	5.75	1.065	1	2	5.36	1.206	2	24	0.406
TAFF3	4.86	1.339	6	32	5.38	1.244	3	29	4.88	1.025	6	31	4.45	1.293	7	41	0.237
TAFF4	4.80	1.472	7	36	5.19	1.289	6	38	5.06	1.063	5	28	4.64	1.629	6	39	0.606
TAFF5	5.16	1.280	2	21	5.33	1.390	4	30	5.56	1.413	4	9	5.27	1.104	4	27	0.621
TAFF6	4.94	1.506	5	29	5.10	1.411	7	41	4.88	1.204	7	33	5.00	1.789	5	35	0.936
TAFF7	5.37	1.410	1	13	5.76	1.179	1	16	5.69	1.250	3	5	5.73	1.009	1	4	0.713
TANX1	4.53	1.276	9	43	5.24	1.411	4	36	4.44	1.209	10	43	4.27	0.905	8	42	0.095
TANX2	4.86	1.291	2	31	5.00	1.549	8	42	4.56	1.031	8	41	4.00	1.342	10	44	0.287
TANX3	4.94	1.281	1	28	4.86	1.493	10	44	5.00	1.317	6	30	4.09	1.375	9	43	0.379
TANX4	4.86	1.486	3	33	5.19	1.569	6	39	4.81	1.471	7	35	5.27	1.737	1	29	0.653
TANX5	4.59	1.240	6	40	4.95	1.596	9	43	5.13	1.628	4	27	4.73	1.794	6	37	0.458
TANX6	4.59	1.606	7	41	5.29	1.554	3	33	5.06	1.652	5	29	5.09	1.221	4	34	0.303
TANX7	4.67	1.519	5	39	5.24	1.609	4	37	5.19	1.424	2	25	4.64	1.502	7	38	0.302
TANX8	4.76	1.315	4	37	5.33	1.494	2	32	5.13	1.544	3	26	5.18	1.471	2	30	0.257
TANX9	4.29	1.414	10	44	5.33	1.426	1	31	4.50	1.155	9	42	4.82	1.779	5	36	0.044*

(*Continued*)

Appendix A-7 (Continued)

Factors Affecting Attitude Towards Technologies

Impact	25 to 34 Years old				35 to 44 years old				45 to 54 years old				55 to 64 years old				KW
	Mean	SD	RW	RA	Mean	SD	RW	RA	Mean	SD	RW	RA	Mean	SD	RW	RA	
TANX10	4.57	1.486	8	42	5.19	1.662	6	40	5.25	1.653	1	23	5.18	1.601	3	32	0.274
PU1	5.61	1.204	1	3	5.90	1.044	1	8	5.50	1.155	2	11	5.73	1.009	1	4	0.764
PU2	5.51	1.023	6	9	5.81	1.078	6	15	5.25	1.125	6	22	5.36	1.286	6	25	0.490
PU3	5.53	1.340	5	8	5.86	1.062	4	12	5.44	1.153	3	15	5.64	1.362	3	13	0.744
PU4	5.61	1.222	2	4	5.81	0.981	5	14	5.31	1.195	5	20	5.55	1.036	4	15	0.634
PU5	5.57	1.225	3	5	5.86	1.014	3	11	5.50	0.966	1	10	5.73	1.104	2	8	0.749
PU6	5.55	1.174	4	6	5.86	0.964	2	10	5.38	1.088	4	17	5.55	1.214	5	17	0.599
PEU1	5.27	1.319	3	16	6.00	1.000	4	5	5.31	1.352	7	21	5.64	1.120	3	12	0.165
PEU2	5.39	1.469	2	12	6.10	0.995	2	3	5.75	1.065	1	2	5.82	1.168	1	3	0.296
PEU3	5.18	1.349	7	20	5.81	1.250	7	13	5.50	1.211	4	12	5.18	1.471	7	30	0.289
PEU4	5.27	1.351	4	17	6.10	0.944	1	2	5.50	1.265	6	14	5.55	1.214	4	17	0.118
PEU5	5.20	1.274	6	19	5.95	1.024	5	6	5.56	1.153	3	8	5.45	1.214	6	21	0.139
PEU6	5.22	1.327	5	18	5.90	1.261	6	9	5.50	1.211	4	12	5.55	1.214	4	17	0.186
PEU7	5.53	1.260	1	7	6.05	1.071	3	4	5.69	1.138	2	7	5.73	1.272	2	10	0.439
PIT1	5.41	1.322	2	11	5.67	1.197	2	20	5.38	1.455	2	18	5.09	0.944	3	33	0.659
PIT2	5.12	1.333	4	25	5.43	1.326	4	26	5.44	1.209	1	16	4.55	1.036	4	40	0.273
PIT3	5.31	1.530	3	14	5.71	1.347	1	17	5.338	1.455	3	19	5.45	1.864	1	22	0.805
PIT4	5.41	1.306	1	10	5.62	1.161	3	21	5.19	1.223	4	24	5.36	1.120	2	23	0.743

Significance (Applicability × Impact)

	25 to 34 Years old				35 to 44 years old				45 to 54 years old				55 to 64 years old				KW
	Mean	SD	RW	RA	Mean	SD	RW	RA	Mean	SD	RW	RA	Mean	SD	RW	RA	
TSE1	32.80	14.006	1	1	36.43	9.953	2	4	36.38	14.486	1	1	33.64	15.082	2	7	0.747
TSE2	25.73	11.109	8	31	30.86	11.337	5	25	24.75	8.714	6	36	33.18	11.098	4	9	0.067
TSE3	24.27	10.998	10	37	29.52	10.591	9	33	25.19	10.291	4	34	32.27	12.523	7	16	0.077
TSE4	27.92	10.986	4	24	31.76	10.554	4	22	24.94	8.473	5	35	33.55	9.994	3	8	0.065
TSE5	26.88	11.599	6	26	29.76	11.184	8	32	23.19	7.232	9	39	32.55	10.792	6	13	0.115
TSE6	27.31	10.520	5	25	30.24	10.473	6	27	23.75	7.912	8	38	29.00	9.338	10	29	0.245
TSE7	25.53	10.073	9	33	29.81	11.635	7	31	26.00	8.501	3	31	32.82	9.347	5	12	0.122
TSE8	25.80	12.605	7	30	28.00	11.349	10	41	23.06	10.459	10	40	30.06	7.968	8	24	0.316
TSE9	32.33	14.538	2	4	40.24	11.371	1	1	34.88	12.753	2	2	29.55	15.623	9	28	0.111
TSE10	29.41	11.781	3	16	33.90	11.260	3	19	23.81	8.503	7	37	36.36	12.388	1	1	0.030*
TAFF1	28.88	11.923	4	18	31.86	12.121	2	21	34.81	12.051	1	3	31.55	13.888	4	20	0.316
TAFF2	28.92	11.622	3	17	29.90	11.920	4	30	34.44	11.905	2	4	31.73	12.815	3	18	0.411
TAFF3	25.90	12.096	6	29	29.33	11.599	5	35	2.94	8.306	7	44	19.82	8.328	7	41	0.221
TAFF4	25.71	12.680	7	32	28.38	12.290	6	37	26.25	8.836	5	30	22.73	11.525	6	39	0.667
TAFF5	29.90	11.913	2	13	31.00	13.236	3	24	33.81	13.600	3	5	32.45	9.353	2	14	0.595
TAFF6	26.37	13.905	5	28	28.33	13.937	7	38	25.38	11.638	6	33	25.73	13.410	5	35	0.933
TAFF7	31.98	14.234	1	7	34.19	12.440	1	16	33.25	13.309	4	8	35.09	10.616	1	3	0.901
TANX1	22.41	11.885	8	42	28.29	14.065	6	39	20.94	10.504	9	42	18.09	5.700	8	42	0.104
TANX2	25.10	12.797	2	35	27.14	14.333	8	42	21.13	10.164	8	41	17.00	8.854	10	44	0.179
TANX3	24.84	13.136	3	36	25.76	13.778	10	44	25.94	13.005	7	32	17.55	8.299	9	43	0.286
TANX4	25.51	13.939	1	34	29.38	14.868	4	34	27.25	13.931	6	29	28.73	17.596	2	30	0.794
TANX5	21.37	12.332	9	43	28.10	14.515	7	40	27.31	15.143	5	28	25.18	16.964	5	36	0.275
TANX6	23.94	14.757	6	40	30.14	14.161	3	29	28.12	14.408	4	27	26.36	12.372	4	34	0.356
TANX7	24.27	14.006	4	38	30.76	14.866	1	26	29.50	14.000	2	24	24.00	13.660	7	38	0.212
TANX8	24.18	13.159	5	39	30.19	13.692	2	28	29.00	14.579	3	25	29.64	14.746	1	27	0.226
TANX9	19.90	10.654	10	44	26.48	13.552	9	43	19.56	11.063	10	43	24.82	15.098	6	37	0.254

(Continued)

Appendix A-7 (Continued)

Factors Affecting Attitude Towards Technologies

Significance (Applicability × Impact)

	25 to 34 Years old				35 to 44 years old				45 to 54 years old				55 to 64 years old				KW
	Mean	*SD*	*RW*	*RA*	*Mean*	*SD*	*RW*	*RA*	*Mean*	*SD*	*RW*	*RA*	*Mean*	*SD*	*RW*	*RA*	
TANX10	22.80	12.824	7	41	28.57	14.685	5	36	29.75	15.382	1	23	28.45	15.902	3	31	0.286
PU1	32.49	11.386	2	3	36.05	11.456	1	5	30.63	11.667	4	19	34.73	10.432	2	4	0.545
PU2	31.61	10.136	6	9	35.24	11.661	6	12	30.06	11.246	6	22	30.82	10.971	6	22	0.492
PU3	31.94	12.531	5	8	36.00	11.459	2	6	31.94	10.810	1	12	34.64	12.002	3	5	0.618
PU4	32.55	11.869	1	2	35.38	11.088	5	11	30.19	12.597	5	20	32.36	11.775	5	15	0.684
PU5	32.10	12.329	4	6	35.48	11.906	4	10	30.88	11.944	2	16	35.55	11.953	1	2	0.598
PU6	32.18	12.036	3	5	35.95	11.156	3	7	30.81	11.280	3	17	33.91	10.968	4	6	0.520
PEU1	28.37	12.799	6	22	34.90	12.864	5	13	30.69	13.573	7	18	30.73	11.594	5	23	0.299
PEU2	30.86	13.701	2	11	37.00	11.118	1	2	33.38	11.775	2	7	33.00	12.066	2	11	0.362
PEU3	28.51	13.037	5	20	34.43	12.303	6	14	32.12	12.706	4	10	28.36	13.017	7	33	0.346
PEU4	28.63	13.705	4	19	35.52	11.682	4	9	32.31	13.250	3	9	31.36	12.209	4	21	0.224
PEU5	28.02	12.750	7	23	34.33	12.940	7	15	32.00	12.377	5	11	29.91	12.128	6	26	0.256
PEU6	29.45	12.934	3	15	35.57	12.216	3	8	31.38	12.816	6	14	32.00	11.790	3	17	0.301
PEU7	31.41	12.530	1	10	36.48	11.250	2	3	33.44	12.066	1	6	33.09	12.136	1	10	0.499
PIT1	29.90	12.808	2	14	34.14	12.807	1	17	31.19	15.079	2	15	28.36	9.036	3	32	0.556
PIT2	26.84	12.226	4	27	31.10	14.057	4	23	30.12	12.873	3	21	22.00	8.367	4	40	0.334
PIT3	28.39	13.862	3	21	34.14	14.827	2	18	31.50	14.560	1	13	31.64	16.512	1	19	0.374
PIT4	30.31	12.979	1	12	32.48	12.722	3	20	28.81	12.740	4	26	29.91	10.644	2	25	0.882

KW, Kruskal Wallis; RA, rank across components; RW, rank within component; SD, standard deviation.

* Significant at *p*-value <0.05.

Appendix A-8 Summary of Results for the Perceived Importance of Knowledge in Project Management by Respondent's Education Level

Project Management Knowledge Areas	Diploma and Below			Bachelor			Postgraduate			Kruskal Wallis
	Mean	Standard Deviation	Rank	Mean	Standard Deviation	Rank	Mean	Standard Deviation	Rank	
Project Integration Management	6.18	1.259	6	6.23	1.165	4	6.26	0.852	6	0.886
Project Scope Management	5.91	1.269	9	6.00	1.155	9	6.11	0.900	7	0.947
Project Schedule Management	6.32	1.287	2	6.15	1.369	8	6.40	0.976	3	0.835
Project Cost Management	6.27	1.279	3	6.28	1.240	3	6.34	1.027	5	0.986
Project Quality Management	6.27	1.279	3	6.15	1.312	7	6.40	0.812	2	0.895
Project Resource Management	5.95	1.327	1	6.20	1.137	5	5.94	1.027	8	0.332
Project Communication Management	6.27	1.279	3	6.40	1.105	1	6.46	1.010	1	0.849
Project Risk Management	6.09	1.306	8	6.33	1.141	2	6.37	1.060	4	0.608
Project Procurement Management	5.68	1.359	10	5.90	1.236	10	5.86	0.944	10	0.770
Project Stakeholder Management	6.14	1.356	7	6.18	1.107	6	5.94	1.162	9	0.500

Appendix A-9 Summary of Results for the Perceived Importance of Skills to Manage Projects by Respondent's Education Level

Skills to Manage Projects	Diploma and Below			Bachelor			Postgraduate			Kruskal Wallis
	Mean	Standard Deviation	Rank	Mean	Standard Deviation	Rank	Mean	Standard Deviation	Rank	
Technical and Operational Technology Skills	6.18	0.958	8	5.83	1.083	11	5.83	1.175	8	0.434
Project Management Skills	6.32	0.945	3	6.05	1.176	4	6.11	0.832	2	0.551
Information Management Skills	6.27	1.162	5	6.20	1.265	3	6.09	1.095	3	0.545
Planning and Organising Skills	6.32	0.995	4	6.00	1.132	6	5.97	0.985	5	0.324
Communication Skills	6.45	0.858	1	6.30	1.018	1	5.77	1.239	9	0.033*
Social Awareness	5.77	1.066	14	5.30	1.363	17	5.09	1.222	18	0.109
Cultural Awareness	5.59	1.054	18	5.08	1.289	19	5.03	0.985	19	0.123
Organisational Awareness	5.73	1.077	17	5.38	1.234	16	5.54	1.120	13	0.506
Creativity	5.82	1.097	11	5.98	0.947	7	5.66	1.083	11	0.478
Problem-Solving Skills	6.41	0.908	2	6.25	0.899	2	6.26	0.817	1	0.620
Ethical Awareness	5.77	1.193	15	5.63	1.372	14	5.46	1.268	14	0.561
Strategic Planning Skills	6.18	1.181	9	6.05	1.197	5	5.83	0.891	7	0.146
Active Learning Skills	5.18	1.220	19	5.20	1.285	18	5.43	1.119	15	0.582
Conflict Management Skills	6.23	1.232	7	5.85	1.122	10	5.89	1.022	6	0.202
Decision-Making Skills	6.27	1.162	5	5.95	1.339	9	6.06	1.136	4	0.625
Delegation Skills	5.82	1.140	12	5.63	1.234	13	5.40	1.117	16	0.361
Motivation Skills	5.82	1.500	13	5.70	1.305	12	5.74	1.010	10	0.670
Negotiation Skills	5.77	1.193	15	5.55	1.260	15	5.23	1.087	17	0.116
Team-Building Skills	6.18	1.259	10	5.98	1.250	8	5.66	1.162	12	0.120

* Significant at p-value <0.05.

Appendix A-10 Summary of Results for Perceived Differences in Factors Affecting Attitude towards Technologies by Respondent's Education Level

Factors Affecting Attitude Towards Technologies

Applicability

	Diploma and Below				Bachelor				Postgraduate				KW
	Mean	SD	RW	RA	Mean	SD	RW	RA	Mean	SD	RW	RA	
TSE1	6.00	1.024	1	1	5.77	1.527	2	2	5.43	1.313	1	14	0.190
TSE2	5.55	0.858	4	23	5.17	1.059	6	24	5.11	1.022	9	28	0.227
TSE3	5.32	0.894	7	28	4.93	1.163	10	38	5.03	1.014	10	30	0.308
TSE4	5.45	0.912	6	26	5.23	1.074	4	21	5.31	0.993	4	21	0.623
TSE5	5.27	0.883	8	29	5.13	1.223	7	25	5.23	0.973	6	25	0.879
TSE6	5.18	0.853	10	32	5.20	1.043	5	22	5.34	0.968	3	20	0.758
TSE7	5.59	0.959	3	22	5.08	1.207	9	29	5.17	0.891	8	27	0.121
TSE8	5.23	0.612	9	30	5.10	1.172	8	27	5.26	.1010	5	24	0.829
TSE9	6.00	1.480	2	2	5.82	1.662	1	1	5.23	1.497	6	26	0.058
TSE10	5.50	0.859	5	25	5.47	1.240	3	14	5.40	1.090	2	16	0.852
TAFF1	5.68	1.129	4	20	5.53	1.176	3	11	5.54	0.980	2	10	0.870
TAFF2	5.91	1.109	1	5	5.48	1.109	4	12	5.40	1.035	4	15	0.192
TAFF3	4.82	1.181	6	39	5.08	1.269	6	30	5.00	0.939	5	31	0.884
TAFF4	4.82	1.332	7	40	5.28	1.198	5	20	4.94	0.938	6	35	0.316
TAFF5	5.91	1.231	2	7	5.60	1.277	2	10	5.54	1.221	3	11	0.408
TAFF6	4.95	1.463	5	38	5.05	1.552	7	32	4.94	1.305	7	36	0.784
TAFF7	5.73	1.316	3	18	5.73	1.339	1	5	5.66	1.235	1	2	0.901
TANX1	4.77	1.110	8	42	4.75	1.316	8	42	4.49	1.095	8	42	0.603
TANX2	4.82	1.402	7	41	4.95	1.413	4	37	4.40	1.288	9	43	0.218
TANX3	4.41	1.652	9	44	4.95	1.377	3	36	4.66	1.305	6	40	0.442
TANX4	5.14	2.077	2	33	5.05	1.518	1	31	4.89	1.388	4	38	0.517
TANX5	5.05	1.838	6	37	4.62	1.580	9	43	4.60	1.397	7	41	0.498
TANX6	5.09	1.601	3	34	4.87	1.488	7	41	4.97	1.272	1	33	0.784
TANX7	5.09	1.659	4	35	5.02	1.510	2	35	4.91	1.442	3	37	0.805
TANX8	5.23	1.716	1	31	4.93	1.421	5	39	4.97	1.294	2	34	0.519
TANX9	4.41	1.593	9	43	4.52	1.536	10	44	4.26	1.379	10	44	0.726
TANX10	5.09	1.743	5	36	4.92	1.366	6	40	4.71	1.250	5	39	0.481
PU1	5.77	1.193	3	13	5.70	1.091	4	7	5.66	0.938	1	1	0.867
PU2	5.68	1.086	6	19	5.70	0.966	3	6	5.63	0.843	2	3	0.909
PU3	5.82	1.097	2	9	5.73	1.154	2	4	5.63	1.114	4	5	0.808
PU4	5.73	1.202	5	17	5.68	1.118	6	9	5.60	1.143	5	6	0.932
PU5	5.77	1.152	3	12	5.70	1.285	5	8	5.57	1.145	6	7	0.838
PU6	5.82	1.053	1	8	5.75	1.006	1	3	5.63	1.087	3	4	0.863
PEU1	5.82	1.220	4	10	5.10	1.614	6	28	5.40	1.265	5	18	0.215
PEU2	5.91	0.971	2	4	5.48	1.261	1	13	5.57	1.267	2	9	0.454
PEU3	5.73	1.077	7	16	5.38	1.254	4	18	5.3	1.239	7	22	0.330
PEU4	5.91	1.109	3	5	5.13	1.453	5	26	5.40	1.193	5	17	0.094
PEU5	5.77	1.232	6	14	5.05	1.568	7	33	5.46	1.172	4	13	0.161
PEU6	5.77	1.066	5	11	5.43	1.238	3	17	5.51	1.245	3	12	0.558
PEU7	5.95	0.999	1	3	5.45	1.218	2	15	5.57	1.145	1	7	0.279
PIT1	5.73	1.032	1	15	5.35	1.388	2	19	5.37	1.140	1	19	0.532
PIT2	5.41	1.054	4	27	5.02	1.423	4	34	4.97	1.224	4	32	0.429
PIT3	5.64	1.706	2	21	5.20	1.400	3	23	5.11	1.530	3	29	0.234
PIT4	5.55	1.011	3	24	5.43	1.130	1	16	5.29	1.226	2	23	0.731

(Continued)

Appendix A-10 (Continued)

Factors Affecting Attitude Towards Technologies

Impact

	Diploma and Below				Bachelor				Postgraduate				KW
	Mean	SD	RW	RA	Mean	SD	RW	RA	Mean	SD	RW	RA	
TSE1	5.82	1.296	2	8	6.03	1.349	1	1	5.54	1.358	2	7	0.194
TSE2	5.50	0.859	4	23	4.97	1.387	8	32	4.94	1.211	7	28	0.160
TSE3	5.27	1.077	8	29	4.93	1.439	9	38	5.06	1.136	5	25	0.563
TSE4	5.50	1.058	5	24	5.25	1.235	4	21	5.20	1.183	4	20	0.624
TSE5	5.27	0.827	7	28	5.10	1.464	6	27	5.03	1.150	6	26	0.812
TSE6	5.23	0.869	9	30	5.20	1.285	5	25	4.94	1.211	7	28	0.617
TSE7	5.59	0.854	3	14	4.97	1.368	7	31	4.89	1.105	9	33	0.054
TSE8	5.09	1.192	10	33	4.75	1.565	10	42	4.89	1.430	10	34	0.702
TSE9	5.95	1.290	1	4	5.95	1.260	2	2	5.57	1.335	1	3	0.318
TSE10	5.36	1.049	6	26	5.40	1.317	3	13	5.26	1.221	3	19	0.822
TAFF1	5.55	1.262	3	21	5.23	1.441	4	24	5.31	1.105	2	16	0.721
TAFF2	5.55	1.184	2	19	5.25	1.391	3	23	5.17	1.071	4	23	0.553
TAFF3	4.77	1.232	7	40	4.95	1.484	6	34	5.00	1.057	5	27	0.852
TAFF4	4.82	1.368	6	38	5.10	1.533	5	28	4.74	1.221	7	38	0.475
TAFF5	5.45	1.299	4	25	5.25	1.335	2	22	5.20	1.279	3	21	0.710
TAFF6	5.05	1.588	5	34	4.95	1.568	7	36	4.94	1.259	6	30	0.916
TAFF7	5.64	1.329	1	12	5.50	1.359	1	10	5.54	1.221	1	5	0.917
TANX1	4.73	1.120	9	43	4.70	1.436	9	43	4.51	1.222	8	41	0.790
TANX2	4.82	1.435	6	39	5.00	1.301	2	30	4.40	1.265	10	44	0.140
TANX3	4.32	1.555	10	44	5.13	1.244	1	26	4.83	1.272	5	37	0.177
TANX4	5.05	1.786	3	35	4.95	1.568	5	36	4.94	1.305.	1	31	0.862
TANX5	4.91	1.823	5	37	4.82	1.318	7	40	4.63	1.352	7	40	0.724
TANX6	4.95	1.676	4	36	4.82	1.615	8	41	4.89	1.491	3	35	0.932
TANX7	4.77	1.716	8	42	4.95	1.518	4	35	4.86	1.438	4	36	0.944
TANX8	5.14	1.699	1	31	4.95	1.358	3	33	4.94	1.305	1	31	0.676
TANX9	4.77	1.541	7	41	4.63	1.580	10	44	4.49	1.292	9	43	0.702
TANX10	5.14	1.807	2	32	4.87	1.539	6	39	4.74	1.482	6	39	0.598
PU1	5.64	1.093	1	10	5.75	1.127	1	3	5.60	1.193	1	2	0.819
PU2	5.55	1.224	5	20	5.60	1.057	6	8	4.50	1.035	6	42	0.824
PU3	5.55	1.299	6	22	5.75	1.276	3	5	5.46	1.197	5	10	0.499
PU4	5.64	1.093	1	10	5.65	1.122	4	6	5.51	1.222	4	9	0.963
PU5	5.59	1.098	3	15	5.75	1.127	1	3	5.54	1.146	2	4	0.692
PU6	5.59	1.182	4	16	5.65	1.122	4	6	5.51	1.095	3	8	0.903
PEU1	5.82	1.053	6	7	5.38	1.353	5	16	5.37	1.262	4	13	0.387
PEU2	6.00	0.976	2	2	5.55	1.413	1	9	5.54	1.336	2	6	0.470
PEU3	5.77	1.152	7	9	5.40	1.411	4	15	5.09	1.292	7	24	0.156
PEU4	5.95	0.950	3	3	5.40	1.392	3	14	5.37	1.262	4	13	0.229
PEU5	5.86	1.037	4	5	5.38	1.353	5	16	5.29	1.126	6	17	0.199
PEU6	5.86	1.037	4	5	5.28	1.396	7	20	5.40	1.288	3	12	0.270
PEU7	6.05	0.844	1	1	5.50	1.396	2	11	5.69	1.132	1	1	0.417
PIT1	5.55	1.101	3	18	5.38	1.462	2	18	5.40	1.168	1	11	0.923
PIT2	5.36	1.049	4	26	5.05	1.431	4	29	5.20	1.279	4	21	0.766
PIT3	5.64	1.706	1	13	5.42	1.375	1	12	5.29	1.545	3	18	0.522
PIT4	5.55	1.011	2	17	5.37	1.275	3	19	5.37	1.330	2	15	0.839

Factors Affecting Attitude Towards Technologies

Significance

	Diploma and Below				Bachelor				Postgraduate				KW
	Mean	SD	RW	RA	Mean	SD	RW	RA	Mean	SD	RW	RA	
TSE1	35.86	12.422	2	4	36.12	13.740	1	1	31.14	13.240	1	9	0.185
TSE2	31.14	.951	4	24	26.75	12.289	7	33	26.14	10.544	8	29	0.140
TSE3	28.82	10.626	7	31	25.53	12.615	10	41	26.06	9.846	9	31	0.369
TSE4	30.77	9.812	5	25	28.50	11.839	4	22	27.17	9.724	4	24	0.562
TSE5	28.36	8.867	8	32	27.60	13.079	6	27	26.94	9.795	6	26	0.848
TSE6	27.55	8.227	9	34	27.98	11.228	5	26	27.06	9.959	5	25	0.983
TSE7	31.77	8.815	3	22	26.25	11.664	8	36	25.86	8.892	10	33	0.078
TSE8	27.00	8.165	10	36	25.68	13.244	9	40	26.60	11.640	7	28	0.733
TSE9	37.05	13.820	1	1	35.77	13.603	2	2	30.46	14.141	2	12	0.138
TSE10	30.27	10.218	6	26	30.93	12.789	3	12	29.46	11.625	3	20	0.840
TAFF1	32.77	13.158	4	20	30.15	13.012	3	17	30.31	10.830	2	14	0.764
TAFF2	33.77	12.336	2	12	29.88	12.484	4	18	28.77	10.652	4	22	0.385
TAFF3	24.23	10.212	7	41	26.67	13.190	7	34	25.77	9.472	5	35	0.946
TAFF4	24.86	11.886	6	40	28.20	13.135	5	25	24.31	10.093	7	38	0.471
TAFF5	33.36	12.381	3	15	30.68	12.294	2	13	30.09	12.011	3	16	0.534
TAFF6	27.00	13.874	5	37	27.08	14.357	6	30	25.69	12.092	6	36	0.908
TAFF7	33.82	13.661	1	11	32.95	13.899	1	8	32.60	12.439	1	2	0.904
TANX1	23.64	11.454	8	42	24.08	13.199	8	42	21.23	10.762	8	42	0.662
TANX2	25.00	12.479	7	39	26.30	13.430	4	35	20.66	11.261	9	43	0.204
TANX3	21.09	12.463	10	44	26.80	13.107	3	32	23.71	12.676	6	40	0.296
TANX4	28.95	16.910	3	30	27.13	14.594	1	29	25.63	12.781	4	37	0.745
TANX5	27.50	16.402	5	35	23.90	13.703	9	43	22.57	12.601	7	41	0.691
TANX6	27.59	15.187	4	33	25.75	14.861	7	39	25.97	13.542	2	32	0.890
TANX7	26.64	14.796	6	38	27.05	14.685	2	31	25.80	13.767	3	34	0.931
TANX8	29.41	15.414	1	28	26.23	13.828	5	37	26.09	12.763	1	30	0.630
TANX9	22.68	13.007	9	43	22.78	13.069	10	44	20.20	10.426	10	44	0.740
TANX10	29.09	16.639	2	29	25.88	14.002	6	38	23.74	12.566	5	39	0.465
PU1	33.45	12.235	2	14	33.67	11.583	3	5	32.51	10.711	1	3	0.933
PU2	32.55	12.335	6	21	32.65	10.625	6	9	31.06	10.003	6	11	0.744
PU3	33.23	12.398	4	17	34.18	12.283	1	3	31.86	11.420	5	8	0.659
PU4	33.23	12.348	3	16	33.07	11.687	5	7	32.09	11.743	2	6	0.987
PU5	33.18	12.145	5	18	33.93	12.595	2	4	31.89	11.694	4	7	0.864
PU6	33.50	12.098	1	13	33.45	11.624	4	6	32.09	11.464	2	5	0.894
PEU1	34.27	11.281	5	8	28.50	13.633	6	23	30.23	12.737	5	15	0.239
PEU2	36.09	10.551	2	3	31.53	13.297	1	10	32.31	13.321	2	4	0.360
PEU3	33.91	11.670	7	10	30.58	13.563	3	15	27.91	12.589	7	23	0.223
PEU4	35.77	10.801	3	5	29.05	14.193	5	21	30.34	12.866	4	13	0.100
PEU5	34.27	11.281	5	8	28.43	14.189	7	24	29.83	11.562	6	18	0.169
PEU6	34.64	11.232	4	6	29.83	13.044	4	19	31.11	13.067	3	10	0.321
PEU7	36.50	9.797	1	2	31.33	13.137	2	11	32.80	12.097	1	1	0.325
PIT1	32.77	11.860	2	19	30.60	14.313	1	14	29.94	11.717	1	17	0.717
PIT2	30.05	11.504	4	27	27.20	13.631	4	28	26.94	11.963	4	27	0.650
PIT3	34.36	15.426	1	7	29.80	13.879	3	20	28.91	14.460	3	21	0.325
PIT4	31.73	11.407	3	23	30.43	12.949	2	16	29.77	12.948	2	19	0.882

KW, Kruskal Wallis; RA, rank across components; RW, rank within component; SD, standard deviation.

Appendix A-11 Summary of Results for the Perceived Importance of Knowledge in Project Management by Respondent's Years of Experience in the Construction Industry

Project Management Knowledge Areas	Less than 3 years			3 to 10 years			More than 10 years			Kruskal Wallis
	Mean	Standard Deviation	Rank	Mean	Standard Deviation	Rank	Mean	Standard Deviation	Rank	
Project Integration Management	6.35	0.870	4	5.91	1.311	3	6.29	1.131	6	0.347
Project Scope Management	6.12	0.956	9	5.87	1.254	4	6.00	1.155	9	0.466
Project Schedule Management	6.30	1.166	6	5.96	1.364	2	6.48	1.151	2	0.964
Project Cost Management	6.44	0.934	3	5.74	1.453	7	6.52	1.122	1	0.865
Project Quality Management	6.35	1.044	5	5.96	1.296	1	6.39	1.145	5	0.927
Project Resource Management	6.23	0.868	7	5.61	1.406	10	6.13	1.204	7	0.700
Project Communication Management	6.60	0.760	1	5.87	1.424	5	6.48	1.151	2	0.645
Project Risk Management	6.47	0.855	2	5.78	1.445	6	6.42	1.177	4	0.983
Project Procurement Management	5.93	1.033	10	5.65	1.301	9	5.84	1.241	10	0.691
Project Stakeholder Management	6.19	1.006	8	5.70	1.424	8	6.10	1.221	8	0.799

Appendix A-12 Summary of Results for the Perceived Importance of Skills to Manage Projects by Respondent's Years of Experience in the Construction Industry

Skills to Manage Projects	Less than 3 years			3 to 10 years			More than 10 years			Kruskal Wallis
	Mean	Standard Deviation	Rank	Mean	Standard Deviation	Rank	Mean	Standard Deviation	Rank	
Technical And Operational Technology Skills	5.86	1.014	12	5.70	1.146	8	6.13	1.147	3	0.537
Project Management Skills	6.16	0.924	5	5.91	1.164	4	6.26	0.999	2	0.244
Information Management Skills	6.26	1.115	3	6.09	1.203	2	6.13	1.258	4	0.304
Planning and Organising Skills	6.09	1.087	9	5.96	0.928	3	6.10	1.106	5	0.281
Communication Skills	6.35	0.870	2	5.91	1.276	5	6.03	1.224	6	0.140
Social Awareness	5.58	1.159	16	5.00	1.446	18	5.23	1.230	18	0.478
Cultural Awareness	5.35	1.044	18	4.87	1.217	19	5.16	1.214	19	0.729
Organisational Awareness	5.70	1.081	14	5.35	1.229	14	5.39	1.202	17	0.207
Creativity	5.86	0.990	11	5.57	1.037	10	5.97	1.080	10	0.424
Problem-Solving Skills	6.35	0.842	1	6.22	0.850	1	6.26	0.930	1	0.611
Ethical Awareness	5.70	1.145	15	5.35	1.496	15	5.65	1.330	12	0.708
Strategic Planning Skills	6.16	0.924	5	5.61	1.196	9	6.00	1.181	9	0.106
Active Learning Skills	5.28	1.098	19	5.09	1.125	17	5.42	1.409	16	0.718
Conflict Management Skills	6.12	0.931	8	5.52	1.201	11	6.03	1.224	6	0.840
Decision-Making Skills	6.21	1.081	4	5.83	1.435	6	6.03	1.251	8	0.393
Delegation Skills	5.79	1.059	13	5.26	1.214	16	5.55	1.261	13	0.644
Motivation Skills	5.91	1.151	10	5.78	1.242	7	5.48	1.363	15	0.252
Negotiation Skills	5.47	1.120	17	5.48	1.344	13	5.52	1.208	14	0.421
Team-Building Skills	6.14	0.941	7	5.52	1.473	12	5.87	1.335	11	0.075

Appendix A-13 Summary of Results for the Perceived Differences in Factors Affecting Attitude towards Technologies by Respondent's Years of Experience in the Construction Industry

Factors Affecting Attitude Towards Technologies

Applicability

	Less than 3 years				3 to 10 years				More than 10 years				KW
	Mean	SD	RW	RA	Mean	SD	RW	RA	Mean	SD	RW	RA	
TSE1	5.56	1.436	1	12	5.87	1.486	2	2	5.77	1.146	1	14	0.408
TSE2	5.12	1.028	6	27	5.52	0.947	5	5	5.19	1.014	7	29	0.745
TSE3	4.84	1.090	10	36	5.22	1.043	10	21	5.23	0.990	6	27	0.097
TSE4	5.12	1.051	7	28	5.74	0.915	3	3	5.26	0.930	5	26	0.303
TSE5	5.07	1.142	9	31	5.52	0.846	4	4	5.13	1.056	9	35	0.595
TSE6	5.23	1.065	4	23	5.39	0.891	8	11	5.16	0.898	8	30	0.433
TSE7	5.07	1.009	8	29	5.39	1.196	9	13	5.32	1.013	4	25	0.333
TSE8	5.16	1.153	5	25	5.43	0.945	6	9	5.03	0.795	10	38	0.915
TSE9	5.49	1.594	2	13	6.00	1.706	1	1	5.61	1.476	2	19	0.446
TSE10	5.42	1.180	3	17	5.43	1.080	6	10	5.52	1.029	3	22	0.638
TAFF1	5.44	0.908	4	15	5.39	1.305	2	14	5.87	1.118	3	9	0.410
TAFF2	5.47	0.935	3	14	5.26	1.251	3	20	5.87	1.118	3	9	0.644
TAFF3	5.07	1.078	6	30	5.04	1.261	5	28	4.84	1.128	7	40	0.726
TAFF4	5.16	1.067	5	24	4.83	1.267	7	40	5.06	1.181	5	36	0.549
TAFF5	5.60	1.158	2	9	5.22	1.506	4	22	6.03	1.048	1	3	0.855
TAFF6	4.98	1.439	7	33	5.00	1.651	6	32	5.00	1.291	6	39	0.681
TAFF7	5.65	1.395	1	7	5.48	1.377	1	7	5.94	1.031	2	5	0.302
TANX1	4.65	1.110	7	41	4.83	1.302	7	41	4.55	1.234	8	42	0.813
TANX2	4.81	1.314	4	38	5.00	1.279	3	31	4.39	1.498	10	44	0.116
TANX3	4.77	1.288	5	39	4.70	1.608	8	42	4.68	1.492	7	41	0.875
TANX4	4.95	1.344	1	34	4.91	1.857	6	38	5.16	1.772	5	34	0.730
TANX5	4.49	1.420	9	43	4.65	1.668	9	43	5.06	1.692	6	37	0.330
TANX6	4.84	1.379	3	37	4.91	1.564	5	36	5.16	1.416	3	32	0.667
TANX7	4.86	1.355	2	35	5.04	1.770	1	30	5.16	1.530	4	33	0.546
TANX8	4.72	1.260	6	40	5.04	1.665	1	29	5.39	1.453	1	24	0.085
TANX9	4.28	1.420	10	44	4.48	1.344	10	44	4.52	1.691	9	43	0.397
TANX10	4.63	1.176	8	42	4.91	1.649	4	37	5.23	1.499	2	28	0.144
PU1	5.72	1.054	1	2	5.35	0.982	3	15	5.94	1.063	4	6	0.685
PU2	5.70	0.914	3	3	5.30	0.876	4	16	5.90	0.978	5	7	0.226
PU3	5.58	1.239	6	11	5.48	1.039	1	6	6.06	0.929	1	1	0.283
PU4	5.67	1.128	4	4	5.30	1.020	5	17	5.90	1.193	6	8	0.516
PU5	5.58	1.200	5	10	5.30	1.146	6	18	6.06	1.153	2	2	0.280
PU6	5.72	1.008	1	1	5.39	1.158	2	12	5.97	0.948	3	4	0.535
PEU1	5.42	1.367	5	19	4.91	1.411	6	35	5.65	1.473	7	18	0.742
PEU2	5.67	1.229	1	5	5.17	1.154	2	24	5.84	1.157	2	12	0.952
PEU3	5.42	1.220	4	18	4.96	1.186	5	34	5.71	1.160	5	16	0.707
PEU4	5.30	1.319	7	22	5.09	1.240	3	26	5.77	1.309	4	15	0.501
PEU5	5.37	1.291	6	20	4.87	1.359	7	39	5.71	1.442	6	17	0.413
PEU6	5.63	1.196	3	8	4.96	1.065	4	33	5.84	1.186	3	13	0.781
PEU7	5.65	1.193	2	6	5.17	1.029	1	23	5.87	1.118	1	9	0.842
PIT1	5.35	1.173	2	21	5.48	1.442	1	8	5.55	1.150	1	20	0.549
PIT2	5.00	1.195	4	32	5.17	1.370	3	25	5.16	1.344	4	31	0.508
PIT3	5.16	1.290	3	26	5.09	1.782	4	27	5.55	1.609	2	21	0.625
PIT4	5.44	1.076	1	16	5.30	1.396	2	19	5.42	1.025	3	23	0.771

Factors Affecting Attitude Towards Technologies

Impact

	Less than 3 years				3 to 10 years				More than 10 years				KW
	Mean	SD	RW	RA	Mean	SD	RW	RA	Mean	SD	RW	RA	
TSE1	5.70	1.406	1	2	6.30	1.105	2	2	5.58	1.361	2	19	0.921
TSE2	4.86	1.246	7	32	5.48	1.201	6	10	5.10	1.193	7	32	0.314
TSE3	4.74	1.293	9	38	5.43	1.121	8	12	5.19	1.223	4	28	0.330
TSE4	5.12	1.179	4	22	5.78	1.043	3	3	5.16	1.186	5	30	0.526
TSE5	4.88	1.384	6	31	5.52	0.898	5	6	5.13	1.147	6	31	0.600
TSE6	5.02	1.244	5	26	5.39	1.076	9	13	5.03	1.140	9	35	0.360
TSE7	4.79	1.146	8	36	5.61	1.158	4	5	5.10	1.193	7	32	0.259
TSE8	4.72	1.485	10	39	5.13	1.424	10	28	4.90	1.375	10	40	0.295
TSE9	5.67	1.322	2	4	6.39	1.118	1	1	5.58	1.285	1	18	0.663
TSE10	5.26	1.217	3	20	5.48	1.275	6	11	5.35	1.199	3	21	0.771
TAFF1	5.12	1.258	3	23	5.22	1.347	2	24	5.71	1.216	3	14	0.284
TAFF2	5.05	1.174	4	24	5.17	1.370	3	27	5.71	1.131	2	13	0.109
TAFF3	5.00	1.291	6	28	4.83	1.403	7	36	4.90	1.193	7	39	0.993
TAFF4	4.86	1.424	7	34	4.91	1.443	5	33	4.97	1.329	5	37	0.599
TAFF5	5.16	1.153	2	21	5.00	1.624	4	31	5.65	1.170	4	15	0.337
TAFF6	5.02	1.456	5	27	4.87	1.576	6	35	4.97	1.402	6	38	0.979
TAFF7	5.35	1.343	1	18	5.52	1.410	1	7	5.84	1.098	1	8	0.197
TANX1	4.58	1.239	9	43	4.83	1.435	3	37	4.58	1.259	9	43	0.782
TANX2	4.79	1.301	5	37	4.91	1.443	2	33	4.55	1.312	10	44	0.206
TANX3	4.95	1.174	1	29	4.70	1.521	7	41	4.77	1.477	8	42	0.527
TANX4	4.88	1.366	2	30	4.70	1.717	8	42	5.29	1.553	1	24	0.663
TANX5	4.67	1.107	6	40	4.65	1.722	9	43	5.00	1.653	6	36	0.493
TANX6	4.65	1.510	7	42	4.78	1.757	6	40	5.26	1.483	2	25	0.232
TANX7	4.86	1.390	3	33	4.61	1.803	10	44	5.10	1.491	5	34	0.783
TANX8	4.81	1.239	4	35	5.00	1.624	1	31	5.23	1.470	4	27	0.349
TANX9	4.35	1.412	10	44	4.83	1.527	4	38	4.81	1.470	7	41	0.338
TANX10	4.65	1.412	7	41	4.83	1.825	5	39	5.26	1.570	3	26	0.301
PU1	5.70	1.124	1	1	5.35	1.191	1	15	5.87	1.088	2	5	0.652
PU2	5.60	0.979	6	8	5.22	1.043	4	21	5.61	1.230	6	17	0.473
PU3	5.63	1.346	5	7	5.22	1.085	5	22	5.84	1.186	3	9	0.275
PU4	5.67	1.149	2	3	5.22	1.126	6	23	5.77	1.117	4	11	0.500
PU5	5.65	1.152	3	5	5.26	1.137	3	19	5.90	1.012	1	3	0.401
PU6	5.63	1.113	4	6	5.30	1.063	2	17	5.74	1.154	5	12	0.523
PEU1	5.40	1.312	3	13	5.17	1.154	4	25	5.81	1.223	6	10	0.554
PEU2	5.53	1.386	2	10	5.30	1.329	3	18	6.06	1.063	1	1	0.282
PEU3	5.35	1.289	6	17	5.04	1.364	6	29	5.65	1.330	7	16	0.769
PEU4	5.37	1.328	5	15	5.26	1.137	3	19	5.90	1.221	3	4	0.292
PEU5	5.30	1.245	7	19	5.17	1.193	5	26	5.87	1.118	4	6	0.312
PEU6	5.40	1.237	3	12	5.00	1.446	7	30	5.87	1.147	5	7	0.522
PEU7	5.58	1.277	1	9	5.48	1.082	1	9	6.00	1.155	2	2	0.580
PIT1	5.35	1.232	3	16	5.70	1.428	1	4	5.32	1.222	2	23	0.246
PIT2	5.05	1.194	4	25	5.39	1.500	3	14	5.19	1.276	4	29	0.290
PIT3	5.37	1.254	2	14	5.35	1.873	4	16	5.55	1.567	1	20	0.564
PIT4	5.42	1.159	1	11	5.52	1.473	2	8	5.32	1.166	2	22	0.497

(Continued)

Appendix A-13 (Continued)

Factors Affecting Attitude Towards Technologies

Significance

	Less than 3 years				3 to 10 years				More than 10 years				KW
	Mean	SD	RW	RA	Mean	SD	RW	RA	Mean	SD	RW	RA	
TSE1	33.14	13.465	1	5	37.52	13.242	2	2	33.42	13.299	1	17	0.587
TSE2	25.53	10.561	7	32	31.26	11.234	4	6	27.52	11.222	7	33	0.408
TSE3	23.86	10.803	10	40	29.00	10.749	9	19	28.19	11.634	4	30	0.141
TSE4	27.02	10.769	5	26	33.57	9.539	3	3	28.03	10.419	5	31	0.444
TSE5	25.93	11.947	6	30	30.96	8.921	5	8	27.23	10.776	8	35	0.611
TSE6	27.12	10.797	4	25	29.61	9.505	8	14	26.61	9.503	9	37	0.573
TSE7	25.09	9.579	9	35	30.83	10.978	6	9	27.94	10.279	6	32	0.264
TSE8	25.53	12.552	8	33	28.78	11.824	10	20	25.55	9.999	10	39	0.401
TSE9	32.51	13.653	2	11	39.52	14.798	1	1	32.42	13.261	2	20	0.553
TSE10	29.70	12.003	3	21	30.78	11.824	7	10	30.61	11.627	3	22	0.770
TAFF1	28.58	10.804	4	23	29.70	13.050	2	12	34.71	12.877	3	12	0.273
TAFF2	28.21	10.318	5	24	28.70	12.840	3	22	34.58	12.347	4	13	0.253
TAFF3	26.49	11.403	7	28	25.83	12.576	6	37	24.81	10.196	7	40	0.970
TAFF4	2619	12.236	1	1	25.26	12.527	7	40	26.42	11.135	6	38	0.617
TAFF5	29.84	10.641	3	19	28.17	14.721	4	26	34.94	11.445	2	11	0.543
TAFF6	26.67	13.601	6	27	26.65	15.150	5	29	26.62	11.934	5	36	0.801
TAFF7	31.77	13.685	2	12	31.87	14.204	1	5	35.61	11.724	1	6	0.297
TANX1	22.60	11.648	8	42	24.91	13.270	7	41	21.97	11.485	9	43	0.924
TANX2	24.44	12.758	4	37	26.22	13.038	2	32	21.65	12.071	10	44	0.158
TANX3	24.67	12.301	3	36	24.09	13.800	9	43	24.23	13.368	7	41	0.816
TANX4	25.79	13.049	1	31	25.61	15.400	6	39	29.71	15.612	2	25	0.724
TANX5	21.98	11.659	9	43	24.13	14.876	8	42	27.45	15.906	6	34	0.485
TANX6	24.42	14.029	5	38	25.83	15.299	5	38	29.0	14.058	4	27	0.389
TANX7	25.42	13.236	2	34	25.87	16.170	4	36	28.48	14.362	5	28	0.673
TANX8	24.14	12.351	6	39	27.65	14.920	1	28	30.16	14.365	1	23	0.161
TANX9	19.91	10.926	10	44	22.83	11.384	10	44	23.74	14.017	8	42	0.277
TANX10	22.93	12.166	7	41	26.22	15.676	3	33	29.58	15.046	3	26	0.205
PU1	33.44	11.335	1	2	29.43	10.599	2	15	35.68	11.426	3	5	0.600
PU2	32.63	10.205	6	10	28.35	9.623	6	25	34.00	11.827	6	15	0.312
PU3	32.91	12.645	4	6	29.39	10.483	3	17	36.19	11.400	2	3	0.324
PU4	33.30	11.841	2	3	28.57	10.211	5	24	35.10	12.210	5	9	0.558
PU5	32.72	12.304	5	8	28.70	10.890	4	21	36.65	11.837	1	1	0.374
PU6	33.19	11.587	3	4	29.65	11.219	1	13	35.13	11.615	4	7	0.516
PEU1	30.40	12.659	4	15	26.35	12.294	5	31	33.52	13.137	6	16	0.690
PEU2	32.65	13.397	2	9	28.61	11.773	2	23	36.26	11.855	1	2	0.489
PEU3	30.37	12.836	5	16	26.35	12.261	4	30	33.35	12.955	7	18	0.807
PEU4	30.02	13.785	6	17	27.65	11.734	3	27	34.97	12.656	4	10	0.464
PEU5	29.74	12.737	7	20	26.09	11.973	6	34	34.06	12.564	5	14	0.467
PEU6	31.63	12.636	3	13	25.91	12.203	7	35	35.10	12.004	3	8	0.694
PEU7	32.91	12.761	1	7	29.13	10.507	1	18	36.10	11.825	2	4	0.757
PIT1	29.88	12.133	2	18	32.74	14.824	1	4	30.81	12.362	2	21	0.471
PIT2	26.49	11.581	4	29	29.43	13.823	4	16	28.26	12.992	4	29	0.371
PIT3	28.98	12.305	3	22	30.00	17.315	3	11	33.03	15.050	1	19	0.768
PIT4	30.63	12.218	1	14	31.00	14.120	2	7	29.90	12.043	3	24	0.657

KW, Kruskal Wallis; RA, rank across components; RW, rank within component; SD, standard deviation.

Appendix A-14 Summary of Results for the Perceived Importance of Knowledge in Project Management by Respondent's Experience in Smart Technologies

Project Management Knowledge Areas	No			Yes			Mann-Whitney
	Mean	Standard Deviation	Rank	Mean	Standard Deviation	Rank	
Project Integration Management	6.28	0.998	3	6.16	1.174	7	0.835
Project Scope Management	6.04	1.009	7	6.00	1.195	9	0.820
Project Schedule Management	6.26	1.231	5	6.30	1.206	4	0.749
Project Cost Management	6.30	1.127	2	6.30	1.225	5	0.752
Project Quality Management	6.22	1.144	6	6.33	1.149	2	0.407
Project Resource Management	5.96	1.098	9	6.16	1.194	8	0.157
Project Communication Management	6.43	1.057	1	6.35	1.173	1	0.986
Project Risk Management	6.28	1.123	4	6.30	1.186	3	0.700
Project Procurement Management	5.72	1.123	10	5.98	1.205	10	0.179
Project Stakeholder Management	6.00	1.197	8	6.19	1.160	6	0.339

Appendix A-15 Summary of Results for the Perceived Importance of Skills to Manage Projects by Respondent's Experience in Smart Technologies

Skills to Manage Projects	No			Yes			Mann-Whitney
	Mean	Standard Deviation	Rank	Mean	Standard Deviation	Rank	
Technical and Operational Technology Skills	5.74	1.085	11	6.12	1.074	8	0.070
Project Management Skills	6.06	1.089	3	6.23	0.895	3	0.515
Information Management Skills	5.98	1.325	4	6.42	0.906	1	0.136
Planning and Organising Skills	5.94	1.123	6	6.21	0.940	4	0.296
Communication Skills	6.17	1.145	2	6.12	1.051	7	0.683
Social Awareness	5.31	1.315	16	5.35	1.213	18	0.868
Cultural Awareness	5.07	1.163	19	5.30	1.124	19	0.382
Organisational Awareness	5.41	1.190	15	5.65	1.110	15	0.307
Creativity	5.78	0.984	10	5.88	1.096	10	0.452
Problem-Solving Skills	6.24	0.889	1	6.35	0.842	2	0.577
Ethical Awareness	5.50	1.240	14	5.72	1.351	14	0.306
Strategic Planning Skills	5.85	1.089	9	6.19	1.075	5	0.076
Active Learning Skills	5.17	1.077	18	5.42	1.349	17	0.214
Conflict Management Skills	5.87	1.082	8	6.05	1.154	9	0.324
Decision-Making Skills	5.96	1.288	5	6.19	1.139	6	0.426
Delegation Skills	5.54	1.077	13	5.65	1.289	16	0.521
Motivation Skills	5.63	1.322	12	5.88	1.138	11	0.396
Negotiation Skills	5.30	1.143	17	5.72	1.221	13	0.046*
Team-Building Skills	5.93	1.242	7	5.88	1.219	12	0.799

* Significant at *p*-value <0.05.

Appendix A-16 Summary of Results for the Factors Affecting Attitude towards Technologies by Respondent's Experience in Smart Technologies

Factor Affecting Attitude Towards Technologies

Applicability

	No				Yes				MW
	Mean	SD	RW	RA	Mean	SD	RW	RA	
TSE1	5.67	1.360	2	3	5.74	1.364	1	8	0.743
TSE2	5.09	0.996	8	29	5.42	1.006	5	19	0.117
TSE3	4.94	0.960	10	33	5.19	1.160	10	34	0.221
TSE4	5.19	0.973	5	25	5.47	1.032	4	16	0.162
TSE5	5.06	1.071	9	30	5.37	1.024	6	23	0.163
TSE6	5.24	1.027	4	24	5.26	0.902	8	29	0.862
TSE7	5.15	0.979	6	27	5.33	1.149	7	25	0.295
TSE8	5.15	0.1035	7	28	5.23	0.972	9	30	0.762
TSE9	5.74	1.507	1	1	5.53	1.681	2	14	0.590
TSE10	5.43	1.191	3	17	5.49	0.985	3	15	0.883
TAFF1	5.48	1.161	4	15	5.67	0.993	2	10	0.486
TAFF2	5.61	1.140	3	5	5.47	1.032	4	16	0.391
TAFF3	4.96	1.243	5	31	5.02	0.988	7	40	0.985
TAFF4	4.96	1.288	6	32	5.16	0.949	5	35	0.486
TAFF5	5.69	1.329	1	2	5.60	1.137	3	12	0.450
TAFF6	4.91	1.593	7	34	5.09	1.1211	6	38	0.721
TAFF7	5.67	1.360	2	3	5.74	1.197	1	7	0.949
TANX1	4.44	1.192	8	42	4.93	1.142	8	42	0.029*
TANX2	4.59	1.486	6	40	4.88	1.219	9	43	0.331
TANX3	4.54	1.514	7	41	4.95	1.272	7	41	0.139
TANX4	4.78	1.586	1	35	5.30	1.597	3	28	0.092
TANX5	4.39	1.595	9	43	5.12	1.467	6	37	0.022*
TANX6	4.74	1.443	3	37	5.23	1.377	4	32	0.087
TANX7	4.72	1.522	4	38	5.35	1.429	1	24	0.042*
TANX8	4.76	1.427	2	36	5.33	1.410	2	26	0.055
TANX9	4.24	1.529	10	44	4.6	1.417	10	44	0.275
TANX10	4.67	1.387	5	39	5.16	1.413	5	36	0.089
PU1	5.52	1.128	4	13	5.93	0.910	4	4	0.072
PU2	5.52	0.947	3	12	5.86	0.915	6	6	0.071
PU3	5.56	1.208	1	8	5.91	0.971	5	5	0.169
PU4	5.43	1.191	6	17	5.95	0.999	1	1	0.029*
PU5	5.44	1.239	5	16	5.95	1.090	3	3	0.046*
PU6	5.54	0.985	2	9	5.95	1.068	2	2	0.025*
PEU1	5.35	1.416	6	21	5.40	1.450	7	22	0.840
PEU2	5.52	1.193	3	14	5.72	1.221	1	9	0.386
PEU3	5.39	1.172	4	19	5.42	1.277	5	20	0.896
PEU4	5.37	1.336	5	20	5.44	1.297	4	18	0.817
PEU5	5.33	1.346	7	22	5.40	1.433	6	21	0.780
PEU6	5.54	1.145	2	10	5.53	1.279	3	13	0.925
PEU7	5.56	1.127	1	6	5.67	1.190	2	11	0.569
PIT1	5.56	1.144	1	7	5.30	1.319	1	27	0.372
PIT2	5.16	1.289	4	26	5.05	1.272	4	39	0.879
PIT3	5.31	1.540	3	23	5.21	1.505	3	33	0.675
PIT4	5.54	1.161	2	11	5.23	1.088	2	31	0.283

Factor Affecting Attitude Towards Technologies

Impact

	No				Yes				MW
	Mean	*SD*	*RW*	*RA*	*Mean*	*SD*	*RW*	*RA*	
TSE1	5.81	1.388	1	1	5.79	1.301	2	9	0.809
TSE2	4.93	1.211	9	30	5.28	1.241	5	26	0.126
TSE3	4.94	1.220	8	29	5.19	1.296	7	30	0.248
TSE4	5.26	1.200	5	23	5.33	1.149	4	22	0.770
TSE5	5.07	1.211	6	27	5.16	1.252	8	33	0.681
TSE6	5.26	1.049	4	22	4.93	1.298	9	39	0.241
TSE7	5.00	1.182	7	28	5.19	1.220	6	29	0.387
TSE8	4.89	1.369	10	31	4.86	1.521	10	43	0.962
TSE9	5.76	1.317	2	2	5.88	1.276	1	7	0.628
TSE10	5.28	1.280	3	21	5.42	1.139	3	18	0.702
TAFF1	5.26	1.348	3	24	5.42	1.200	2	19	0.630
TAFF2	5.30	1.268	2	20	5.28	1.202	4	25	0.784
TAFF3	4.80	1.392	6	33	5.09	1.109	6	35	0.389
TAFF4	4.80	1.571	7	34	5.05	1.112	7	36	0.528
TAFF5	5.26	1.376	3	25	5.30	1.206	3	23	0.994
TAFF6	4.85	1.559	5	32	5.12	1.313	5	34	0.423
TAFF7	5.48	1.356	1	7	5.63	1.215	1	11	0.673
TANX1	4.46	1.229	9	43	4.86	1.246	9	42	0.073
TANX2	4.76	1.331	2	36	4.72	1.351	10	44	0.905
TANX3	4.76	1.386	3	37	4.93	1.316	7	40	0.547
TANX4	4.78	1.574	1	35	5.21	1.424	2	27	0.188
TANX5	4.61	1.433	7	41	4.98	1.456	6	38	0.219
TANX6	4.63	1.582	6	40	5.19	1.516	4	32	0.085
TANX7	4.61	1.559	8	42	5.21	1.424	2	27	0.050
TANX8	4.70	1.369	5	39	5.35	1.395	1	21	0.023*
TANX9	4.37	1.445	10	44	4.91	1.444	8	41	0.059
TANX10	4.76	1.553	2	38	5.05	1.603	5	37	0.384
PU1	5.41	1.108	1	9	6.00	1.091	1	1	0.005*
PU2	5.31	1.079	6	18	5.77	1.043	6	10	0.028*
PU3	5.33	1.318	5	17	5.93	1.078	3	3	0.023*
PU4	5.35	1.135	3	15	5.91	1.087	4	4	0.011*
PU5	5.39	1.071	2	10	5.95	1.112	2	2	0.006*
PU6	5.35	1.084	3	14	5.88	1.096	5	5	0.016*
PEU1	5.37	1.248	5	12	5.60	1.275	3	12	0.341
PEU2	5.50	1.314	2	5	5.84	1.271	2	8	0.178
PEU3	5.31	1.241	7	19	5.44	1.436	7	17	0.616
PEU4	5.48	1.270	3	6	5.56	1.278	5	14	0.788
PEU5	5.37	1.202	4	11	5.56	1.240	4	13	0.459
PEU6	5.37	1.263	6	13	5.56	1.333	6	15	0.440
PEU7	5.54	1.239	1	3	5.88	1.138	1	6	0.151
PIT1	5.43	1.191	2	8	5.42	1.384	2	20	0.905
PIT2	5.17	1.255	4	26	5.19	1.350	4	31	0.690
PIT3	5.35	1.481	3	16	5.51	1.549	1	16	0.532
PIT4	5.50	1.209	1	4	5.30	1.264	3	24	0.519

(*Continued*)

Appendix A-16 (Continued)

	Factor Affecting Attitude Towards Technologies								

Significance

	No				Yes				MW
	Mean	*SD*	*RW*	*RA*	*Mean*	*SD*	*RW*	*RA*	
TSE1	34.41	13.489	2	2	34.09	13.322	1	9	0.867
TSE2	26.06	11.055	9	30	29.37	10.900	5	27	0.129
TSE3	25.30	10.775	10	33	27.93	11.667	8	33	0.185
TSE4	28.20	10.910	5	25	29.77	10.279	4	25	0.365
TSE5	26.74	11.141	6	27	28.53	10.864	6	30	0.428
TSE6	28.28	10.053	4	24	26.63	10.123	9	38	0.616
TSE7	26.57	9.903	7	28	28.35	10.812	7	31	0.287
TSE8	26.31	11.717	8	29	26.30	11.548	10	40	0.994
TSE9	34.44	13.629	1	1	33.77	14.588	2	10	0.831
TSE10	29.94	12.331	3	20	30.63	11.058	3	19	0.845
TAFF1	30.07	12.629	4	18	31.72	11.770	2	13	0.514
TAFF2	30.76	11.891	3	11	29.86	11.926	4	23	0.614
TAFF3	25.26	11.996	7	34	26.47	10.292	7	39	0.540
TAFF4	25.33	13.073	6	32	26.93	10.190	6	37	0.390
TAFF5	31.28	12.677	2	8	30.81	11.601	3	18	0.814
TAFF6	25.93	14.509	5	31	27.35	11.840	5	35	0.548
TAFF7	32.65	13.794	1	3	33.49	12.567	1	11	0.804
TANX1	21.15	11.850	9	43	25.21	11.767	8	42	0.048*
TANX2	23.63	12.914	6	40	24.40	12.333	9	43	0.797
TANX3	23.28	13.051	7	41	25.79	12.663	7	41	0.270
TANX4	24.98	14.338	1	35	29.53	14.330	3	26	0.143
TANX5	21.93	13.772	8	42	27.14	13.806	6	36	0.055
TANX6	24.07	14.268	4	38	28.98	14.124	4	28	0.093
TANX7	23.89	14.021	5	39	29.79	13.996	2	24	0.043*
TANX8	24.19	13.041	2	36	30.30	14.020	1	21	0.035*
TANX9	20.20	11.980	10	44	23.86	12.078	10	44	0.097
TANX10	24.15	13.717	3	37	27.95	14.554	5	32	0.208
PU1	30.70	11.373	2	13	36.35	10.562	2	2	0.009*
PU2	30.06	10.453	6	19	34.56	10.667	6	6	0.023*
PU3	30.89	12.202	1	9	35.93	11.070	5	5	0.033*
PU4	30.15	11.799	5	17	36.02	10.951	4	4	0.016*
PU5	30.35	11.784	4	16	36.37	11.735	1	1	0.016*
PU6	30.52	11.030	3	14	36.05	11.625	3	3	0.016*
PEU1	29.74	12.395	6	22	31.30	13.550	5	15	0.658
PEU2	31.52	12.690	2	6	34.51	12.751	1	7	0.243
PEU3	29.80	12.157	5	21	31.09	13.834	6	16	0.753
PEU4	30.76	13.296	4	12	31.40	13.113	4	14	0.759
PEU5	29.69	12.425	7	23	30.98	13.222	7	17	0.727
PEU6	30.83	12.134	3	10	32.07	13.437	3	12	0.647
PEU7	31.94	12.076	1	4	34.40	12.181	2	8	0.379
PIT1	31.43	12.430	2	7	30.14	13.359	2	22	0.642
PIT2	27.96	13.225	4	26	27.49	11.718	4	34	0.828
PIT3	30.46	14.799	3	15	30.58	14.181	1	20	0.894
PIT4	31.80	13.187	1	5	28.84	11.545	3	29	0.300

MW, Mann-Whitney; RA, rank across components; RW, rank within component; SD, standard deviation.

* Significant at *p*-value <0.05.

Appendix A-17 Summary of Results for the Perceived Importance of Knowledge in Project Management by Organisation Domain

Project Management Knowledge Areas	Consultant			Contractor			Developer			Kruskal Wallis
	Mean	Standard Deviation	Rank	Mean	Standard Deviation	Rank	Mean	Standard Deviation	Rank	
Project Integration Management	6.35	0.834	6	6.14	1.353	4	6.14	0.990	4	0.762
Project Scope Management	6.10	0.928	9	5.94	1.305	8	6.00	1.024	7	0.955
Project Schedule Management	6.42	1.083	4	6.20	1.368	3	6.14	1.207	5	0.433
Project Cost Management	6.53	0.877	3	6.23	1.352	1	6.00	1.272	9	0.250
Project Quality Management	6.40	0.928	5	6.14	1.353	4	6.23	1.152	2	0.865
Project Resource Management	6.20	0.911	7	5.91	1.337	9	6.00	1.195	8	0.803
Project Communication Management	6.63	0.740	1	6.20	1.346	2	6.27	1.202	1	0.337
Project Risk Management	6.58	0.813	2	6.03	1.361	6	6.18	1.220	3	0.116
Project Procurement Management	5.95	1.037	10	5.71	1.296	10	5.82	1.181	10	0.813
Project Stakeholder Management	6.18	1.035	8	6.00	1.350	7	6.05	1.174	6	0.938

Appendix A-18 Summary of Results for the Perceived Importance of Skills to Manage Projects by Organisation Domain

Skills to Manage Projects	Consultant			Contractor			Developer			Kruskal Wallis
	Mean	Standard Deviation	Rank	Mean	Standard Deviation	Rank	Mean	Standard Deviation	Rank	
Technical and Operational Technology Skills	5.83	0.958	11	5.91	1.173	7	6.05	1.214	8	0.537
Project Management Skills	6.13	0.966	6	6.09	1.197	4	6.23	0.752	1	0.978
Information Management Skills	6.28	1.154	2	6.14	1.240	2	6.05	1.133	6	0.533
Planning and Organising Skills	6.00	1.109	9	6.11	1.078	3	6.09	0.921	4	0.866
Communication Skills	6.25	1.056	3	6.06	1.259	5	6.09	0.921	4	0.637
Social Awareness	5.32	1.095	17	5.23	1.395	17	5.50	1.371	17	0.687
Cultural Awareness	5.30	0.966	18	4.94	1.305	19	5.32	1.171	19	0.426
Organisational Awareness	5.68	1.095	13	5.26	1.197	15	5.64	1.177	15	0.334
Creativity	5.88	0.939	10	5.80	1.079	8	5.77	1.152	11	0.989
Problem-Solving Skills	6.35	0.770	1	6.26	0.980	1	6.23	0.869	2	0.882
Ethical Awareness	5.70	1.043	12	5.37	1.457	14	5.77	1.412	13	0.525
Strategic Planning Skills	6.10	0.900	8	5.80	1.368	9	6.14	0.889	3	0.752
Active Learning Skills	5.28	0.960	19	5.14	1.332	18	5.50	1.406	18	0.492
Conflict Management Skills	6.20	0.853	4	5.77	1.352	10	5.77	1.066	10	0.251
Decision-Making Skills	6.15	1.122	5	5.97	1.382	6	6.05	1.174	7	0.920
Delegation Skills	5.68	1.095	13	5.40	1.218	13	5.73	1.241	14	0.521
Motivation Skills	5.68	1.309	15	5.74	1.358	11	5.86	0.941	9	0.943
Negotiation Skills	5.62	1.030	16	5.23	1.330	16	5.64	1.217	16	0.377
Team-Building Skills	6.13	0.966	6	5.74	1.421	12	5.77	1.307	12	0.558

Appendix A-19 Summary of Results for the Factors Affecting Attitude Towards Technologies by Organisation Domain

---

Factors Affecting Attitude Towards Technologies

Applicability

	Consultant				Contractor				Developer				KW
	Mean	SD	RW	RA	Mean	SD	RW	RA	Mean	SD	RW	RA	
TSE1	5.60	1.482	1	11	5.91	1.173	2	2	5.55	1.405	2	15	0.565
TSE2	5.25	1.080	8	24	5.20	0.868	4	25	5.27	1.120	8	25	0.980
TSE3	5.07	1.095	10	33	5.00	0.939	8	30	5.09	1.192	10	33	0.948
TSE4	5.40	1.057	4	18	5.14	0.912	6	27	5.41	1.054	4	20	0.517
TSE5	5.30	1.137	7	23	5.00	0.907	7	29	5.32	1.129	7	23	0.327
TSE6	5.35	1.027	5	21	4.97	0.785	10	34	5.50	1.058	3	17	0.117
TSE7	5.30	1.018	6	22	5.20	0.994	5	26	5.14	1.246	9	30	0.906
TSE8	5.25	1.127	9	25	4.97	0.747	9	33	5.41	1.098	5	21	0.203
TSE9	5.50	1.664	3	14	6.00	1.435	1	1	5.36	1.620	6	22	0.182
TSE10	5.55	1.197	2	12	5.23	0.877	3	24	5.64	1.217	1	12	0.231
TAFF1	5.65	1.001	3	7	5.49	1.314	4	21	5.55	0.858	3	14	0.868
TAFF2	5.65	.0975	2	6	5.51	1.292	2	18	5.41	0.959	4	19	0.688
TAFF3	5.17	1.083	6	30	4.71	1.100	7	40	5.09	1.231	6	34	0.165
TAFF4	5.23	1.121	5	26	4.83	1.224	6	38	5.09	1.065	5	32	0.352
TAFF5	5.68	1.141	1	5	5.50	1.397	3	19	5.68	1.211	2	10	0.989
TAFF6	5.12	1.362	7	32	4.91	1.401	5	36	4.86	1.642	7	42	0.719
TAFF7	5.65	1.312	4	8	5.66	1.305	1	8	5.86	1.246	1	4	0.793
TANX1	4.63	1.102	8	42	4.49	1.197	9	43	5.00	1.309	5	36	0.353
TANX2	4.70	1.363	6	40	4.57	1.399	8	42	5.00	1.380	6	37	0.611
TANX3	4.68	1.366	7	41	4.60	1.479	7	41	5,00	1.447	7	38	0.673
TANX4	4.98	1.423	1	34	5.00	1.847	3	32	5.09	1.571	4	35	0.877
TANX5	4.55	1.501	9	43	4.77	1.767	6	39	4.91	1.411	8	41	0.674
TANX6	4.88	1.362	4	37	4.86	1.556	5	37	5.27	1.352	2	27	0.521
TANX7	4.90	1.464	2	35	4.97	1.562	4	35	5.23	1.541	3	28	0.659
TANX8	4.88	1.305	3	36	5.00	1.663	2	31	5.27	1.316	1	26	0.576
TANX9	4.40	1.482	10	44	4.29	1.426	10	44	4.59	1.623	10	44	0.741
TANX10	4.75	1.335	5	39	5.03	1.599	1	28	4.91	1.269	8	40	0.548
PU1	5.70	1.018	1	1	5.57	1.119	4	10	5.91	1.019	1	1	0.518
PU2	5.68	0.888	3	3	5.54	1.010	6	12	5.86	0.941	3	3	0.475
PU3	5.70	1.067	2	2	5.66	1.083	2	6	5.82	1.296	6	7	0.658
PU4	5.63	1.030	5	9	5.57	1.195	5	11	5.86	1.246	4	4	0.491
PU5	5.60	1.172	6	10	5.60	1.241	3	9	5.91	1.192	2	2	0.520
PU6	5.68	0.888	3	3	5.69	1.051	1	3	5.86	1.283	5	5	0.524
PEU1	5.18	1.517	6	29	5.49	1.269	6	20	5.55	1.503	5	16	0.532
PEU2	5.53	1.240	1	13	5.66	1.083	2	6	5.68	1.359	3	11	0.820
PEU3	5.40	1.257	4	19	5.46	1.146	7	23	5.32	1.287	7	24	0.976
PEU4	5.18	1.357	5	28	5.54	1.291	3	15	5.59	1.260	4	13	0.382
PEU5	5.15	1.511	7	31	5.51	1.245	5	17	5.50	1.336	6	18	0.525
PEU6	5.45	1.260	3	16	5.51	1.147	4	16	5.73	1.202	1	8	0.700
PEU7	5.48	1.198	2	15	5.69	1.078	1	4	5.73	1.202	1	8	0.638
PIT1	5.38	1.148	2	20	5.69	1.231	1	5	5.18	1.332	1	29	0.295
PIT2	4.85	1.189	4	38	5.54	1.221	3	14	4.82	1.368	4	43	0.033*
PIT3	5.23	1.330	3	27	5.49	1.755	4	22	5.00	1.447	3	38	0.258
PIT4	5.43	1.035	1	17	5.54	1.146	2	13	5.14	1.283	2	31	0.530

(Continued)

Appendix A-19 (Continued)

Factors Affecting Attitude Towards Technologies

Impact

	Consultant				Contractor				Developer				KW
	Mean	SD	RW	RA	Mean	SD	RW	RA	Mean	SD	RW	RA	
TSE1	5.88	1.436	2	2	5.86	1.353	2	2	5.59	1.182	2	15	0.423
TSE2	5.08	1.269	7	30	4.94	1.162	6	29	5.32	1.287	5	18	0.639
TSE3	5.07	1.207	8	31	4.86	1.264	8	32	5.32	1.323	6	19	0.392
TSE4	5.40	1.128	4	18	5.09	1.147	3	24	5.41	1.297	4	17	0.444
TSE5	5.25	1.296	6	25	4.91	1.011	7	30	5.18	1.402	8	27	0.389
TSE6	5.35	1.122	5	19	4.77	1.031	9	34	5.23	1.378	7	26	0.068
TSE7	5.05	1.260	9	32	5.09	1.147	3	24	5.14	1.207	9	31	0.960
TSE8	4.95	1.449	10	34	4.63	1.285	10	41	5.14	1.612	10	34	0.336
TSE9	5.73	1.396	3	3	5.97	1.248	1	1	5.73	1.202	1	9	0.599
TSE10	7.42	1.238	1	1	5.09	1.147	3	24	5.59	1.260	2	16	0.191
TAFF1	5.48	1.132	3	12	5.23	1.477	4	23	5.23	1.232	3	24	0.739
TAFF2	5.53	1.012	2	9	5.23	1.437	3	22	4.95	1.214	5	38	0.227
TAFF3	5.10	1.150	7	29	4.63	1.285	7	41	5.09	1.444	4	36	0.209
TAFF4	5.15	1.331	5	27	4.66	1.413	6	40	4.86	1.424	7	41	0.329
TAFF5	5.30	1.265	4	21	5.26	1.442	2	21	5.27	1.162	2	21	0.994
TAFF6	5.15	1.406	6	28	4.77	1.477	5	35	4.95	1.527	6	39	0.514
TAFF7	5.65	1.189	1	5	5.37	1.477	1	20	5.64	1.177	1	14	0.775
TANX1	4.55	1.239	9	43	4.46	1.221	10	44	5.09	1.411	6	35	0.128
TANX2	4.75	1.235	7	41	4.71	1.274	7	38	4.77	1.631	10	44	0.986
TANX3	4.80	1.305	4	38	4.69	1.388	8	39	5.14	1.390	5	32	0.595
TANX4	4.83	1.375	2	36	5.00	1.698	2	28	5.18	1.500	2	28	0.583
TANX5	4.78	1.165	5	39	4.77	1.716	5	36	4.77	1.510	9	43	0.994
TANX6	4.78	1.510	6	40	4.80	1.677	4	33	5.18	1.532	3	29	0.595
TANX7	4.83	1.448	3	37	4.74	1.597	6	37	5.18	1.563	4	30	0.510
TANX8	4.90	1.257	1	35	4.91	1.616	3	31	5.27	1.352	1	22	0.563
TANX9	4.48	1.450	10	44	4.63	1.374	9	43	4.82	1.651	8	42	0.646
TANX10	4.65	1.460	8	42	5.09	1.772	1	26	5.00	1.447	7	37	0.373
PU1	5.70	0.992	1	4	5.49	1.173	4	15	5.91	1.306	1	2	0.270
PU2	5.45	0.959	6	14	5.43	1.195	6	19	5.77	1.110	4	6	0.383
PU3	5.60	1.215	2	6	5.53	1.221	3	11	5.68	1.393	6	12	0.797
PU4	5.55	1.061	4	8	5.53	1.094	2	10	5.77	1.378	5	8	0.468
PU5	5.58	1.059	3	7	5.57	1.008	1	6	5.86	1.390	3	5	0.288
PU6	5.53	1.037	5	10	5.49	1.173	4	15	5.86	1.167	2	3	0.360
PEU1	5.27	1.301	6	24	5.51	1.222	5	12	5.77	1.232	3	7	0.338
PEU2	5.50	1.377	1	11	5.77	1.087	2	4	5.73	1.486	5	11	0.656
PEU3	5.20	1.363	7	26	5.60	1.193	3	5	5.32	1.460	7	20	0.462
PEU4	5.33	1.269	3	20	5.51	1.269	5	13	5.86	1.246	2	4	0.259
PEU5	5.28	1.261	4	22	5.54	1.221	4	9	5.64	1.136	6	13	0.441
PEU6	5.28	1.339	5	23	5.49	1.197	7	17	5.73	1.352	4	10	0.370
PEU7	5.45	1.319	2	17	5.80	1.079	1	3	5.95	1.133	1	1	0.276
PIT1	5.45	1.176	3	16	5.51	1.358	3	14	5.23	1.343	2	25	0.712
PIT2	4.98	1.209	4	33	5.57	1.170	1	7	4.91	1.509	4	40	0.086
PIT3	5.48	1.301	1	13	5.46	1.738	4	18	5.27	1.518	1	23	0.765
PIT4	5.45	1.154	2	15	5.54	1.172	2	8	5.14	1.457	3	33	0.577

Factors Affecting Attitude Towards Technologies

Significance

	Consultant				Contractor				Developer				KW
	Mean	SD	RW	RA	Mean	SD	RW	RA	Mean	SD	RW	RA	
TSE1	34.55	13.994	1	3	35.80	13.521	2	2	31.32	11.890	2	15	0.396
TSE2	27.75	12.190	7	29	26.40	9.586	6	30	28.91	11.385	6	25	0.732
TSE3	26.73	11.518	10	33	25.26	10.942	7	33	27.91	11.322	9	30	0.536
TSE4	30.12	11.341	4	18	27.00	9.905	5	28	29.68	10.357	4	19	0.498
TSE5	29.07	12.223	6	23	25.17	8.783	8	34	28.50	11.653	8	27	0.361
TSE6	29.40	10.736	5	21	24.26	7.901	9	37	29.41	10.944	5	21	0.099
TSE7	27.65	10.729	8	30	27.31	9.964	4	27	26.91	10.488	10	37	0.973
TSE8	27.38	12.897	9	32	23.63	8.994	10	39	28.64	12.389	7	26	0.322
TSE9	33.05	14.390	2	6	37.17	13.0874	1	1	31.32	13.145	3	16	0.246
TSE10	32.38	12.403	3	8	27.49	10.354	3	26	32.59	12.195	1	13	0.177
TAFF1	31.85	11.292	3	11	30.37	14.361	3	23	29.59	10.395	3	20	0.781
TAFF2	32.00	10.607	2	10	30.37	13.903	3	22	27.36	10.284	5	35	0.279
TAFF3	27.42	10.848	7	31	22.86	10.276	7	42	27.50	12.831	4	33	0.116
TAFF4	28.17	12.034	5	26	23.69	11.465	6	38	25.91	11.924	7	39	0.280
TAFF5	31.10	11.553	4	15	31.09	13.714	2	21	31.00	11.071	2	17	0.987
TAFF6	28.07	13.258	6	28	25.14	13.084	5	35	26.05	14.228	6	38	0.524
TAFF7	33.30	13.035	1	4	31.94	14.332	1	13	34.23	12.047	1	8	0.841
TANX1	22.35	11.502	9	43	21.09	11.009	10	44	27.00	13.586	6	36	0.145
TANX2	23.78	11.592	5	39	23.09	12.344	8	41	25.73	14.996	8	41	0.883
TANX3	23.77	12.376	6	40	23.14	12.431	7	40	27.50	14.481	5	34	0.590
TANX4	25.68	13.141	1	34	27.69	16.009	2	25	28.32	14.545	4	29	0.830
TANX5	22.98	12.491	8	42	25.03	15.682	6	36	25.27	14.062	9	43	0.905
TANX6	25.18	13.581	4	38	25.69	15.156	5	32	29.09	14.648	3	24	0.598
TANX7	25.65	13.990	2	35	25.71	14.241	4	31	29.32	15.003	2	23	0.601
TANX8	25.42	12.600	3	36	27.00	15.137	3	29	29.41	13.727	1	22	0.567
TANX9	21.28	11.649	10	44	21.20	11.717	9	43	23.82	13.769	10	44	0.772
TANX10	23.75	13.299	7	41	28.17	15.779	1	24	25.91	12.887	7	40	0.446
PU1	33.13	10.006	3	5	31.60	12.250	5	16	35.91	12.043	2	2	0.354
PU2	31.45	9.367	6	12	31.09	11.828	6	20	34.68	11.290	5	6	0.433
PU3	32.90	11.026	4	7	32.43	12.291	1	7	34.64	13.272	6	7	0.675
PU4	92.00	10.551	2	2	32.03	12.256	4	11	35.27	13.116	4	4	0.487
PU5	92.05	11.170	1	1	32.26	12.301	2	9	36.00	13.359	1	1	0.321
PU6	32.08	10.202	5	9	32.26	12.241	2	8	35.73	12.881	3	3	0.390
PEU1	28.12	12.468	7	27	31.31	12.797	7	19	33.23	13.575	5	12	0.317
PEU2	31.45	12.930	1	13	33.60	11.730	2	4	34.18	14.215	2	9	0.619
PEU3	29.38	12.883	4	22	31.80	12.672	4	15	29.91	13.561	7	18	0.716
PEU4	28.87	13.464	5	24	31.86	13.218	3	14	33.68	12.396	4	11	0.361
PEU5	28.22	13.061	6	25	31.60	12.751	5	17	31.82	12.152	6	14	0.361
PEU6	29.98	12.861	3	20	31.34	12.438	6	18	34.00	12.877	3	10	0.453
PEU7	31.13	12.562	2	14	33.89	11.512	1	3	35.14	12.299	1	5	0.428
PIT1	30.35	11.668	2	17	32.91	13.874	1	5	28.50	13.074	1	28	0.470
PIT2	25.25	11.419	4	37	32.11	12.676	3	10	25.36	12.819	4	42	0.046*
PIT3	30.03	13.524	3	19	32.74	16.112	2	6	27.86	13.375	3	32	0.414
PIT4	30.60	11.873	1	16	32.00	12.882	4	12	27.86	13.282	2	31	0.408

KW, Kruskal Wallis test; RW, rank with component; RA, rank across components; SD, standard deviation.

* Significant at *p*-value <0.05.

Appendix A-20 Summary of Results for the Perceived Importance of Knowledge in Project Management by Organisation's Experience in Smart Technologies

Project Management Knowledge Areas	No			Yes			Mann-Whitney
	Mean	Standard Deviation	Rank	Mean	Standard Deviation	Rank	
Project Integration Management	6.13	1.212	5	6.28	1.008	6	0.675
Project Scope Management	5.78	1.184	7	6.14	1.029	9	0.119
Project Schedule Management	6.16	1.298	4	6.34	1.176	4	0.429
Project Cost Management	6.22	1.263	3	6.34	1.122	3	0.688
Project Quality Management	6.22	1.184	2	6.29	1.128	5	0.725
Project Resource Management	5.69	1.306	9	6.23	1.012	7	0.032*
Project Communication Management	6.25	1.295	1	6.46	1.001	1	0.443
Project Risk Management	6.06	1.343	6	6.40	1.028	2	0.225
Project Procurement Management	5.47	1.191	10	6.02	1.111	10	0.021*
Project Stakeholder Management	5.78	1.431	8	6.23	1.012	7	0.187

* Significant at *p*-value <0.05.

Appendix A-21 Summary of Results for the Perceived Importance of Skills to Manage Projects by Organisation's Experience in Smart Technologies

Skills to Manage Projects	No			Yes			Mann-Whitney
	Mean	Standard Deviation	Rank	Mean	Standard Deviation	Rank	
Technical and Operational Technology Skills	5.78	1.099	11	5.97	1.089	8	0.398
Project Management Skills	6.16	1.167	3	6.12	0.927	4	0.553
Information Management Skills	6.06	1.294	6	6.23	1.115	2	0.710
Planning and Organising Skills	6.19	0.998	2	6.00	1.075	7	0.425
Communication Skills	6.06	1.366	7	6.18	0.950	3	0.811
Social Awareness	5.22	1.313	18	5.38	1.246	17	0.576
Cultural Awareness	5.03	1.257	19	5.25	1.090	19	0.418
Organisational Awareness	5.47	1.295	14	5.54	1.091	16	0.908
Creativity	5.97	0.999	8	5.75	1.046	12	0.370
Problem-Solving Skills	6.31	0.859	1	6.28	0.875	1	0.870
Ethical Awareness	5.50	1.270	13	5.65	1.304	14	0.571
Strategic Planning Skills	5.91	1.146	10	6.05	1.067	5	0.551
Active Learning Skills	5.31	1.120	16	5.26	1.253	18	0.874
Conflict Management Skills	5.97	1.177	9	5.94	1.088	9	0.771
Decision-Making Skills	6.16	1.298	4	6.02	1.192	6	0.427
Delegation Skills	5.41	1.214	15	5.68	1.147	13	0.271
Motivation Skills	5.69	1.554	12	5.77	1.072	11	0.686
Negotiation Skills	5.31	1.256	17	5.57	1.159	15	0.294
Team-Building Skills	6.09	1.279	5	5.82	1.198	10	0.164

Appendix A-22 Summary of Results for the Factors Affecting Attitude Towards Technologies by Organisation's Experience in Smart Technologies

Factors Affecting Attitude Towards Technologies

Applicability

	No				Yes				MW
	Mean	SD	RW	RA	Mean	SD	RW	RA	
TSE1	5.84	1.439	1	1	5.63	1.318	1	11	0.294
TSE2	4.94	0.914	9	30	5.38	1.026	5	22	0.043*
TSE3	4.94	0.914	9	30	5.11	1.120	10	34	0.424
TSE4	5.09	0.963	4	25	5.42	1.014	4	19	0.131
TSE5	4.97	1.092	8	29	5.31	1.030	8	28	0.143
TSE6	5.06	1.014	6	27	5.34	0.940	6	25	0.172
TSE7	5.03	1.031	7	28	5.32	1.062	7	27	0.145
TSE8	5.09	1.027	5	26	5.23	0.996	9	29	0.631
TSE9	5.81	1.554	2	2	5.57	1.600	2	14	0.459
TSE10	5.28	1.170	3	21	5.54	1.062	3	16	0.318
TAFF1	5.47	1.344	4	14	5.62	0.947	3	12	0.726
TAFF2	5.53	1.344	3	9	5.55	0.952	4	15	0.892
TAFF3	4.66	1.260	7	38	5.15	1.034	6	31	0.065
TAFF4	4.81	1.278	5	32	5.17	1.009	5	30	0.188
TAFF5	5.66	1.359	2	4	5.65	1.192	2	10	0.771
TAFF6	4.75	1.566	6	35	5.11	1.359	7	35	0.255
TAFF7	5.78	1.157	1	3	5.66	1.350	1	8	0.836
TANX1	4.22	1.039	9	43	4.88	1.206	9	43	0.016*
TANX2	4.28	1.464	7	41	4.94	1.285	6	40	0.054
TANX3	4.37	1.561	6	40	4.89	1.324	8	42	0.127
TANX4	4.78	1.660	2	34	5.12	1.576	2	33	0.351
TANX5	4.28	1.631	8	42	4.92	1.514	7	41	0.059
TANX6	4.69	1.491	4	37	5.09	1.389	4	37	0.210
TANX7	4.72	1.529	3	36	5.14	1.488	1	32	0.217
TANX8	4.81	1.401	1	33	5.11	1.459	3	36	0.391
TANX9	4.03	1.576	10	44	4.58	1.413	10	44	0.127
TANX10	4.66	1.473	5	39	5.00	1.381	5	39	0.280
PU1	5.41	1.132	5	16	5.85	0.988	1	1	0.065
PU2	5.47	1.047	3	12	5.77	0.880	5	5	0.149
PU3	5.62	1.100	1	5	5.75	1.132	6	6	0.487
PU4	5.41	1.188	6	17	5.78	1.097	3	3	0.124
PU5	5.44	1.268	4	15	5.78	1.152	4	4	0.174
PU6	5.56	1.014	2	6	5.80	1.049	2	2	0.194
PEU1	5.31	1.330	5	20	5.40	1.477	6	21	0.605
PEU2	5.50	1.244	2	11	5.66	1.189	1	7	0.563
PEU3	5.34	1.260	4	19	5.43	1.199	5	18	0.800
PEU4	5.28	1.326	7	23	5.46	1.312	4	17	0.522
PEU5	5.28	1.276	6	22	5.40	1.434	6	20	0.536
PEU6	5.47	1.164	3	13	5.57	1.224	3	13	0.646
PEU7	5.53	1.135	1	8	5.65	1.165	2	9	0.629
PIT1	5.56	1.216	1	7	5.38	1.234	1	23	0.517
PIT2	5.22	1.362	4	24	5.03	1.237	4	38	0.571
PIT3	5.38	1.699	3	18	5.33	1.431	3	26	0.504
PIT4	5.50	1.218	2	10	5.35	1.096	2	24	0.690

Factors Affecting Attitude Towards Technologies

Impact

	No				Yes				MW
	Mean	*SD*	*RW*	*RA*	*Mean*	*SD*	*RW*	*RA*	
TSE1	5.84	1.462	1	1	5.78	1.293	2	3	0.651
TSE2	4.72	1.143	10	32	5.26	1.241	5	23	0.039*
TSE3	4.84	1.221	7	28	5.15	1.265	8	30	0.192
TSE4	5.06	1.243	5	26	5.40	1.129	4	19	0.176
TSE5	4.91	1.146	6	27	5.22	1.256	7	27	0.195
TSE6	5.12	1.040	3	24	5.11	1.239	9	33	0.895
TSE7	4.81	1.203	8	29	5.22	1.179	6	26	0.112
TSE8	4.75	1.368	9	30	4.94	1.467	10	40	0.491
TSE9	5.66	1.382	2	2	5.89	1.252	1	1	0.412
TSE10	5.09	1.254	4	25	5.46	1.187	3	16	0.163
TAFF1	5.34	1.428	4	14	5.32	1.213	2	22	0.825
TAFF2	5.38	1.408	3	12	5.25	1.146	3	24	0.455
TAFF3	4.47	1,436	7	41	5.15	1.135	5	29	0.025*
TAFF4	4.66	1.619	6	35	5.03	1.250	7	36	0.291
TAFF5	5.38	1.385	2	11	5.23	1.260	4	25	0.590
TAFF6	4.69	1.595	5	33	5.11	1.371	6	34	0.201
TAFF7	5.53	1.344	1	4	5.55	1.275	1	12	0.981
TANX1	4.22	1.157	9	43	4.85	1.302	9	43	0.021*
TANX2	4.47	1.295	7	40	4.88	1.341	8	42	0.248
TANX3	4.63	1.497	3	36	4.94	1.273	6	39	0.391
TANX4	4.56	1.664	5	38	5.17	1.409	1	28	0.088
TANX5	4.47	1.545	8	42	4.92	1.384	7	41	0.183
TANX6	4.53	1.606	6	39	5.05	1.535	3	35	0.144
TANX7	4.62	1.581	4	37	5.00	1.490	4	37	0.249
TANX8	4.66	1.405	2	34	5.15	1.395	2	31	0.111
TANX9	4.16	1.417	10	44	4.83	1.442	10	44	0.028*
TANX10	4.75	1.626	1	31	4.95	1.556	5	38	0.565
PU1	5.31	1.176	4	18	5.85	1.079	1	2	0.029*
PU2	5.22	1.184	6	23	5.66	1.003	6	10	0.056
PU3	5.32	1.230	3	16	5.74	1.241	3	7	0.074
PU4	5.38	1.100	2	9	5.71	1.155	5	9	0.128
PU5	5.38	1.008	1	8	5.77	1.156	2	4	0.047*
PU6	5.31	1.091	4	17	5.72	1.111	4	8	0.089
PEU1	5.34	1.234	4	13	5.54	1.276	4	13	0.438
PEU2	5.47	1.459	2	6	5.74	1.215	1	6	0.470
PEU3	5.38	1.289	3	10	5.37	1.353	7	21	0.997
PEU4	5.31	1.330	6	20	5.62	1.234	3	11	0.297
PEU5	5.31	1.281	5	19	5.52	1.187	6	15	0.462
PEU6	5.28	1.250	7	21	5.54	1.312	4	14	0.303
PEU7	5.59	1.214	1	3	5.74	1.203	1	5	0.533
PIT1	5.44	1.268	2	7	5.42	1.286	2	18	0.975
PIT2	5.28	1.301	4	22	5.12	1.293	4	32	0.729
PIT3	5.34	1.638	3	15	5.46	1.448	1	17	0.830
PIT4	5.50	1.244	1	5	5.37	1.232	3	20	0.710

(Continued)

Appendix A-22 (Continued)

Factors Affecting Attitude Towards Technologies

Significance

	No				Yes				MW
	Mean	*SD*	*RW*	*RA*	*Mean*	*SD*	*RW*	*RA*	
TSE1	35.91	14.026	1	1	33.46	13.034	2	9	0.304
TSE2	24.03	9.852	10	36	29.25	11.276	5	25	0.029*
TSE3	24.81	10.709	9	30	27.28	11.423	9	35	0.212
TSE4	26.72	10.863	4	25	29.97	10.398	4	22	0.123
TSE5	25.41	10.839	6	27	28.58	11.007	6	26	0.163
TSE6	26.63	9.814	5	26	28.00	10.232	8	30	0.366
TSE7	25.09	10.126	8	29	28.48	10.276	7	27	0.087
TSE8	25.19	11.183	7	28	26.86	11.819	10	38	0.558
TSE9	34.47	13.765	2	2	33.98	14.206	1	6	0.890
TSE10	28.19	12.114	3	24	31.26	11.494	3	14	0.235
TAFF1	30.91	14.193	4	12	30.75	11.244	2	17	0.972
TAFF2	31.28	13.829	3	9	29.91	10.838	4	23	0.649
TAFF3	22.31	11.378	7	39	27.51	10.651	6	34	0.024*
TAFF4	24.06	13.337	6	35	27.02	11.025	7	37	0.204
TAFF5	31.87	13.290	2	5	30.68	11.638	3	18	0.742
TAFF6	24.44	14.482	5	31	27.60	12.734	5	33	0.255
TAFF7	33.28	13.489	1	3	32.89	13.164	1	11	0.901
TANX1	18.69	9.492	9	43	25.05	12.496	9	43	0.014*
TANX2	20.78	11.339	8	42	25.54	12.972	7	41	0.147
TANX3	22.22	13.377	6	40	25.46	12.587	8	42	0.231
TANX4	24.22	14.628	3	34	28.37	14.259	1	28	0.223
TANX5	21.00	14.115	7	41	25.83	13.711	6	40	0.073
TANX6	23.41	14.212	5	38	27.65	14.305	4	32	0.165
TANX7	24.00	14.197	4	37	27.74	14.215	3	31	0.219
TANX8	24.25	12.971	2	33	28.20	14.039	2	29	0.247
TANX9	18.41	11.503	10	44	23.51	12.115	10	44	0.035*
TANX10	24.28	14.501	1	32	26.60	14.021	5	39	0.405
PU1	29.66	11.834	5	18	34.95	10.724	1	1	0.039*
PU2	29.44	11.461	6	19	33.34	10.198	6	10	0.064
PU3	30.91	12.116	1	11	34.22	11.767	4	4	0.139
PU4	30.13	11.851	4	15	34.05	11.564	5	5	0.138
PU5	30.19	11.966	3	14	34.42	11.983	2	2	0.077
PU6	30.44	11.262	2	13	34.22	11.603	3	3	0.103
PEU1	28.97	11.471	7	23	31.15	13.538	5	15	0.462
PEU2	31.25	13.033	2	10	33.63	12.619	1	7	0.448
PEU3	30.06	12.510	3	16	30.52	13.142	7	19	0.984
PEU4	29.41	13.361	5	20	31.85	13.075	4	13	0.436
PEU5	29.00	12.303	6	22	30.88	12.989	6	16	0.529
PEU6	29.91	11.904	4	17	32.11	13.066	3	12	0.399
PEU7	32.03	11.674	1	4	33.52	12.395	2	8	0.634
PIT1	31.69	13.277	1	6	30.45	12.640	1	20	0.693
PIT2	29.09	13.992	4	21	27.09	11.784	4	36	0.682
PIT3	31.31	16.223	3	8	30.12	13.616	2	21	0.754
PIT4	31.69	13.575	2	7	29.89	12.017	3	24	0.621

MW, Mann-Whitney; RW, rank within component; RA, rank across component; SD, standard deviation.

* Significant at *p*-value <0.05.

Index

For Product Safety Concerns and Information please contact our EU
representative GPSR@taylorandfrancis.com
Taylor & Francis Verlag GmbH, Kaufingerstraße 24, 80331 München, Germany

www.ingramcontent.com/pod-product-compliance
Lightning Source LLC
Chambersburg PA
CBHW060237220326
41598CB00027B/3965